Handbook of structural welding

Processes, materials and methods used in the welding of major structures, pipelines and process plant

Handbook of structural welding

Processes, materials and methods used in the welding of major structures, pipelines and process plant

JOHN LANCASTER

ABINGTON PUBLISHING

Woodhead Publishing Ltd in association with The Welding Institute
Cambridge England

Published by Abington Publishing, Abington Hall, Abington,
Cambridge CB1 6AH, England
www.woodhead-publishing.com

First published 1992, Abington Publishing
Reprinted in paperback 1997
Reprinted 2003

British Library Cataloguing in Publication Data
A catalogue record for this book is available from the British Library.

ISBN 1 85573 343 9

Printed by Lightning Source, Milton Keynes, England

Contents

Preface

So far as the fabrication of heavy equipment is concerned, fusion welding has been a slow starter. Welding with a stick electrode was invented in the 1890s, but it was not applied to large structures to any significant extent for half a century. There were, of course, exceptions. The technique of stovepipe welding for pipelines was developed in the USA during the 1920s, and has survived more or less unchanged to the present day. However, not all of the early welding applications were so successful. A torpedo-boat was fabricated by welding in 1895; unfortunately the material chosen was the aluminium, 6% copper alloy, which is highly susceptible to stress corrosion cracking, and the boat did not survive long enough to demonstrate its military potential.

Structural welding on a large scale began with the construction of Liberty ships and T2 tankers in the Kaiser shipyards during World War II. To many welding technologists these ships epitomize catastrophic brittle fracture, and a picture of the 'Schenectady', which broke in two whilst moored in calm water, is still sometimes used to chill the hearts of aspiring young welding engineers. In fact, the Kaiser shipbuilding effort was a major success, enabling large quantities of supplies to be shipped from the United States across all the oceans of the world. Even some of those ships that suffered catastrophic failure were (unlike Humpty Dumpty) put together again and subsequently provided years of useful service. More important, perhaps, was the fact that fusion welding then became the norm for the fabrication of boilers, pressure vessels, bridges, structural steelwork and, of course, shipping.

In the post-war years such applications have stimulated a remarkable development of welding technology, and processes have multiplied at a considerable rate. Materials have likewise changed, very much for the better as far as welding is concerned. The original incentive for the invention of the basic oxygen converter was to increase steelmaking capacity, but from this and other improvements such as hot metal desulphurisation has emerged a new generation of clean steels that possess the levels of notch-ductility and through-thickness ductility essential for heavy structural fabrication.

It is the purpose of this book to describe the current state of welding technology as applied to large structures, and to say something of the problems that have been encountered in this work, together with the means that are used to assure reliability. Recent steelmaking developments have been included because, as noted earlier, they are highly relevant to the integrity of welded structures.

It seems likely that some major hazards, in particular catastrophic brittle fracture, are now avoidable but others, such as fatigue failure, will continue to present difficulties in the future. It is increasingly necessary for engineers at all levels to be aware of such problems and to take all reasonable steps to avoid them. It is hoped that this book will help to promote such awareness, and, in a small way, contribute to the increasing dependability of welded fabrication.

The author would like to acknowledge the encouragement and help of colleagues and friends in assembling this material, but most particularly wishes to thank his wife, Eileen, for her contribution to the finished manuscript. Thanks are also due to Ralph Yeo for advice on welding processes, to Peter Lane of Lloyds Register and to staff at The Welding Institute (now TWI), in particular Richard Dolby and Trevor Gooch for helpful advice. The librarians at TWI have given most valuable assistance in providing literature. And finally the author acknowledges the help of companies such as Cooperheat and Magnatech, who furnished background materials and illustrations.

John Lancaster

1 Processes

Introduction

The materials with which this book is concerned are used in the fabrication of process plant for power generation, or for structures such as steel buildings, bridges and offshore oil rigs. The joining method appropriate to this field of construction is predominantly fusion welding. Exceptionally a solid-phase process may be practicable, as for example in the longitudinal welding of pipe and tube using the electrical resistance welding process.

Amongst the more industrialized nations competition provides a spur to seek more economical methods of fusion welding. One route to greater productivity is through automation and by the use of robots, and to this end the process must be capable of operating consistently over as long a period of time as possible. When automatic techniques are not practicable the use of gas metal arc (GMA) welding, where wire is fed continuously, gives higher deposition rates than welding with coated electrodes as shown for the case of fixed position welding on an offshore oil rig in Table 1.1.[1]

Consequently there is a trend towards increased use of GMA welding and a corresponding decrease in that of coated electrodes. At the same time variants of the GMA process have been developed so as to widen the current range over which it can be used, make it applicable to all-position welding and generally to increase its versatility. One of the most important contributions to such improvements has been through changes in the design of welding power sources. This has been made possible by the availability of high power switching devices;

Table 1.1 Typical deposition rates for welding of carbon steel in offshore fabrication

Process	Position	Deposition rate, kg/hr (lb/hr)
Manual metal arc 7018	All	1.2–1.8 (2.6–4.0)
Self-shielded flux-cored	All	1.5–2.0 (3.3–4.4)

firstly the thyristor and later, high power transistors. So before discussing fusion welding processes, it will be convenient to describe these power source developments, particularly as they have also proved to be applicable to welding and cutting operations other than GMA.

Welding power sources

Simple machines consist of a transformer with, typically, a three-phase AC output with an open-circuit voltage of 65–100 V: a level which is usually considered low enough to be safe. Current control is obtained either by taking different tappings on the secondary coil or by moving the transformer core. Alternatively for site work, a rotating power source may be driven by an internal combustion engine. In the past, AC was more commonly used in Europe and Japan but US fabricators preferred DC for quality work. Almost all welding was carried out using stick electrodes and the machines were traditionally designed to have a drooping characteristic. In this way, the change in current produced by any inadvertent change in arc length (with corresponding change in arc voltage) is minimised, so helping to maintain steady welding conditions (Fig. 1.1).[2] The availability of solid state rectifiers has led to a more general use of DC, and simple transformer-rectifier sets are often employed for welding with coated electrodes.

More recent developments owe much to a need to improve the operating characteristics of pulsed arc welding (see later in this chapter). The first pulsed arc machines were only capable of pulse frequencies of either mains frequency or multiples thereof, and were dynamically slow, such that the process was difficult to use (Fig. 1.2a).[3] Variable frequency sources were required, and these became available as a result of the development of a transistor chopper controlled regulator in the early 1970s. Figure 1.2b illustrates a typical circuit. Such machines were however large and expensive, and largely used for research and development. An improved version, which became commercially available, utilised solid state power devices on the output side of the transformer to generate pulses.

The third generation of pulsed arc machines (Fig. 1.2c) is a radical departure. The three-phase power supply is first rectified, and then inverted to high frequency AC. The AC is fed to a transformer, the output of which is rectified and provides either a pulsed or steady DC source. For pulsed current, the switching is carried out using power transistors or thyristors on the primary side of the transformer, and this makes possible a nearly square wave shape for the pulse. When the switch is on the secondary side, however, the wave form is trapezoidal (Fig. 1.3).[4] The high frequency transformer requires a smaller core than a normal mains frequency type, such that the weight and volume of an inverter power source is about half that of a

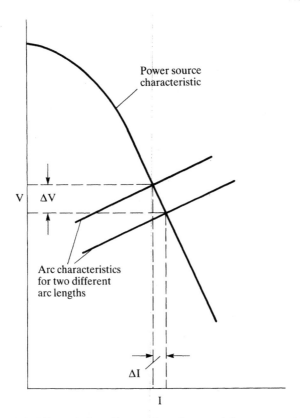

1.1 Effect of changing arc length on welding current.[2]

conventional type, whilst the price, although higher, may well be justified by improved performance.

Apart from pulsed arc welding, current pulsing may be useful in gas tungsten arc (GTA) or plasma welding, in short circuiting and flux-cored GMA welding, and in high speed plasma cutting. Inverter power sources may also be used primarily for portability, for example in low current (below 50 A) plasma cutting machines and for site welding.

Thus the number of types of welding power source has multiplied remarkably, as has the number of welding processes. Figure 1.4 shows a classified list of power sources used in Japan, in which the listing relates to the control system.[5] Within such general categories, there are various types adapted to particular processes or to a range of processes; for example, in 1989 one company listed seven types of inverter source[3], whilst the old moving core or stepped voltage AC welding set was at that time still available. The general tendency with all power sources however is towards electronic control both on the input side (to compensate for fluctuations in mains voltage) and on the

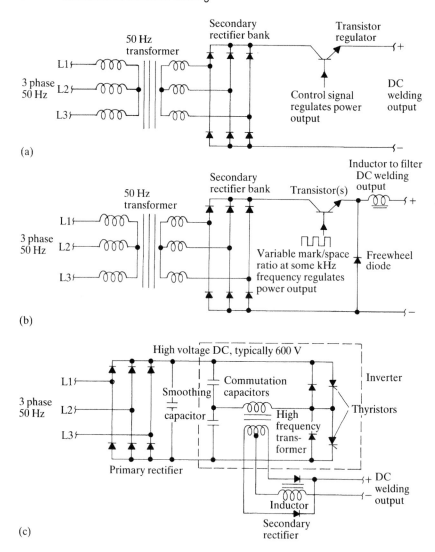

1.2 Various types of pulsed GMA welding power sources:[3] a) 50 Hz secondary transistor regulator; b) Secondary transistor switched mode regulator; c) Primary switched inverter.

output side to provide a minimum stepless control of both current and voltage.

Gas tungsten arc (GTA) welding

This process became established in industry after the Second World War, mainly for welding austenitic chromium-nickel steels and

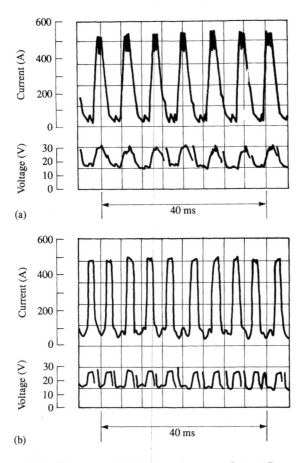

1.3 Oscillograms of voltage and current for: a) Conventional pulsed power source (trapezoidal wave); b) Inverter type (rectangular or square wave) (source: Daihen Corporation).[4]

non-ferrous metals, particularly aluminium. It provides a heat source only and filler metal must be added separately. In this and other respects it resembled the oxyacetylene and carbon arc welding processes which it rapidly displaced. Originally called tungsten inert gas (TIG) welding, it was redesignated gas tungsten arc welding by the American Welding Society because for some purposes argon-hydrogen, argon-nitrogen or pure nitrogen gas shields are used. However in the UK and in Europe the title tungsten inert gas welding is still employed.

An argon gas shield containing 2–5% hydrogen may be employed in welding austenitic stainless steels and nickel alloys to reduce oxidation and obtain higher welding speeds. Nitrogen and argon-nitrogen mixtures may be used for copper and provide a higher heat input rate for any given current. Helium-argon mixtures (typically 76He 25Ar) also

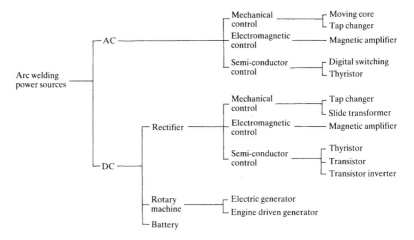

1.4 Classification of arc welding power sources.[5]

increase the heat input rate and are applicable to most metals, particularly aluminium, stainless steel and nickel-base alloys. Pure (99.95%) argon is the most commonly used shielding gas.

The process is operated by striking an arc between the tungsten electrode and the workpiece, with shielding gas flowing through a nozzle disposed circumferentially around the electrode. In the majority of applications the current is DC, electrode negative (straight polarity, as opposed to reverse polarity which is electrode positive). For aluminium it is normal to use AC because during the electrode positive part of the cycle the cathode removes oxide from the surface of the weld pool. However if a helium or argon-helium mixture is used for the gas shield it is possible to obtain good fusion of aluminium with electrode negative. This process is used for the automatic seam welding of irrigation tubes, and also for welding seams in vessels where a high quality weld free from porosity is required, as in the nuclear industry. With DC the heat input to the workpiece per ampere is substantially greater than with AC so that the DC automatic process is faster and capable of welding thicker material: up to 25 mm. For descriptive purposes it is convenient to consider the process in three parts: the electrode, the arc column and the weld pool.

The tungsten electrode

In DC operation the electrode is typically a rod 1.5–3.5 mm diameter ground to a conical tip. Pure tungsten is not used because the tip of the cone melts and the cathode spot moves over the molten drop, making the arc difficult to control. The most common electrode material is thoriated tungsten; this contains 1.5–2.0% ThO_2 and operates (at normal welding currents of say 200 A and below) without melting. At the high temperature of the electrode tip the thorium oxide is reduced by tungsten to form thorium, which diffuses to the electrode surface. Thorium-coated tungsten has a lower electron work function than the pure metal and therefore can emit the required electron current density (which is of the order of 10^7 A/m^2) at a lower temperature.

In automatic welding operations electrodes may be required to operate for extended periods of time and the thoria content may be reduced due to evaporation, so that the arc may become unstable. To improve electrode life various alternative additions have been made to tungsten, notably the oxides of cerium, lanthanum and yttrium. Such rare additions react with tungsten to form tungstates or oxy-tungstates, which also diffuse to the electrode surface. Here they have two effects; firstly to reduce the electron work function and secondly to increase the thermal emissivity.[6] Table 1.2 shows the results of measurements suggesting that rare earth additions could indeed improve GTA electrode performance, and that in particular lanthanated electrodes should perform well.

Other factors that may contribute to longer electrode life are tungsten grain size and tip shape. Diffusion of both thorium and rare earths is largely intergranular, and both are exhausted more slowly in coarse-grained material. The cone angle, if too small, may result in overheating and disintegration. A cone angle of 60° appears to be a good choice.

These considerations do not necessarily apply to welding with AC which is normally used for aluminium. For any given arc current the rate of heat liberation at the electrode is greater when it is anodic than when it is a cathode; consequently all these materials melt at the tip although thoriated tungsten may still have some advantage. Using say a 6 mm diameter electrode gives better cooling and normally permits stable operation but there is a limiting current above which drops of tungsten may be emitted by the electrode, giving rise to tungsten inclusions in the weld.

Table 1.2 Tip temperature, emissivity value and work function for several electrodes after 30 min of arcing at 150 A in pure argon[6]

	Electrical material		
	W/ThO$_2$	W/CeO$_2$	W/La$_2$O$_3$
Tip temperature, ° C	3340	2800	2440
Emissivity	0.18	0.22	0.30
Work function, eV	2.38	2.14	2.0

The GT arc column

Axial flow rate in the GT arc column are of the order of 100 m/s for currents in excess of 100 A, and lend the quality of stiffness to the arc. The flow is from electrode to workpiece as shown in Fig. 1.5 and is jet-like.[7] It is caused by the electromagnetic forces induced by the interaction between the arc current and its own magnetic field. Impingement of this jet on the weld pool results in a stagnation pressure. Figure 1.6 gives the pressure distribution on the anode for a 200 A argon-shielded TIG arc.[8] The total force due to this condition is about 5×10^{-3} N, and it is not sufficient to cause the deep depression of the weld pool surface (the crater) that is characteristic of submerged-arc or manual metal arc welding (MMA). The arc force may be reduced by using helium as a shielding gas or by increasing the cone angle of the electrode tip. It may be increased (to obtain greater stiffness) by using a pulsed current in the range 5–10 kHz. This effect may be useful for welding at currents below 100 A.

Except for welding in an enclosed chamber there is always some degree of oxidation of the weld pool in argon-shielded GTA welding.

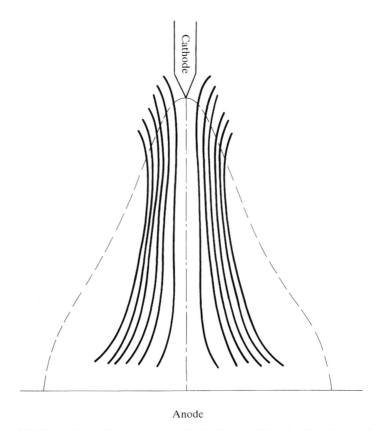

Anode

1.5 Streamlines of electromagnetically induced flow in the column of a GT arc.[7]

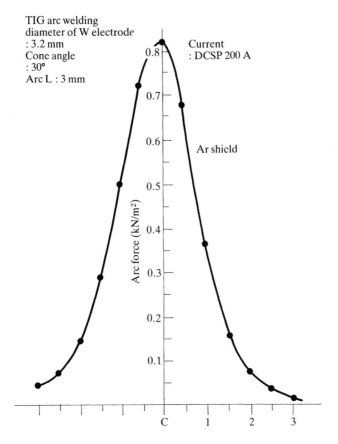

1.6 The pressure distribution on the anode of a 200 A arc in argon.[8]

Such oxidation is probably due to the action of the plasma jet in drawing in contaminated argon from the outer region of the gas shield.

The weld pool

The pool is subject to electromagnetic, buoyancy and surface tension forces, as well as shear forces due to the outward spread of the plasma jet. For short (say 2–4 mm) arcs used in welding the surface tension force is potentially the strongest. In a pure metal surface, tension decreases with increasing temperature, and in a weld pool with a negative temperature gradient from centre to edge, this could result in a shear force at the surface acting in an outward direction. This in turn may generate an outward flow of metal across the surface, giving a relatively wide, shallow pool (Fig. 1.7 illustrates the mechanism).

In practice weld metals are not pure and in particular are contaminated

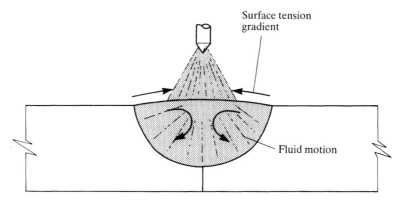

(a)

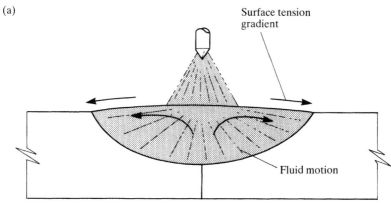

(b)

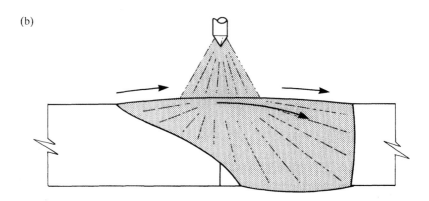

(c)

1.7 Flow in a GTA weld pool induced by a gradient of surface tension:[9] a) Inwardly-directed surface flow, associated with the presence of surface-active elements; b) Outwardly-directed surface flow, as in a pure metal; c) A combination of normal sulphur steel on the left with low sulphur steel on the right.

with oxygen (from the plasma jet) and, in steel, with sulphur. In steel both these elements are surface-active, and in their presence the surface tension gradient with temperature becomes zero or positive. Normally, therefore, the flow in the weld pool is relatively sluggish and the profile of the fusion boundary is more or less semicircular, as would be expected in the absence of flow.

Abnormal conditions, however, may prevail in welding austenitic stainless steels. These alloys are aluminium killed and when the residual aluminium content is 0.01% by mass or greater, the oxygen content of the liquid metal is very low, even when there is oxide on the surface. Thus, if the sulphur is also low (say below 50 ppm) the alloy behaves like a pure metal, and there is an outward flow across the surface of the weld pool which convects heat in a radial direction. In this way the depth/width ratio of the fused zone may be reduced from a normal value of about 0.5 to as low as 0.2. Also, welding a normal-sulphur to a low-sulphur steel can give an asymmetric fused zone, as shown in Fig. 1.7.[9]

Such conditions can be avoided by specifying a lower limit for the sulphur content. Too much sulphur makes the alloy susceptible to hot-cracking during welding, and it has been proposed that the sulphur content of austenitic stainless steel be maintained between 0.01% and 0.02% by mass.

Similar effects have (exceptionally) been observed with ferritic steel. Using the oxygen converter it is practicable to manufacture steel of very low impurity content, and although this may improve the notch-ductility of the unwelded material it may not benefit welded joints. Apart from loss of control of penetration, such steels may suffer greater hardening in the heat affected zone. Calcium-treated steel is particularly subject to variable penetration because calcium reacts with sulphur and removes it from solution, as does aluminium with oxygen.

Except for AC welding of aluminium, the weld pool forms the anode of the arc. In argon or helium this is normally a diffuse region and, unlike the cathode, is not fixed to any point on the metal surface. At low currents and at high welding speeds, however, a fixed anode spot may form and cause the arc to move forward discontinuously. This defect can be overcome by applying a pulse of 5–10 kHz to the arc current.

Anode spots may also be formed around patches of oxide on the weld pool surface; these move along with the weld pool but may distort the heat flux distribution and modify the fused zone profile.

AC welding

During the electrode positive part of the current cycle the workpiece becomes the cathode and emits electrons. The mechanism of electron emission from such a cathode is very different from that of a tungsten thermionic cathode. There is a layer of oxide on the metal surface which normally has a very low electrical conductivity. Where positive ions condense on the oxide surface an electric field of up to 10^9 V/m may be set up, and by one mechanism or another, the oxide locally conducts and emits electrons, at the same time vaporising or blowing away surrounding material. The diameter of an emitting site is very small, in the region of 10^{-7} m, and its lifetime is in the range of 1×10^{-9} to 1×10^{-6} s. Concentrations of these active spots move over the metal surface within the area of the arc column and in so doing remove the surface oxide layer.

The ease with which a non-thermionic cathode spot moves over a metal surface may be assessed by moving the arc column using a magnetic field and observing the motion of the cathode arc root. In pure argon or helium it moves in jumps, with a relatively long interval between each jump. In argon containing a small amount of contaminant such that a thin surface layer is formed on the metal the cathode spot moves smoothly and freely. With thick surface layers however the movement once again becomes slower and more erratic. Under welding conditions the shielding gas is always contaminated to some degree with air, and the conditions are such as to produce a thin oxide layer and a freely moving cathode spot.

Oxide removal is very beneficial in fusion welding aluminium, where oxide films may persist and have detrimental effects on the integrity and the acceptability of the joint. In a 50 cycle sinusoidal AC operation there are however some difficulties. Because of deionisation the mechanism of the cathode is disrupted after the current falls to zero and the arc must restrike more or less from cold. Thus there is a high restriking voltage and a corresponding delay, such that the workpiece negative half cycle is shorter than electrode positive. In effect this results in a DC component of current which can overload the transformer and it may be necessary to run a high frequency igniter continuously, with a risk of radio interference and damage to computer-controlled installations.

A solution to this problem is found with electronically controlled power sources which are capable of generating a square wave AC. The recovery from current zero is rapid enough to avoid deionisation and there is no delay in arc reignition. Moreover it is possible to vary the relative duration of the electrode negative and electrode positive pulses. With longer values of this ratio the heat load on the torch is reduced, whilst with smaller values (longer periods of electrode positive) the degree of oxide removal is greater.

Equipment

Torches

Figure 1.8 illustrates typical GTA torches for the middle and higher DC current range. Above 200 A, and for AC, water-cooling is necessary but for the majority of DC manual applications an air-cooled type is adequate. The nozzle is usually ceramic for DC torches but may be metallic and water-cooled for heavy duty or AC. The tungsten electrode is held by a collet and projects upwards. A plastic cover protects this upward projection from accidental arc strikes. There is a diffuser or gas lens that spreads the shielding gas stream and promotes laminar flow. This consists typically of a disk or cylinder of porous metal. Flexible torches are available that can be bent to provide access to confined spaces. Normally there is a control switch which connects with the power source and activates power, gas and water supplies but when the torch is used with a simple power source it may have a built-in gas valve. In manual operation filler rod is usually fed into the weld pool by hand, but there are torches with an integral wire feed, making one handed operation possible. Automatic equipment is described later in this chapter.

Power sources

The power source characteristic required for GTA welding is similar to that for SMA (MMA) welding, namely one which minimises the change in arc current with arc length. Thus it is possible to use a simple transformer/rectifier set for DC welding, as indicated above. However for quality applications a specialised or multi-purpose machine is normally employed. Such machines incorporate gas valves with pre-welding flow and a delayed cut-off to protect the finish of the weld from oxidation, an arc striking device, and a controlled reduction of current such that

1.8 Typical GTA welding torches (reproduced by courtesy of ESAB Ltd).

the end crater of the weld solidifies gradually to avoid cracks. They may also incorporate a low frequency pulsed current output.

One method of striking the arc is to use a high frequency (HF) spark generator. In principle it is possible to start by applying a high voltage DC pulse but this may be hazardous. The alternative is to use a train of high voltage sparks to ionise and break down the gap. HF does not penetrate the body and is therefore safe. On the other hand it causes radio interference and may be unacceptable in built-up areas. The alternative is to touch the electrode to the workpiece. GTA power sources may be programmed such that the risk of tungsten inclusions or electrode contamination due to this operation are minimised; for example by providing an initial small warming up current with a gradual rise of current after the electrode is lifted and the arc is established. This type of arrangement also avoids the formation of a non-thermionic cathode on the cold electrode.

For DC the power source itself is usually a thyristor controlled mains frequency transformer/rectifier set. In addition, units incorporating an inverter may be designed for multiple use, including GTA welding. AC sources are also either thyristor controlled or inverter types with a square wave output and a facility for adjusting the relative durations of electrode positive and electrode negative pulses, as described earlier.

Applications

Materials
One of the outstanding virtues of GTA welding is its wide range of applicability to different metals. Copper that is used for electrical conductors presents some difficulties due to porosity and, in thick sections due to its high thermal conductivity, which may cause lack of fusion defects. There may also be problems with the reactive metals such as titanium because of absorption of oxygen, and in some cases it may be necessary to weld in an argon-filled container or 'glove box'. Aluminium and magnesium alloys are normally welded with AC, but DC is used for other metals.

Because of the lack of flux and generally clean operation, GTA welding is often used for stainless steel and nickel-base alloys in equipment for the food, pharmaceutical, nuclear and aerospace industries; also in the chemical and petrochemical industry where there are special requirements for cleanliness (e.g. high pressure oxygen piping).

Product forms
One of the earliest applications for the process was in making the root pass in pipe welds. When the pipe can be rotated it is relatively easy to control penetration. Dams (which may be of paper, plastic or metal)

are placed inside the pipe on either side of the joint and argon is fed into the space in between to prevent oxidation of the underside of the weld. In Continental Europe a mixture of hydrogen and nitrogen (often called Formier gas, originally cracked ammonia) may be used for this purpose. When the pipe is fixed it may be helpful to apply a low frequency (within the range 0.1–100 Hz) pulsed current to make, in effect, a sequence of overlapping spot welds. An ideal weld preparation for a root pass with no filler is shown in Fig. 1.9.

Automatic GTA welding equipment is available for tube and pipe joints made in the fixed position; an operation known as orbital welding. Such machines may be used in conjunction with a programmable power source so as to obtain good control in all positions. One object of automatic welding is to obtain a smooth hygienic root pass for stainless steel food and pharmaceutical equipment. Another, of course, is to improve productivity. In such cases much depends on the set-up time of the machine and its reliability. Figure 1.10 shows an orbital system set up for butt welding thick-wall pipe.

Another automatic application is the welding of heat exchanger tubes to tubesheets. This may be an end weld on the outside of the tubesheet or a butt weld to a projection on the inside of the tubesheet. Figure 1.11 shows some typical joint configurations. Large numbers of identical welds are required and there is a good case for an automatic technique. Figure 1.12 shows a head for welding on the face of a tubesheet; this is equipped for filler wire addition and is water cooled (see also Chapter 5).

Heads for rolled pipe joints or for flat position welding (for example longitudinal welds in tubing) generally take the form of a vertical cylinder and are equipped with some method of maintaining constant arc length. In such operations it is essential to establish strict control over all details of the set-up, such as nozzle-to-work distance, stick-out of electrode, electrode cone angle as well as the welding variables. In some applications it is possible to control penetration by for example an optical device in the bore of the tube. Flow in the GTA weld pool is

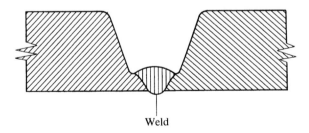

Weld

1.9 Preparation for an autogenous (no added filler) GTA root pass in pipe welding.

1.10 Automatic (orbital) GTA welding of thick-wall pipe in the fixed position (reproduced by courtesy of Magnatech).

not positively determined, as it is in many other welding processes, and may be sensitive to relatively small changes in the physical set-up.

A major disadvantage of GTA welding in some applications (surfacing for example) is its low deposition rate. When using a cold filler wire this is usually less than 1 kg/hr. The situation is improved by passing a current through the filler wire and thereby heating it – the hot-wire GTA process. The current is supplied to the wire by an independent power source and is normally AC. The wire is fed close behind the tungsten electrode at a steep angle (30 ° to 40 ° from vertical) and conditions are adjusted so that it is molten when it feeds into the pool. In this way deposition rates can be increased to 2.5 kg/hr whilst preserving the good quality of GTA welding; in particular, freedom from porosity and slag inclusions.

An advantage is also obtained by operating so that the arc burns below the plate surface. This technique is known as buried gas tungsten arc welding and for thick plate typical welding variables are a voltage setting of 10 V and current 600 A. This means 19 mm plate can be welded in two passes. The same process was used experimentally by

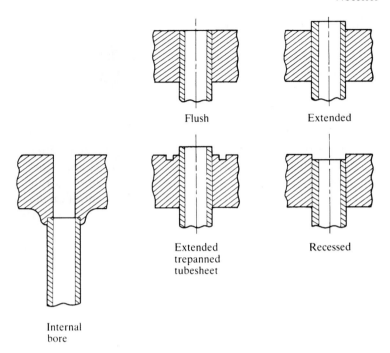

Flush   Extended

Extended
trepanned
tubesheet

Recessed

Internal
bore

1.11 Some typical configurations for tube-to-tubesheet welded joints.

1.12 A gun for making automatic tube-to-tubesheet GTA welds. The projecting rod is inserted in the tube to centre the device. Filler wire mounted on the reel at the back is fed in as the tungsten electrode rotates within the gas chamber. A different design is used for internal bore welds.

the author in about 1950 for welding 2 mm thick austenitic chromium-nickel sheet. The resulting weld was completely oxide free, but with tramlines of condensed metal vapour on either side. It appeared to be a keyhole type weld with metal vaporisation providing an additional shield.

Health hazards
GTA welding is a clean arc process; that is to say, the amount of particulate fume generated is small. However, because of the lack of fume, ultraviolet radiation from the arc column is at a high level and ozone may be generated in a region some distance from the arc. As a rule, good ventilation is sufficient to maintain the ozone concentration at a safe level but where high currents are used and for automatic welding it may be necessary to monitor the ozone level, and to consider the use of extraction equipment.

Thorium is slightly radioactive and is considered by some to be a potential problem. This reinforces the trend towards the use of alternative additives such as lanthanum. Regardless of which additive is used, a health risk is most likely to arise due to grinding electrodes. Grinders fitted with integral dust extractors are now available.

In all arc welding processes other than submerged-arc, the ultraviolet radiation from the arc is damaging to the eye and welders' helmets must be fitted with a suitable filter. It is also necessary to ensure that welding areas are free from reflecting surfaces that can project arc radiation into the back of the helmet, whence it can be reflected from the filter. Normally the welder sights the position of the electrode with the helmet up, and drops the helmet and strikes the arc at the same time. In restricted areas this may not be practicable, and here it is possible to use a filter that is normally clear, but which is fitted with a circuit that darkens the filter as soon as the arc is struck.

Filters are standardised in BS 679, and there is now an 'R' grade that is resistant to impact.

Plasma welding and cutting

General

Plasma welding is usually employed for thin sections and microwelding and only comes marginally within the scope of this book. Plasma cutting on the other hand is employed to an increasing extent in metal fabrication. Figure 1.13 shows typical ranges of current and gas flow rates for both plasma welding and cutting.[10] It was at one time supposed that constricting the arc through an orifice would raise the plasma temperature and increase the power density above that normally available in arc welding. Measurements of temperatures in

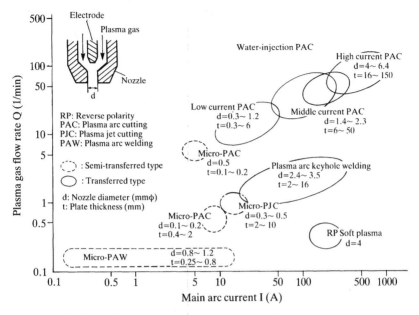

1.13 Operating ranges for plasma arc cutting and welding.[10]

plasma jets used for spraying do not show such augmented temperatures whilst the conditions indicated in Fig. 1.13 correspond to power densities in the range 2×10^7 to 2×10^8 W/m², which are very similar to those for GTA arcs. It would seem that plasma welding is really very similar to GTA welding but with an increased plasma jet flow, giving sufficient force to penetrate the weld pool in keyhole welding and cutting. The other advantage of the process is that for non-keyhole welding it has better directionality than the GT arc, particularly at low currents.

Torches

Figure 1.14 shows the two basic designs of torch.[10] In the non-transferred type the current flows between electrode and nozzle and the hot plasma is blown through the orifice. This configuration is used primarily for cutting non-metallic materials such as plastic. In the second, transferred type, the arc burns between electrode and workpiece but is constricted and directed by the orifice. For microwelding and also for many general applications a pilot arc of a few amperes is maintained between electrode and nozzle so that the main arc can be struck without the use of a high frequency spark generator. For welding, the plasma gas is argon and there is an outer nozzle to provide a gas shield. The shielding gas is normally argon but argon/hydrogen and

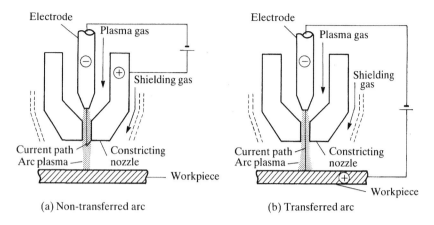

(a) Non-transferred arc   (b) Transferred arc

1.14 Basic modes of plasma arc operation.[10]

argon/helium mixtures may be used in the same way as in GTA welding.

The electrodes are the same as for GT welding, usually thoriated tungsten and, being protected from contamination, have a longer life than in GTA welding. They operate with DC electrode negative, except for the 'soft plasma' torches used in surfacing, which have electrode positive. Aluminium welding is carried out with AC. Pulsed current is advantageous for high speed welding and for positional work.

A cutting torch is sketched in Fig. 1.15. The electrode consists of a cylinder of hafnium, ruthenium or zirconium embedded in copper.[11] Early torches used nitrogen or an argon/hydrogen mixture for the plasma gas; performance was improved by water injection but for low and medium currents (up to 150 A) air is now commonly used for both the plasma and the shielding gas. In some applications there may be concern about nitrogen contamination of the cut edge, and where this is so the air is replaced by oxygen.

Techniques and applications

One of the attractions of plasma welding is its capability of making single pass welds of square edge close butt joints in thicknesses up to 6 mm and exceptionally up to 10 mm. The technique is known as 'keyhole' welding. This condition is intermediate between normal welding and cutting: the plasma jet penetrates the joint forming a circular hole, but the liquid metal is not blown away and it flows together behind the keyhole to form a full penetration weld. In cross-section the fused zone has a shape somewhat resembling a wine glass. The fluid mechanics of this process are not well understood, nor is it easy to use in production. Recent surveys of plasma welding indicate that industrial applications of keyhole welding are still rather limited.

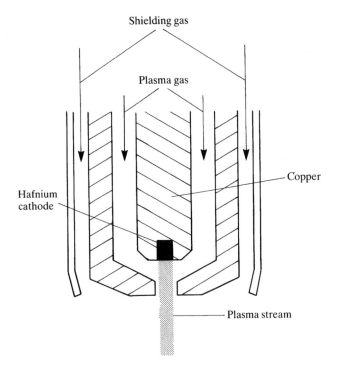

1.15 Diagrammatic section of a plasma cutting torch.

Normally the plasma torch is used as a heat source for welding sheet material without addition of filler; e.g. for corner welds or butt welds in stainless steel sheet. With pulsed arc operation it is possible to achieve high welding speeds, of the order of 1 m/min.

Cutting torches are used for carbon steel, stainless steel and aluminium and can operate at high speed, up to 5 m/min on thin material. It is not always easy to obtain a good square cut and development work on this aspect is continuing.

Manual metal arc (MMA) welding

General

Early attempts to weld steel with a fusible electrode were made with bare wire, typically fencing wire, but the weld quality was poor due mainly to the high nitrogen content. Subsequent improvements included wrapping short lengths of wire with paper bonded, with sodium silicate or wrapping with asbestos string. The next step was to coat the wire with a paste made of a mixture of clay, mica and sodium or potassium silicate together with minerals, organic material, ferro-alloys and other materials. Such coatings were at one time applied by dipping the wire into a slurry, but now they are extruded coaxially on to the wire in a press and dried by baking.

Cellulosic electrodes (BS class C and AWS 6010) have coatings which contain a mixture of cellulose, rutile (TiO_2) and magnesium silicate with ferro-manganese. Rutile coatings (BS class R and AWS 6013) contain a smaller quantity of cellulose with rutile, carbonate and silica, again with ferro-manganese as a deoxidant. These two types of electrode are baked after extrusion at 100–150 °C; higher baking temperatures would destroy the cellulose. As a result the coating has a substantial water content.

Basic electrodes (BS class B and AWS 7015, 7016 and 7018) have coatings that contain a mixture of calcium carbonate and calcium fluoride together with ferro-manganese, ferro-silicon and ferro-titanium. There is no organic content and such electrodes are baked at 400–450 °C, which drives off the bulk of the water content.

There are of course electrode coatings that do not fall into these three categories, and in particular the coatings for stainless steel and non-ferrous electrodes are individually formulated.

Iron powder may be added to all three types of coating to increase the metal deposition rate per ampere of welding current. Such electrodes have a thicker coating and, when large additions are made, may only be suitable for welding in the flat position. Figure 1.16 illustrates the advantage that can be obtained under ideal conditions.[12]

Ferro-alloys may also be added to the coating to produce a ferritic alloy steel deposit with a normal rimming steel core wire. This is a common practice with, for example, low alloy chromium-molybdenum steel electrodes, although some electrode manufacturers prefer to use an alloy core wire on the grounds that damage to the coating could result in a deposit of incorrect composition.

Basic coatings tend to pick up moisture from the atmosphere and are usually stored prior to use in a heated oven, and held by welders in a heated quiver. Some manufacturers have overcome this deficiency by firstly using a vacuum pack so as to increase the shelf life of packed electrodes, and secondly by formulating the coating such that the moisture pick-up rate is low. This type of electrode can be left in the open at room temperature for a specified period without a significant increase in moisture content.

For basic electrodes that are not so formulated it is necessary to establish a drying procedure before use. Packets of electrodes should be stored unopened in a dry atmosphere, preferably at 20 °C with a relative humidity of less than 60%. Immediately prior to welding they are removed from the packets and dried in an oven according to the manufacturer's instructions as to temperature and time of drying. Electrodes

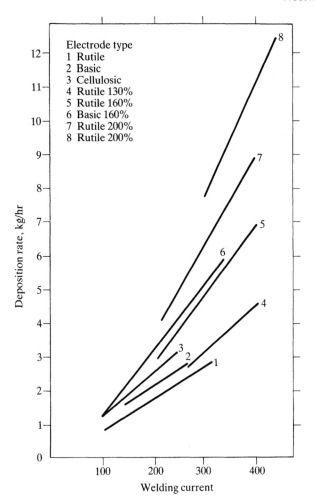

1.16 Deposition rate as a function of current for various electrodes. The coating of types 4–8 contain iron powder and the percentage figures represent the ratio between mass of deposited metal and mass of core wire.[2]

for welding stainless steel, copper, nickel and aluminium alloys should be treated in the same way. Rutile electrodes may be dried but this is not normal practice; cellulosic electrodes *must not* be dried since they depend on the presence of moisture in the coating to function correctly.

Functions of the coating

The electrode coating serves a number of functions. It provides gas and slag shields that prevent nitrogen pick-up from the atmosphere. It generates the arc flow, ensures directional metal transfer, stabilises the arc and promotes a favourable weld metal chemistry by, for example, deoxidation or alloy addition.

The gas generated by cellulosic and rutile electrodes is a mixture of carbon monoxide, carbon dioxide and water vapour with a high proportion of hydrogen. The gas produced by basic coatings is largely carbon monoxide plus carbon dioxide with small amounts of hydrogen and water vapour, and other things being equal, the volume is less than with rutile or cellulosic coatings.

There are various ways in which directionality of metal transfer is achieved in welding with coated electrodes. Figure 1.17 illustrates the results obtained from high speed X-ray films of fully killed carbon steel electrodes.[13] In one case a drop is extruded from the cup formed by the coating at the electrode tip by the stagnation pressure of the gas flow; in another the drop is detached by electromagnetic forces. Others have observed bubbles which form in the liquid metal and eventually burst, producing a scatter of small drops which are carried into the weld pool. In some cases (notably in welding with basic coated electrodes when a short arc is maintained) the transfer occurs as the result of a short circuit. Gas flow undoubtedly plays a part in the transfer process, because if an all-position electrode is completely dried out by heating at 700 °C it is no longer capable of welding in the overhead position; drops which form at the electrode tip simply fall down the side of the rod.

The arc characteristics may be modified by the addition of easily ionisable elements. For example, reignition in AC welding is facilitated by the presence of potassium, which has a low ionisation potential. Therefore potassium silicate is preferred as a binder for electrodes that are designed for AC work, whilst sodium silicate is suitable for DC.

The weld pool

X-ray films of downhand bead-on-plate welds made with coated electrodes have shown that there is a cavity below the electrode at the front of the weld pool. It may be surmised that the arc force is sufficient to maintain such a cavity and to force the liquid metal to the rear of the weld pool, giving a circulation rather like that in submerged-arc welding. However, such a condition does not necessarily apply in positional or fillet welding, where it is necessary to maintain a small weld pool supportable by surface tension forces. Also in making the root pass in a fixed pipe, the state of the weld pool must necessarily change as it moves around the circumference.

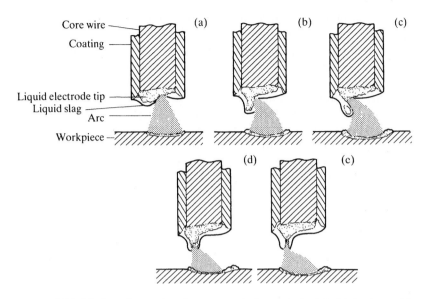

1.17 Modes of transfer from coated electrodes: a) Initial phase; b) Deformation of liquid metal due to pressure at the arc root; c) Separation of drop due to surface tension; d) Alternatively; arc moves to tip of projection; e) Separation of drop by electromagnetic force.[13]

Equipment

Power sources

The tendency for stick electrode welding to be replaced by automatic or semi-automatic processes is most advanced for work in permanent shops, but less in the case of site welding. This factor influences the choice of welding source.

In more remote areas where mains power is lacking motor-generator sets must be used and these may provide AC or DC. Such conditions are increasingly uncommon, and most MMA power sources are transformers or transformer/rectifiers. Current control may be of the moving core or tap changer type, but thyristor controlled machines are available which provide a constant current regardless of voltage (a vertical static characteristic). These have an advantage for welding on a pipe rack where long leads may be required. Another solution to this problem is provided by inverter sources which may be as light as 8 lb and can therefore be carried to the job location.

Health hazards

Fume may be a problem in confined spaces (including workshops) and some form of protection against inhaling fumes is normally provided. Light respirators are available but are not popular with welders and most organisations rely on the removal of fume at source. An overhead hood is sometimes used for fixed welding positions, but a more efficient device is a portable hose that can be fixed (for example by magnets or using a cantilever arm) so as to draw fume directly away from the welding location. Some welding guns for GMA welding are equipped with integral fume extraction but these are inevitably more bulky and difficult to manipulate. Most developed countries have legal requirements that place limitations on the exposure of individuals to dusts, and welding fume comes into this category. Thus in addition to providing extraction equipment it is necessary to monitor the degree of exposure and if necessary set up control systems.

Applications

Welding with coated electrodes remains the most widely used of welding processes. It is the most versatile, since it can be used in all positions, in most confined spaces, and for the great majority of welding applications, with relatively simple equipment. It is not applicable to metals such as titanium but such limitations are confined to a small minority of joining requirements. Its major disadvantage is that of being a manual and discontinuous process with limited prospects for improvement in productivity, so as noted earlier, there is a trend towards more continuous manual techniques such as GMA welding, and towards automatic and robotic operation. Nevertheless in many areas the MMA process holds its own. Most of the world's gas and oil pipelines are still welded with coated electrodes, in spite of the fact that automatic equipment is available for this purpose.

The problems associated with hydrogen in welded joints and the means of avoiding such problems will be discussed in the sections concerned with carbon and low alloy steels. It is worth noting at this point that low hydrogen weld deposits are not exclusively associated with basic coatings. By eliminating the organic content in the coating formulation it is possible to produce a rutile coating which can be baked at high temperature and thus be of the low hydrogen type.

Rutile and cellulosic electrodes are general purpose rods for carbon steel fabrication whilst basic coated rods are used where there is a requirement for improved notch-ductility in the deposit, or where there is a known risk of hydrogen-induced cracking (except for stovepipe welding of girth joints in pipelines, where cellulosic electrodes are used even for high tensile steel). Rutile coatings are favoured in Europe (including the UK) whilst in the USA, and for work engineered by US companies, cellulosic rods are usually specified. It is sometimes thought that because basic electrodes are specified for some quality work they are invariably superior to other types. This is not the case;

for example the root pass of a pipe weld made with basic electrodes usually contains porosity whilst for work where no special mechanical properties are required, rutile-coated rods give a much better weld profile. For MMA welding operations, the best electrode is the one most suited to the job.

Gas metal arc (GMA) welding

General

GMA welding was originally developed in the early 1950s by the Air Reduction Company and marketed under the trade name 'Sigma'. This process used a continuously fed solid wire with an inert gas shield, and later became known as metal inert gas (MIG) welding. With the introduction of CO_2 and mixed gas shields it acquired the name gas metal arc (GMA) or MIG/MAG, where MAG stands for metal active gas. The term GMA will be used here.

Solid wires with continuous current

At low electrode current densities in argon with electrode positive the arc melts the wire progressively until the drop so formed falls under its own weight. Increasing the current density results in decreasing drop size, and above a critical value the drop no longer falls but is projected across the arc. With 1.2 mm diameter steel wire the critical current for such free flight metal transfer is 200–250 A, and the weld pool so formed is too large for positional welding or for controlled penetration in a root pass. In the case of aluminium the critical current is lower – 135 A for a 1.6 mm diameter wire – and positional welding is practicable.

The other problem with carbon and alloy steel is that when using an argon shield the cathode, which forms on the plate, wanders over too wide an area and causes the arc likewise to wander. Addition of 2% oxygen to the argon thickens the oxide layer on the plate surface and restricts the cathode to an acceptable degree. Argon/CO_2 mixtures may be used for the same purpose.

With such mixed gas shields welding carbon, low alloy and stainless steel is possible in the downhand position, with the advantage that the wire feed is continuous, there is little spatter and no slag.

One of the features of GMA welding using a conventional type of transformer/rectifier power source is the self-adjusting property of the arc. Suppose that due to a movement of the gun the arc length is shortened, then the arc voltage decreases, the current increases and so does the burnoff rate. Thus the arc returns to its original length. The opposite effect occurs if the gun is moved away from the workpiece. Thus, once the arc voltage is set on the power source, the arc length will remain more or less constant independent of involuntary movements of the gun relative to the weld pool. This self-adjusting tendency is reinforced by using a power source with a flat characteristic, thereby obtaining the maximum current change for any momentary change of voltage.

Such considerations do not necessarily apply to electronically controlled power sources, with which it is possible, for example, to maintain a constant current independent of the arc voltage. Such sources may nevertheless be designed to produce volt-ampere characteristics that mimic those of conventional machines and can be switched from a drooping characteristic for coated electrodes to a flat characteristic for GMA welding.

When welding steel any gross displacement of the gun will result in a change in the stick-out length and this will also change the arc current due to an alteration of the resistance of the welding circuit. One advantage of open arc welding is that the arc current may be less sensitive to such changes, and the process is therefore more adaptable to positional work than welding with solid wire.

Equipment

Figure 1.18 shows the general layout for GMA welding processes. The gas is usually bottled; exceptionally it may be economic to install a bulk supply which is piped around the workshop. From a pressure let-down valve it passes through a flowmeter to control valves, usually located in the wire feed machine, and thence to the torch. Wire is normally pushed through a guide tube to the gun, but if the leads are exceptionally long a push-pull system may be used, with pneumatically driven or electronically driven rolls on the gun itself. The power source may range from a flat characteristic transformer-rectifier to an electronically controlled inverter type. Some power sources incorporate a circulation system for water cooling, complete with filter.

Controls usually include a start-up arrangement which feeds wire slowly until it touches the workpiece, provides a high current when the short circuit occurs, followed by a modulation of the current until steady conditions are established. Likewise there may be a finishing sequence which (for solid wire) ensures that the last drop is detached, the wire does not freeze in the puddle and the crater is filled. The operation of the gas valves is synchronised with such programmes.

Welding guns are either air-cooled or water-cooled. In either case one of their most important components is the contact tube (Fig. 1.19). The welding current is picked up by gliding contact between wire and the hardened copper tube, the tip being replaceable because it can be damaged by spatter. Natural curvature of the wire normally ensures good current pick-up but some manufacturers supply a spirally formed tube to make sure. An important variable in GMA welding of steel is the stick-out; that is the distance between the point of current pick-up

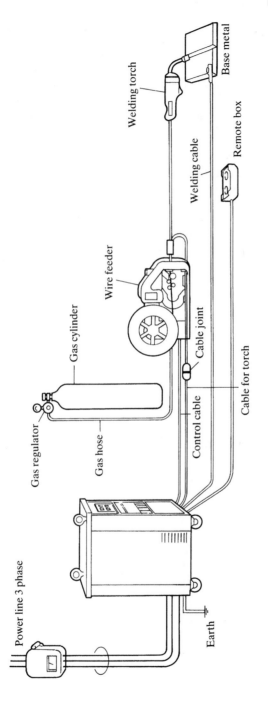

1.18 Typical GMA welding layout.

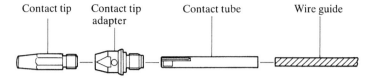

1.19 Typical contact tube arrangement for a GMA welding torch.

and wire tip. This portion of the wire experiences a significant amount of Joule heating which in turn can influence the burnoff rate. Stick-out is one of the variables to be controlled in the welding procedure.

Shielding gas
Pure argon is used for AC GMA welding of aluminium since the presence of any oxidising gas would cause unacceptable oxidation; at the same time, the degree of cathode wander with argon shielding is not excessive. Argon-helium mixtures are also used where it is required to obtain a higher heat input rate, and pure helium has been employed in DC electrode negative automatic welding.

For steel the required oxidising additions to argon are O_2 or CO_2 or both. Oxygen alone has the effect of constricting the arc and making the process less tolerant of poor fit-up. For general purposes an argon $5CO_2$ $2O_2$ may be used for thin material and argon $15CO_2$ $2O_2$ for thicker sections. Argon-helium mixtures with CO_2 additions can be useful in welding austenitic stainless steel to produce a flatter weld bead than is possible with argon based mixtures. Helium based mixtures are also used for some purposes.

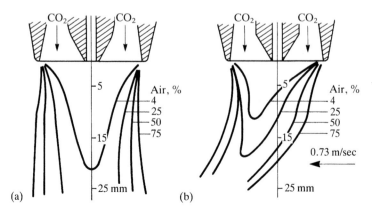

1.20 Contamination profiles of CO_2 shielding gas flow of 15 litre/mm (about 0.8 m/s at the tip of the nozzle): a) No wind; b) Side wind of 0.73 m/s.[15]

Table 1.3 GMA welding costs in DM/m for Al 3 Mg with various shielding gases[14]

	Shielding gas		
	Argon	75 Ar 25 He	50 Ar 50 He
Gas cost	0.52	0.94	1.04
Wire cost	2.74	2.48	1.98
Labour and overheads	1.75	1.46	1.12
Total costs	5.01	4.88	4.14

Pure CO_2 is still employed as a shield for short circuiting welding of carbon steel in relatively thin sections, and recent developments in power source technology have made possible a reduction of the spatter normally associated with CO_2 welding.

It is worth noting that the cost of gas is not a major proportion of the total cost of making a weld. Therefore a relatively small improvement in productivity can pay for a significant increase in gas costs. Table 1.3 shows the advantage gained in GMA welding of aluminium 3% magnesium cylindrical shells by substituting argon-helium mixtures for pure argon. This improvement resulted from the higher welding speed associated with helium mixtures.[14]

Contamination of the shielding gas due to draughts is always a potential hazard. The rate of flow through the nozzle is of the order 1 m/s (about 2 mph) so that very light winds can blow it away. Figure 1.20 shows the degree of contamination of CO_2 downstream of a GMA welding nozzle in still air and with a wind of 0.73 m/s (1.63 mph).[15] GMA welding should be carried out in an enclosed shop or welding booth, and fume extractors must be oriented so as to minimise air flow around the torch itself. Welding guns with integral fume extractor are designed to overcome this problem.

Health hazards
Some GMA welding processes, notably open arc welding, produce large amounts of fume. Open arc or self-shielded welding is unique in that there is no external gas shield and therefore fume extraction systems similar to those used for coated electrodes are applicable. At the other extreme, solid wire welding generates little fume and presents the same problem as GTA welding, namely ozone formation due to ultraviolet radiation (see earlier section on GTA welding).

It has been suggested that adding small amounts (0.03%) of nitric oxide (NO) to the gas can reduce ozone formation by the reaction:

$$NO + O_3 = NO_2 + O_2$$

and gases containing such additions are commercially available. Others, however, claim that NO additions cause a slight increase in ozone formation and that NO_2 is itself an irritant and may be toxic. Ozone concentrations are normally in the hazardous range close to the arc and not in the vicinity of the welder, but this may not always be the case. The best solution to this problem is to ensure adequate ventilation at all times, with fume extractors for automatic welding where an accumulation of ozone is possible.

Solid wire with short circuiting

One way to overcome the problem of excessive current and arc force is to employ a short circuiting mode of metal transfer. In the early days of GMA welding there was also another incentive for such a development, namely the possibility of using the relatively cheap CO_2 as a shielding gas. This is practicable provided the wire contains strong deoxidants, but the drop at the electrode tip is repelled and detached in an erratic manner. Such irregular behaviour is brought under control by short circuiting. Early equipment used a choke in series with a normal transformer-rectifier set to limit the short circuit current and its rate of rise, and used the arc voltage as a means of controlling the short circuit frequency. Amongst other applications short circuiting GMA proved to be especially suitable for pipe welding, because it was possible to make a controlled penetration root pass and then weld out with the same equipment. The process suffers however from a tendency to lack of sidewall fusion, which is a particularly serious defect because it is difficult to detect by radiography. CO_2 welding also generates quantities of spatter, some of which sticks to the nozzle of the welding gun so that the operation must be interrupted periodically to remove it, whilst the rest sticks to the workpiece as well as to jigs and fixtures, requiring subsequent cleaning operations.

Short circuiting GMA welding is a cyclic process operating in the range 20–250 Hz. The weld pool surface oscillates, and during its upward movement collides with the electrode, causing a short circuit. It then moves downwards and, when the connection with the electrode breaks and the arc is reignited, a miniature explosion occurs. This drives the centre of the weld pool downwards and maintains the oscillation.

The explosion is almost certainly an electrical explosion of the liquid metal bridge between the electrode and the descending surface of the weld pool. Such explosions occur when liquid metal is superheated by Joule heating to a temperature where it contains sufficient energy for instantaneous vaporisation. This results in the formation of a small volume of metal vapour at very high pressure, generating a shock wave which decays rapidly into a sharp-fronted burst of sound. At the same time it exerts a force in the weld pool, the strength of which is proportional to the energy of the explosion.

Various modes of oscillation are possible in weld pools. Figure 1.21 illustrates some possibilities for the partial penetration case.[16] Two of these are axially symmetrical, the first and second harmonic respectively. The other two are asymmetric with a diametral mode; these are more likely to occur in elongated weld pools where the bulk of the pool trails behind the electrode. In short circuiting welding the oscillation is frequently irregular, but the axisymmetric mode of Fig. 1.21 has been observed and when the oscillation is violent, spatter detaches from the outer part of the pool as shown in Fig. 1.22.[17] One means of minimising this and other forms of spatter is to reduce the force of the explosion by minimising the current that flows immediately prior to rupture of the liquid metal bridge. With a conventional power source the short circuit current is limited by means of a series choke but it still may be in the region of 300–400 A. An alternative means of current limitation is to use an electronically controlled pulsed source and to set the pulse variables so that both short circuit and arc reignition occur during the low current phase of the cycle. This is possible using an inverter power source that generates a square wave pulse (Fig. 1.23).

A second way to minimise spatter is to control the frequency of short circuiting. With a conventional power source it is possible to obtain a

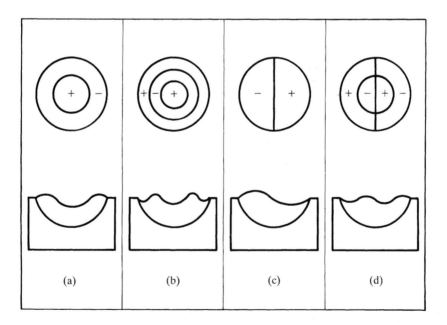

1.21 Possible oscillation modes for a partial penetration weld pool: a) Axi-symmetric, first harmonic; b) Axi-symmetric, second harmonic; c) Asymmetric, fundamental mode; d) Asymmetric, first harmonic.[16]

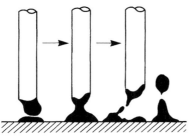

1.22 Spatter in short circuiting GMA welding due to excessive amplitude of oscillation of the weld pool.[17]

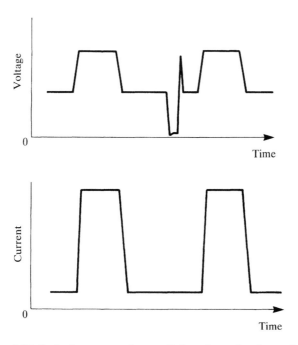

1.23 Reducing spatter by confining short circuits to the low current phase of a pulsed current source: form of typical oscillogram (source: Daihen Corporation).[4]

condition where the current and voltage supplied by the source oscillate at the same frequency as the natural frequency of oscillation of the weld pool, giving a steady condition with minimum spatter (Fig. 1.24).[18] This is only possible for one current setting however. Using an inverter source no such limitation exists, and with a square wave pulse the pulse variables may be programmed such that for any given combination of current and voltage the amount of spatter is minimal.

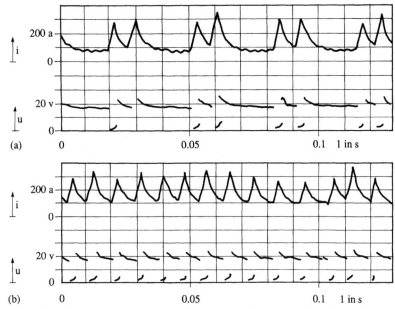

1.24 Voltage and current oscillograms during short circuiting GMA welding with a conventional power source:[18] a) Mean current 120 A, not sufficient to maintain steady conditions: spatter 6.3%; b) Mean current 180 A, steady oscillation: spatter 4.5%.

Ideally, such controls permit the transfer of one drop of metal per pulse, and minimise the amplitude of oscillation of the weld pool.

Short circuiting GMA welding is used primarily for carbon steel, particularly for thin sections, and for pipe welding and positional welding with a CO_2 or mixed gas shield. In high speed (25–30 mm/s) GMA welding, it also becomes necessary to reduce the voltage and use short circuiting metal transfer.

The problem of lack of sidewall fusion limits the use of the process to thinner sections, say up to 18 mm. Spatter is also a limitation, particularly for applications where the final appearance is important and spatter must be removed by costly grinding operations. This defect can be minimised but not eliminated by the use of special power sources, as discussed above, and by using an argon-CO_2 mixture for the shielding gas, keeping the CO_2 content of the mixture as low as practicable.

Solid wire pulsing

It was realised at an early stage of the development of GMA welding that its range of application could be widened by applying a high current pulse to a low background current, such that projected metal transfer could be obtained at a lower average current than that required for constant current operation. The background current melts the wire and forms a drop, whilst the high current pulse causes the drop to elongate and gives it an acceleration in the direction of the weld pool. The degree of elongation increases with the pulse current and with reduced surface tension, and since this is a dynamic condition, it will be affected to some degree by density and viscosity.

In the early stages of development it was (as noted earlier) only possible to generate pulses at mains frequency or multiples thereof, whereas an infinitely variable range of frequency was required. The advent of electronically switched power sources provided this facility and experimental work with such sources established that optimum welding conditions were obtained when one drop was detached per current pulse. The pulse variables are illustrated in Fig. 1.25.[3] These are pulse amplitude I_p and frequency F, pulse duration T_p and base current I_b. An independent variable is the wire feed speed. Investigations have established the relationship between these variables which will produce one drop per pulse for a range of wire diameter, materials and shielding gases. Electronically controlled power sources (both inverter and mains frequency transformer types) may be programmed so that if one variable is set (normally the wire feed speed) the other four are automatically adjusted for one-drop-per-pulse operation. Such machines incorporate settings for wire diameter and type as well as shielding gas.

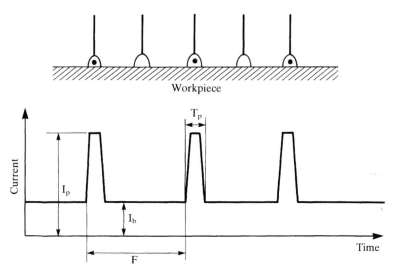

1.25 Ideal conditions for pulsed GMA welding; one drop detaches at each pulse.[3] The lower figure designates pulse variables: I_p = pulse current, T_p = pulse duration, I_b = base current and F = frequency.

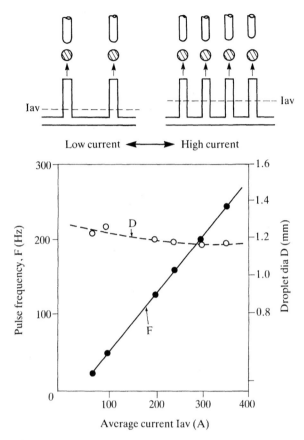

1.26 Pulse frequency and droplet dia (wire: mild steel 1.2 mm dia gas: 20% CO_2 + Ar).[5]

The equations that relate the pulse variables have, in Great Britain, been called 'synergic algorithms' whilst the power sources in question are said to have 'synergic control'. Another term, which is applicable also to other electronically controlled systems such as short circuiting welding is 'one knob control'. Figure 1.26 shows the relationship between mean current and pulse frequency for 1.2 mm diameter carbon steel wire. With a square wave pulse it is possible to ensure that drop detachment takes place during the base current period. This minimises the risk of spatter associated with drop detachment (Fig. 1.27).[5]

One of the applications of pulsed GMA welding is in high speed welding of chassis parts in the automotive industry. In order to avoid humping or undercut it is necessary to use a short arc, and there is a risk of short circuiting, and consequently an unacceptable amount of

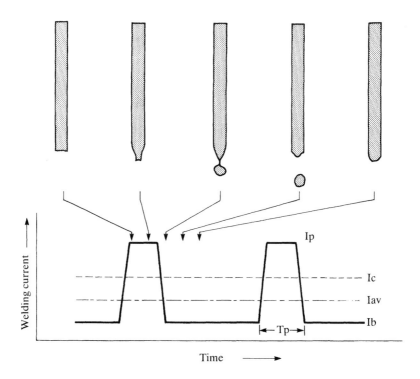

1.27 Drop detachment during the low current period in pulsed GMA welding.[5]

spatter. One way of overcoming this problem is to reduce the size of the transferring drop. Two methods of doing this have been tested. The first is to increase the electromagnetic force by using a square wave pulse, which gives a higher peak current for any given average current than a trapezoidal pulse, as shown in Fig. 1.28. The other is to modify the wire composition so as to decrease the viscosity of the liquid metal. In this way the deformability of the drop is increased and it is possible to achieve one drop per pulse with a shorter pulse duration and hence obtain a smaller drop.[19]

Pulsed GMA welding is characterised by low spatter loss and improved productivity as compared with MMA welding. It is applicable to metals other than the low melting or reactive metals, and is most generally used for carbon steel, stainless steel and aluminium. It is particularly useful for sheet metal work in the automotive industries but also has a wide variety of applications. The shielding gases are as described for GMA welding generally but it cannot be used with a pure CO_2 shield.

Cored wires

Stick electrodes in the form of a bare wire with a flux core were first produced by F Rapatz and others in 1927 and were used, amongst other things, for welding submarine hulls. The weld metal suffered embrittlement due to nitrogen pick-up such that its impact strength was in the region of 2 J. Nevertheless welds made with cored rods performed well and Stular records the case of a submarine hull that was severely damaged by collision but where none of the cored wire welds failed.[20] Production of this type of electrode continued at least into the 1960s. In 1958 the Lincoln Electric Company introduced the Innershield process. This is an automatic or semi-automatic process employing a continuous cored wire with no gas shield. The core contains titanium or aluminium to counter the embrittling effect of nitrogen. Flux-cored wires were developed for use in GMA welding at a somewhat earlier date, whilst at a later stage metal-cored wires appeared.

Cored wire is made by metering the required mixture of flux and ferro-alloys into a grooved strip of metal, then folding the strip to enclose the flux, and finally drawing or rolling to the required diameter and reeling on to spools. It is possible to incorporate a baking operation

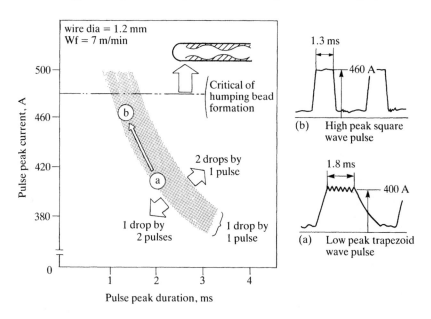

1.28 Reducing the pulse peak duration and drop size for high speed pulsed GMA welding of steel.[19]

for low hydrogen content of the weld deposit. Finished diameters are typically 1.2 mm to 2.4 mm, larger than solid wire. GMA equipment is normally sized to accommodate such thicker wire.

Advantages over solid wire include the ability to incorporate alloying elements and, for some wires, the presence of a flux to aid wetting and penetration and to provide support in positional welding.

Gas-shielded flux-cored wires

These are available with either rutile or basic flux cores. The rutile-cored types are formulated and manufactured so that the weld deposit has low hydrogen content, typically below 5 ml/100g. With rutile type fluxes, metal transfer is in the form of fine drops that are projected across the arc, so that positional welding is possible in the free flight mode. Such wires provide a sound technical alternative to stick electrodes for low hydrogen welding higher-strength steel combined with the advantage of improved productivity. Good impact properties at subzero temperatures can also be achieved.

With a basic flux core the metal transfer is in the form of coarse drops and positional welding is only possible using a short circuiting or pulsed arc technique. For both types of flux the gun is pointed backwards along the joint to prevent flux being trapped in the weld pool, in much the same way as welding with coated electrodes.

Flux-cored wires are also available for low alloy, stainless steel and for hardfacing compositions. In most cases the shielding gas is an argon/CO_2 mixture but straight CO_2 shielding is applicable in some cases. Most flux-cored wires are operated with electrode positive, but some wires can be used with electrode negative. Cored wires are developed to run with the polarity that gives the smoothest transfer. With some wires it is positive, with others negative. A few wires can run with either.

Flux-cored wires are mainly used for fabricating carbon and carbon-manganese steels, particularly for piping and steel structures. The rutile-cored type has been particularly successful because it can be operated in all positions and because the impact and other properties of the weld metal are satisfactory. A typical procedure for welding a heavy-wall pipe would use a metal-cored wire for the root pass, rutile-cored stringer beads for the filler passes, and a full width weave, also rutile-cored, for the capping pass.

Metal-cored wires

These wires contain an iron or alloy powder core with minor additions to promote arc stability but no flux. They are therefore suitable for applications where flux residues are undesirable and may be used for example to provide a higher productivity substitute for GTA welding. Metal-cored wires are not suitable for positional welding in the free flight mode and are usually employed in the flat position, e.g. in robotic welding. A combination of pulsed operation with metal-cored wire has been used to replace a GTA root pass for carbon-manganese pipe and pressure vessel welds. Figure 1.29 illustrates the procedure. The torch position is such as to apply an upward force to the weld pool and make possible a relatively wide and thick root pass. Filler passes are then made either by metal-cored GMA welding or by submerged-arc welding.[21]

Metal-cored wires have the advantage of higher deposition rate, very little spatter, suitability for pulsed current, no slag, low hydrogen (below 5 ml/100g) and satisfactory impact properties. There is little fume but this implies the disadvantage of ozone generation. The shielding gas is an argon-rich CO_2 or 1–5% O_2 mixture.

Self-shielded flux-cored wires
As indicated earlier, the problem that had to be solved with self-shielded or open arc welding is potential porosity and embrittlement of the fused zone due to nitrogen and further embrittlement due to oxygen pick-up. Such embrittlement has been avoided by adding a sufficient amount of deoxidants and nitride-formers to combine with these contaminants. Titanium and silicon have been used for this purpose but the amounts required lead to reduced ductility and their use is

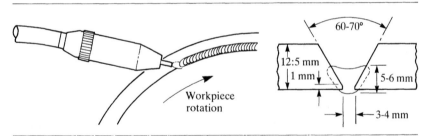

Procedure for root pass welding in C-Mn steel applying to all thicknesses	
Material	: Steel, H11, all thicknesses
Welding wire	: PZ6103, 1.2 mm∅
Shielding gas	: 98/2, Ar+O_2
Welding current	: Pulse, 360 A, 2 m sec Base, 50 A, synergic regulated frequency (about 4.7 m sec)
Average welding current and arc voltage	: 95 A, 18 V
Wire speed	: 3.9 m/min
Welding speed	: 0.20 m/min, workpiece rotation

For larger circumference vessels the average welding current/wire feed rate can be slightly increased to give up to 0.25 m/min, welding speed.

1.29 Procedure for metal-cored GMA root pass welding of a circumferential joint in a pressure vessel.[21]

Table 1.4 Typical chemical composition of weld deposits from various grades of Lincoln Innershield wires[22]

Application	Grade	AWS class	Composition, %						
			C	Mn	Si	P	S	Al	Other
High speed welding of sheet steel	NR1	E70T-3	0.23	1.23	0.5	0.006	0.016	0.076	Ti 0.13
General downhand welding of mild steel	NR202	E71T-7	0.12	0.67	0.2	0.008	0.009	1.09	–
Downhand welding of steel with impact strength 27 J at −29 °C	NR302	E70T-6	0.11	1.75	0.14	0.006	0.009	0.89	–
All-position welding of steel with impact strength of 27 J at −29 °C	NR203 NiC	E61T8-K6	0.06	0.72	0.03	0.004	0.002	0.85	Ni 0.52

restricted to single pass high speed welds in mild steel sheet. For the majority of applications the main deoxidiser and nitride-former is aluminium. In steel, aluminium has the effect of closing the gamma loop of the iron-carbon equilibrium diagram, such that when more than 1.2% is present there is no transformation. The weld metal consists of coarse ferrite grains and is unacceptably brittle. This problem is overcome by adding austenite stabilisers (carbon and manganese) and by restricting the aluminium content. The toughness of the deposit may be enhanced by the addition of nickel. Subject to such controls aluminium can have a beneficial effect in reducing sulphur content (and hence improving resistance to hot-cracking), eliminating any tendency to strain-age embrittlement, and increasing the corrosion resistance, so that for example in sea water exposure the parent metal corrodes preferentially rather than the weld metal.[22] Some typical chemical compositions of weld metals produced by Lincoln self-shielded wires are shown in Table 1.4.

Some degree of vapour shielding of the arc is obtained by adding magnesium and lithium to the core mixture, and this makes it possible to reduce the aluminium content of the deposit. Barium fluoride may also be substituted for part of the calcium fluoride element of the flux, partly to enhance the vapour shield, and partly to aid the formation of a quick freezing slag which helps in positional welding.

The voltage, and hence the arc length, is controlled at the power source. The welding current and voltage drop across the arc may then be modified by the welder through altering the torch-to-workpiece distance, which changes the stick-out length. With self-shielded wires this may be used for controlling penetration in one sided and positional work.

The advantage of self-shielded welding is that the need for a gas supply is eliminated and the welding gun and leads are simplified. Within reasonable limits it is not adversely affected by wind. This makes the process especially useful for site fabrication of structural steelwork. It is also used in shop fabrication, notably for oil rig construction. In the workshop, fume extraction is required, particularly for wires containing barium, but this does not present significant problems because there is no shielding gas to disturb. Disadvantages of the process are, firstly, that the high aluminium content of the deposit makes it unsuitable for mixing with other processes. Thus it cannot be used for root passes with high basicity submerged-arc weldout. Secondly, although hydrogen contents of 5 ml/100g weld metal can be achieved, it is difficult to guarantee such a level. Therefore it is more realistic to assume 10 ml/100g hydrogen, and for some joints preheat may be required.

Submerged-arc welding (SAW)

The submerged-arc process was developed during the 1930s in the USA as (primarily) an automatic technique for joining thick sections of carbon and alloy steel. Manually operated versions have been introduced and have a limited application. Likewise the process has been applied to non-ferrous metals but these are rarely used in the thicknesses where SAW becomes economic.

The essential features of SAW are illustrated in Fig. 1.30. The arc is struck between a continuously fed bare electrode and the work, and the whole is covered with a granular flux fed through a nozzle ahead of the electrode. Part of the flux melts and forms a liquid bubble around the arc which periodically expands, bursts and then reforms. Metal transfer is somewhat irregular but is confined within the flux bubble so that there is no spatter. A cavity forms in the forward part of the weld pool and this generates a strong rearward flow such that the pool is significantly more elongated than would be calculated assuming isotropic heat flow. Unmelted flux may be removed by a vacuum device for reuse, whilst the melted portion is usually easy to detach at a convenient point behind the welding head.

By its nature SAW may only be operated in the downhand or horizontal/vertical positions. Welding currents may be as low as 200 A but are usually within the range 500–1200 A. The large weld pools produced by high currents solidify with coarse-grain and a correspondingly poor notch-ductility, so that at one time it was customary to specify a limit on the weld thickness per pass of say ¾ inch. To keep

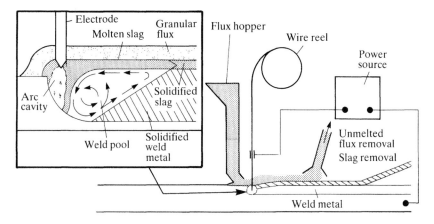

1.30 Typical submerged-arc welding layout.

the weld pool size down, whilst obtaining high deposition rates, two, three or four heads may be operated in tandem. For single head operation the current is normally DC electrode positive. Where tandem heads are used with DC for both electrodes, the arc may be deflected by the magnetic field of its neighbour. This problem is avoided by applying DC to the leading head and AC to the others.

The first submerged-arc fluxes, patented by the Union Carbide Company, were of the calcium silicate type and were made by fusing the constituents and then producing a powder either by crushing or by water quenching. An alternative type of flux (not covered by the Union Carbide patents) was later produced by the Lincoln Electric Company by blending powdered raw materials with a bonding agent to make an 'agglomerated' powder. Fused flux has the advantage of being free from moisture, but the agglomeration process is more flexible and permits a much wider range of additions. American flux formulations are classified into three types: neutral, active and alloy. Neutral fluxes do not contain any metallic elements, and although the slag they produce reacts with the weld metal, the weld metal composition is not much affected by changes in the arc voltage. Active fluxes on the other hand contain manganese and silicon. These deoxidants reduce the risk of porosity but composition of the deposit may differ with arc length and the voltage must be tightly controlled. Alloy fluxes contain ferro-alloys to produce a low alloy weld deposit. This is cheaper than using a low alloy electrode, but again the composition may be affected by welding variables. Use of such fluxes has in the past resulted in some weld failures, and a safer alternative is to use a metal-cored electrode to obtain the required composition.

Requirement for higher impact strength in carbon steel weld deposits led, in Europe, to the production of agglomerated fluxes based on calcium carbonate rather than manganese silicate. Chemically these fluxes are basic so that a second classification arose in which manganese silicate types are acid, calcium carbonate types are basic and intermediate types may be neutral. The chemistry of slag metal reactions will be discussed in Chapter 2. It should be noted here, however, that basic submerged-arc fluxes are, like basic electrode coatings, hygroscopic and therefore require special care in storage and may need to be dried prior to welding.

Apart from the use of twin heads the deposition rate of SAW may be increased by metal powder injection. The powder is delivered to the weld pool from a hopper via a delivery tube and a dosage regulator. In this way deposition rates per ampere may be increased by up to 100%, whilst the properties of the weld metal and in particular the impact strength are not affected. Ferro-alloys may also be introduced in this way but the use of metal-cored wire is a safer method.

Productivity may also be increased by feeding a filler wire into the weld pool behind the arc, as in hot-wire GTA welding. The weld is supplied with AC current and Joule heating brings it close to the melting point as it enters the pool. Close control over wire speed, wire current and stick-out are required to operate this process successfully and the necessary control systems are available.

Submerged-arc welding is the work-horse of heavy metal fabrication. It is capable of producing defect-free welds in carbon, low alloy or stainless steel plate of any thickness, and it is much used for surfacing (see surfacing later in this chapter). It is also employed for making longitudinal welds in pipe and for spirally welded pipe. As applied to pressure vessel manufacture it is normal practice to use a J preparation for the joint and for thick material a very large number of passes may be required. In such applications narrow gap welding (to be discussed in the next section) may be attractive. At the other end of the scale SAW is widely used for sheet metal because it is capable of high travel speeds, low heat input and no spatter.

Power sources for submerged-arc welding (and for other automatic applications) are designed for continuous operation and for high current output. DC units are thyristor controlled transformer-rectifiers with constant voltage or constant current output. There is no need for the sophisticated control systems employed for some GMA power sources but thyristor control makes possible the close regulation of operating voltage. AC machines are transformers with a moving core for current control and a drooping or constant potential characteristic.

There is very little fume and almost no ultraviolet light emission in the submerged-arc process, so that it rarely presents any danger to human health.

Narrow gap welding

One means of improving productivity in welding thick sections is to reduce the volume of weld metal required to make the joint. Figure 1.31 shows the weight of weld metal per metre for a variety of preparations as a function of plate thickness.[23] For a J preparation, reducing the taper from 8° to 1° gives a significant advantage for thick plate. Problems that have to be overcome are lack of sidewall fusion, obtaining good

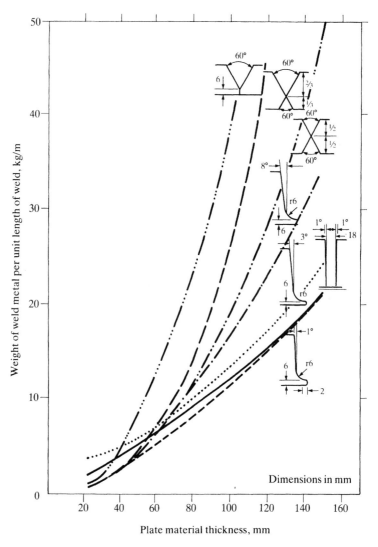

1.31 Mass of weld deposit required per metre of weld per various joint configurations.[23]

(a)

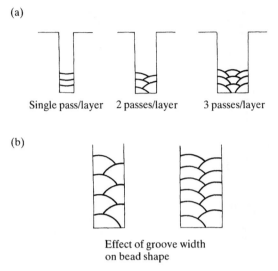

Single pass/layer   2 passes/layer   3 passes/layer

(b)

Effect of groove width
on bead shape

1.32 Various procedures for narrow gap welding.[23]

flux detachment and accurate tracking of the joint. For the circumferential joint of a pressure vessel rotating on rolls, for example, some system of correcting axial drift must be employed.

Both GMA and GTA welding have been proposed for use in narrow gaps. For GMA welding, in particular, many ingenious techniques have been developed for directing the arc alternately to left and right to ensure fusion. Details of such techniques will be found in Ref. 23. GMA narrow gap welding does not appear to have been used to any significant extent in production work except possibly in Japan, but the GTA process has found such application in welding HK40 (high carbon 25Cr20Ni) reformer furnace tubes.

Narrow gap submerged-arc welding has been applied to heavy-wall pressure vessels. With a gap of up to 14 mm it is possible to make the joint in a single pass per layer (Fig. 1.32) but the risk of defects is reduced by making two passes per layer, and this procedure has been employed successfully in practice. Both longitudinal and circumferential seams may be welded in this way. The metallurgical quality of the joint is not significantly different from that of a normal submerged-arc weld. Possibly because of the exacting controls that are necessary to make a successful narrow gap weld, the incidence of defects is very low (see also Chapter 5).

Electroslag welding (ESW)

The electroslag process was invented by H K Hopkins of the M W Kellogg Company in 1936. At that time stainless clad plate was being

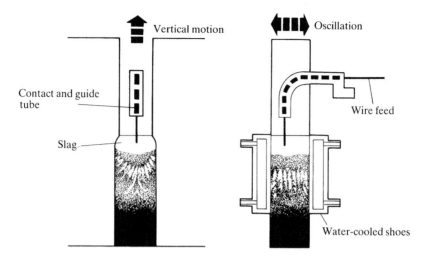

(a)

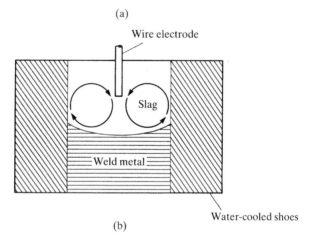

(b)

1.33 The electroslag welding process:[24] a) General arrangement; b) Circulation of slag.

produced by making a submerged-arc weld deposit of the required composition on one side of a slab and then rolling out into plate. The system worked well for 12Cr cladding but with 18Cr8Ni types the weld deposit cracked. Hopkins therefore stood the slab on end and applied the weld metal by traversing to and fro and at the same time moving upwards. The outer edge of the weld pool was restrained by a dam. In the USA this process was later developed for production of high quality alloy ingots, but not for welding. The welding application was due to the Paton Institute in Kiev.

The principle of the welding process is shown in Fig. 1.33. The joint preparation is simply the gap (of 25–30 mm) between two square cut edges of plate. Wire is fed vertically downwards into a pool of electrically conducting liquid slag which rests on top of the weld pool. The liquid metal and slag are held in place by water-cooled dams and the whole assembly moves upwards as the weld is made. The slag is maintained at a temperature of about 2000 °C by resistance heating and this in turn melts the wire and maintains the pool of liquid metal.[24] Electromagnetic forces cause circulation in the slag pool and this melts the edges of the plate, giving a weld that is wider than the original gap and ensuring good fusion. As well as moving upwards the wire feed traverses from side to side of the joint with a dwell at each end. Even so the plan section of the fusion zone is somewhat barrel-shaped.

Wire and flux combinations are similar to those used in submerged-arc welding except that the flux contains a higher percentage of calcium fluoride in order to increase electrical conductivity and reduce viscosity.

The power source is an AC transformer with a flat characteristic rated for continuous operation at 300–600 A with a voltage of about 50 V. Sensors are built into the arrangement such that the rate of upward traverse matches the rate of rise of the slag pool.

The weld run is started on a block of metal fixed below the joint. An arc is struck and flux is fed in until the flux pool is established and the arc extinguished. A similar method must be used to restart the weld after an interruption, and it is difficult to do this without producing a defect.

Because of the large weld pool and the low rate of traverse the thermal cycle of an electroslag weld is relatively protracted as shown in Fig. 1.34. Cooling rates are relatively low and the process could be applied to hardenable steels that would not normally be considered weldable. At the same time there is excessive grain growth in the heat affected zone and the weld metal may be coarse-grained. For application requiring notch-ductility therefore the joint must be normalised after welding. In the case of pressure vessels this limits the use of the process to longitudinal welds of individual strakes.

One variant of electroslag welding is the consumable guide process. A set of parallel wires are fed down tubes in a fixed steel frame. As the weld pool rises up the joint the guide melts and is eventually consumed.

Other vertical-up techniques that require dams and which produce a more or less coarse-grained weld are electrogas welding and the Lincoln consumable guide method. Both these are arc welding processes. Electrogas welding employs a flux-cored wire and either CO_2 or argon-rich gas shielding. In the Lincoln process a self-shielded flux-cored wire (but without aluminium addition) is fed down a consumable tube. In both cases there is provision for draining off excessive amounts of flux. The vertical-up arc welding processes are suitable for a thickness range of 18–100 mm in carbon and carbon-manganese steel, and the

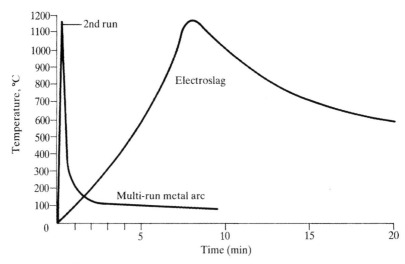

1.34 Thermal cycle in the heat affected zone of an electroslag weld compared with that due to a multipass weld made with coated electrodes.[24]

impact properties are better than those obtained with the electroslag process.

Single pass vertical-up welding processes are best suited to joining heavy steel sections where notch-ductility requirements are modest or non-existent so that a post-weld normalising heat treatment is not required. They have the advantage that the set-up is simple and they can be used for either butt or T joints.

High power density welding

The processes in question are those that develop a beam with a power density of 1×10^{10} W/m² or more, capable of vaporising metals and of maintaining a deep, small diameter keyhole type weld pool; namely electron beam and laser welding. When these processes were first introduced their potential for single pass welding of thick plate was recognised and much development work has been, and is still being done to achieve this end. In practice most welding applications for laser beams are for thin sheet and microjoining, whilst electron beam welding of thick sections is mainly confined to easily manipulated machined parts such as gearwheels.

There are two major obstacles to electron beam welding of large pieces of equipment such as pressure vessels. The first is that of providing a vacuum, or near vacuum, either by means of a chamber or by an enclosed volume local to the electron beam and moving with it. Making a large chamber is not difficult but evacuating it is both difficult and time consuming. Evacuating a local enclosure is not difficult but maintaining the vacuum is. The second problem is that of achieving the required accuracy of joint preparation and joint tracking. Any failure here could result in a severe defect, one which would be hard to repair.

For the immediate future, the use of electron beam, or laser welding for large structures is only a remote possibility, but they may well be employed in fabrication of components such as machine parts or thin-wall tubing.

Surfacing

The oil which was pumped from the first wells in Pennsylvania was sweet; that is to say, its sulphur content was less than 0.7% and it did not corrode mild steel at the temperature used for distillation. Later discoveries in west Texas were not so accommodating; the sulphur content was relatively high and steel for distillation towers required a corrosion allowance of an inch or more. A more secure arrangement was to use steel clad with 12Cr alloy for the crude and vacuum stills. Such clad steel was first made by rolling out slabs with a 12Cr weld deposit on one side, as noted earlier. Subsequent developments have required corrosive fluids to be handled at progressively high temperatures and pressures, and the thickness of steel required may be such that roll-cladding is impracticable, so that the protective layer must be applied by fusion welding.

In all arc welding processes there is a more or less vigorous circulation of the weld pool and some degree of melting of the base plate. As a result, there is dilution of the cladding, which may affect its corrosion resistance. Dilution can be minimised by putting down two layers of weld metal, and especially if the first layer is a highly alloyed steel such as 25Cr12Ni. It may also be controlled by adjusting the composition of the deposit upwards so as to compensate for dilution; for example the carbon content of the consumable is kept to low whilst alloying elements may be added via the flux. A more radical solution is to use electroslag cladding. This is compared with submerged-arc cladding in Fig. 1.35. Using ESW the circulation in the weld pool is weak and dilution low. It is still necessary to adjust the deposit composition by additions to the flux, however. Another process that is capable of surfacing with low dilution is soft plasma arc welding. This employs a plasma torch with a nozzle diameter of 4 mm, and is capable of a dilution level of less than 0.1%.

The most important cladding processes are manual welding with coated electrodes, submerged-arc welding with either wire or strip electrodes, electroslag welding with strip electrodes and GTA welding. In

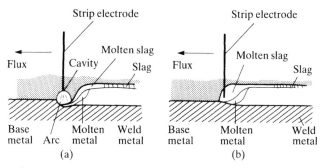

1.35 Submerged-arc strip cladding of steel compared with electroslag cladding: a) SAW process; b) ESW process.[25]

the USA there is some preference for an array of wire electrodes arranged in a wedge shape. European and Japanese practice is to use strip electrodes, usually 50 mm in width. Both techniques result in the formation of a wide weld deposit and it is important that the profile is such that adjacent beads merge smoothly with each other. In strip cladding there is a tendency to undercut due to electromagnetic forces, which act inwards on the liquid metal close to the electrode. This force may be neutralised by means of a permanent magnet arranged as shown in Fig. 1.36.[25]

The electroslag cladding process requires a slag depth greater than 3 mm combined with an operating voltage of about 25 V. At higher voltages an arc forms because the electrode is out of the flux, whilst at low voltages an arc forms due to periodic short circuits.

In internal cladding of pressure vessels, automatic processes are applied as far as practicable and manual welding is used for the less accessible areas, for example around nozzles. Tubesheets are also clad by fusion welding as a general rule, but explosive cladding may be possible. In such cases it is necessary to consider the possibility of brittle fracture due either to low notch-ductility or the presence of notches or both.

Cladding may also be accomplished using one of the variants of GMA welding such as flux-cored arc welding. For high quality work hot-wire TIG welding, whilst not having such a high deposition rate, may still be a good choice because of the low incidence of weld defects.

Solid phase processes

The use of solid phase welding for structures and pressure equipment is almost entirely confined to piping applications. A classic case was the flash butt welding of longitudinal joints in lengths of line pipe, as carried out by the A O Smith Corporation. Pipes were formed and then

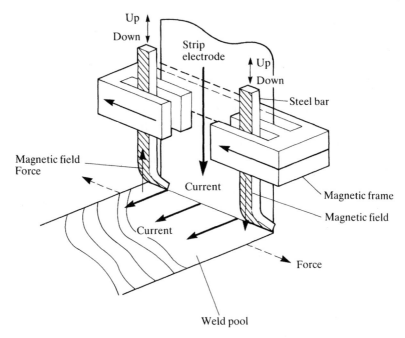

1.36 Electromagnetic control to prevent undercut in SAW strip cladding of steel.[25]

placed in the machine, where the whole weld was flashed and then forged in one operation. After removal of surplus metal the pipes were expanded by operating an internal hydraulic mandrel progressively along the length. The result was to round up the pipe, increase the yield strength and test the weld at the same time. This heroic technique has now been displaced by submerged-arc welding for larger pipe diameters and by electric resistance welding for smaller diameters. Flash welding is still used for welding boiler and superheater tubes. As noted earlier heat exchanger tubesheets are sometimes clad with stainless steel by explosive welding. Most solid phase processes require use of heavy equipment and are therefore employed primarily for shop fabrication.

Electric resistance welded (ERW) tube or pipe is made by forming a round section from strip, and then passing the joint under a pair of high frequency induction coils. The induced electric current flows mainly near the surface so as to heat the skin of the metal. The tube then passes through a pair of forging rolls where surplus metal is extruded and a weld is made. Downstream of these rolls draw knives remove the flash and the joint is reheated by another coil to normalise the weld and heat affected zone. The pipe is then cut into lengths by a flying saw.

Pipe is tested by flattening samples with the weld in the twelve o'clock and three o'clock positions. Normally such welds are ductile but embrittlement can occur, possibly due to the presence of non-metallic films in the joint area.

Explosive welding is a solid phase technique that may be used for cladding heat exchanger tubesheets and for welding tubes to tubesheets (see Chapter 5). In principle, the two metal surfaces to be joined are spaced at a small angle and a charge is exploded close to one surface (that of the thinner material). The surfaces are thus forced together at high velocity. As they collide, a jet of liquid metal is formed; this cleans the two surfaces and allows them to weld together. The welded interface is undulating and liquid metal may be trapped in some of the undulations, forming hard spots. For most applications these hard spots appear to have little effect on the integrity of the joint.

The hazards attending such operations will be self-evident. In addition, the noise associated with cladding means that the work must be carried out at remote locations. Such considerations are likely to limit the use of explosive welding in the future.

REFERENCES

1 Rogers K J and Lockhead J C: 'The use of gas-shielded FCAW for offshore fabrications' *Weld J* 1989 Vol 68 (2) 26–32.

2 Lancaster J F (Ed): 'Physics of Welding'. 2nd Ed, Pergamon, Oxford, 1986.

3 Pixley M: 'Power sources for pulsed MIG welding' *Joining and Materials* 1989 Vol 2 268–271.

4 Yamomoto H, Harada S and Yasuda T: 'The effect of wave shape on spatter generation'. Daihen Corporation Report No R97–T27.

5 Yamomoto H, Nishida Y and Ueguri S: 'Trends and problems of arc welding power source in Japan'. IIW Document XII-960-86.

6 Ushio M, Sadek A A and Matsuda F: 'GTA electrode temperature measurement and its related phenomena'. IIW Document 212–733–89.

7 Seeger G and Tiller W: 'Laser diagnostics on the TIG arc'. in Arc Physics and Weld Pool Behaviour, The Welding Institute, Cambridge, 1979.

8 Yamauchi N and Taka T: 'TIG arc welding with hollow tungsten electrodes'. IIW Document 212–452–79.

9 Tinkler M, Grant I, Mizuno G and Gluck C: 'Welding 340L stainless steel tubing having variable penetration characteristics' in Proc of Conf on the effects of residual, trace and microalloying elements on weldability and weld penetration, The Welding Institute, Cambridge, UK 1983.

10 Nishiguchi K: 'Plasma arc welding and cutting'. in Advanced Joining Technologies, Ed T North, Chapman and Hall, London 1990.

11 Harris I et al: 'Plasma welding and cutting' *Joining and Materials* 1989 Vol 2 326–338.

12 van den Berg R W A and Haverhals J: 'Advantages of high efficiency MMA electrodes' *Metal Construction* 1987 Vol 19 643–644.

13 Becken O: 'Metal transfer from welding electrodes'. IIW Document 212–179–69.

14 Boeme D: 'Welding gases'. IIW Document XII-1197–90.

15 Yeo R B G: 'Cored wires for lower cost welds' *Joining and Materials* 1989 Vol 2 68–73.

16 Xiao Y H and den Ouden G: 'Weld pool oscillation during GTA welding of mild steel'. IIW Document 212-776–90.

17 Yamomoto H, Okazaki K and Harada S: 'The effect of short circuiting current control on the reduction of spatter generation in CO_2 arc welding'. IIW Document 212–649–86.

18 Gupta S R, Gupta P C and Rehfeldt D: 'Process stability and spatter generation during tip transfer in MAG welding' *Welding Review* Nov 1988 232–241.

19 Matsui H and Suzuki H: 'The effects of current wave shape and electrode wire contents on metal transfer in high speed pulsed MAG welding'. IIW Document No XII–1161–90.

20 Stular P: 'Metal transfer with cored electrodes in various shielding atmospheres' in Physics of the Welding Arc, The Institute of Welding, 1966.

21 Sol A M: 'Pulsed MIG metal-cored wire welding replaces TIG root pass' *Joining and Materials* 1989 Vol 2 372–376.

22 Yeo R B G et al: 'Welding with self-shielded flux-cored wire' *Metal Construction* 1986 Vol 18 491–494.

23 Ellis D J: 'Mechanised narrow gap welding of ferritic steel' *Joining and Materials* 1988 Vol 1 80–86.

24 Houldcroft P T: 'Welding Processes'. Cambridge University Press, London, 1967.

25 Anon: 'Overlay welding with strip electrode'. Kobe Steel Ltd Technical Report No. 539.

2   The metallurgical effects of fusion welding

Introduction

Fusion welding subjects the metal to quite a severe thermal treatment, and one of the objects of the welding procedure is to ensure that no significant damage occurs. A strain cycle accompanies the thermal cycle, and it is the combination of embrittlement with tensile strain that represents the most serious hazard. Metallurgical changes take place in the weld pool and in the high, intermediate and low temperature regions of the heat affected zone and it will be convenient to discuss these four regions separately. But first we will consider the nature and effects of the thermal cycle.

The nature and effects of the weld thermal cycle

Thermal effects

It is customary to distinguish between two thermal regimes in fusion welding. The first is three dimensional (3D) and is typified by a weld run on the surface of a thick plate. The second is two dimensional (2D), and this type of heat flow occurs with deep penetration keyhole welds. Multipass welds are usually treated as 3D whilst a single pass weld in thin sheet is 2D. There are intermediate conditions, sometimes known as 2½D.

A quantity that may be used to characterise 3D welds is the heat input rate, q/v, which may be conveniently expressed in kJ/mm. In this expression q is voltage × arc current × arc efficiency, where the efficiency represents the proportion of arc heat that is absorbed by the weld. Another relevant quantity is qv, which in non-dimensional form is known as the operating parameter. At high values of the operating parameter, as found in submerged-arc welds for example, the weld width d is proportional to the square root of the heat input rate

$$d = \text{constant} \times (q/v)^{1/2} \qquad\qquad [2.1]$$

For low values of qv, typical of low current GTA welds, the weld width is simply proportional to the effective arc power q

$$d = \text{constant} \times q \qquad [2.2]$$

The same rules apply to the width of the heat affected zone (HAZ).

These expressions follow from Roberts and Wells equation for the weld width[1]

$$q = {}^{5}\!/_{4}\,\pi d\kappa T_m \left({}^{2}\!/_{5} + \frac{vd}{4\alpha}\right) \qquad [2.3]$$

Where κ = thermal conductivity, α = thermal diffusivity, and T_m is the difference between the melting point of the metal and the initial temperature of the base metal. They are confirmed by the measurements of Christensen et al[2] (Fig. 2.1).

For 2D conditions the corresponding expression is

$$q/w = 8\kappa T_m \left({}^{1}\!/_{5} + \frac{vd}{4\alpha}\right) \qquad [2.4]$$

where w represents plate thickness. In this case for high arc power

$$d = \text{const} \times q/v \qquad [2.5]$$

The weld pool and heat affected zone widths are important in that they affect the grain size. Grain growth is rapid at temperatures near the melting point, and the time spent in the grain growth temperature range is larger for the larger weld pools. For most metals (magnesium is an exception) the grain size in the weld metal is proportional to that of the parent metal at the weld boundary. Thus weld metal solidified from large weld pools tends to be coarse-grained. This is not of much consequence for metals having a face-centred cubic structure, such as austenitic chromium-nickel steel, aluminium and copper, but the notch-ductility of ferritic steels is adversely affected by coarse grain. For such materials it is necessary to control the heat input rate.

The heat input rate also affects the cooling rate of the weldment. In 3D welding the cooling rate at the rear of the weld pool (which is the maximum cooling rate) is

$$dT/dt = 2\pi\kappa T_m^2/(q/v) \qquad [2.6]$$

so that high rates of heat input are associated with slow cooling rates and vice versa. This may be significant in welding of ferritic steel where rapid cooling may increase both the hardness and risk of hydrogen cracking. Thus for the simple case of a multipass weld on thick plate there may be an optimum weld pool size where the best results are obtained. In practice, for welding steel structures, the cooling rate is the dominant consideration.

Equation [2.6] is valid for any point along the weld axis except that

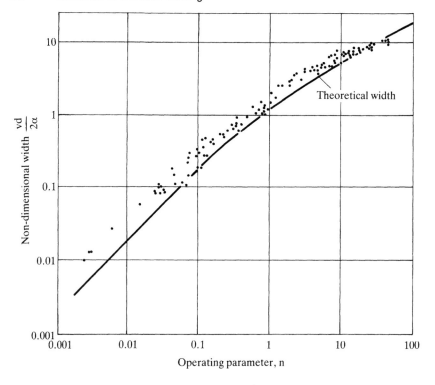

2.1 Measured values of the quantity $\frac{vd}{2d}$ as a function of the operating parameter

$$n = \frac{qv}{4\pi a^2 \rho C_p \ (T_m - T_o)}$$

where q is arc power, v = welding speed, α = thermal diffusivity, ρ = metal density C_p = specific heat, T_m = melting point and T_o = initial plate temperature. [2]

T_m becomes $(T - T_o)$ where T is the temperature at the point in question and T_o is the initial plate temperature. If the plate is preheated, T_o is the preheat temperature T_p and

$$dT/dt = 2\pi\kappa \ (T - T_p)^2/(q/v) \tag{2.7}$$

At high temperatures a preheat in the range of 50–200 °C will, according to [2.7], have a relatively small effect on the cooling rate but at temperatures below 200 °C (which is the hazardous range for hydrogen cracking in steel) the cooling rate will be reduced by the ratio $(T - T_p) \ /(T - T_o)$ and this can be very significant.

The cooling rate for 2D welds is always less than for 3D welds of the same size and for the same welding speed. The ratio of the cooling rates on the weld axis at the rear of the weld pool is, for a line and point heat source respectively

$$\left(\frac{dT}{dt}\right)_{2D} \Big/ \left(\frac{dT}{dt}\right)_{3D} = \frac{vx_1}{2\alpha} \left[\frac{K_1\left(\frac{vx_1}{2\alpha}\right)}{K_0\left(\frac{vx_1}{2\alpha}\right)} - 1 \right]$$ [2.8]

where x_1 is the distance between the heat source and the rear of the weld pool K_0, K_1 are the modified Bessel functions of order zero and one respectively. When $vx_1/2\alpha$ is large the right hand side of [2.8] tends to 0.5, whilst at low values of $\frac{vx_1}{2\alpha}$ it approximates to $- [\ln (\frac{vx_1}{2\alpha})]^{-1}$. In practice the values of $\frac{vx_1}{2\alpha}$ range from 0.01 for low current GTA welds to about 50 for submerged-arc welds and the corresponding 2D/3D cooling rate ratio is 0.2 to 0.5, as plotted in Fig. 2.2. These figures relate to idealised line and point heat sources respectively, but in real welds the same tendency is observed.

Mechanical effects

Distortion
In fusion welding the metal immediately surrounding the weld, being restrained by cooler areas, is first subjected to compressive stress at an elevated temperature. If the metal yields, it will be in a state of residual tensile stress on cooling. The weld metal itself starts to contract as soon as it solidifies, and continues to do so on cooling to room temperature. In thin sheet such contraction results in distortion and buckling. To

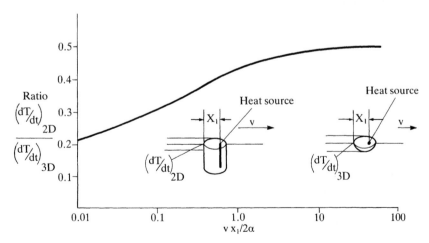

2.2 Ratio of axial cooling rates for two dimensional as compared with three dimensional heat flow.

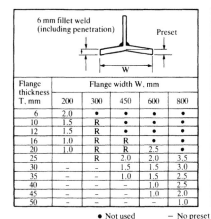

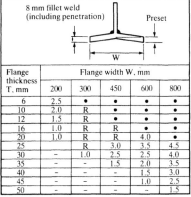

Flange thickness T, mm	Flange width W, mm				
	200	300	450	600	800
6	2.0	•	•	•	•
10	1.5	R	•	•	•
12	1.5	R	•	•	•
16	1.0	R	R	•	•
20	1.0	R	R	2.5	•
25		R	2.0	2.0	3.5
30	–	–	1.5	1.5	3.0
35	–	–	1.0	1.5	2.5
40	–	–	–	1.0	2.5
45	–	–	–	1.0	2.0
50	–	–	–	–	1.0

Flange thickness T, mm	Flange width W, mm				
	200	300	450	600	800
6	2.5	•	•	•	•
10	2.0	R	•	•	•
12	1.5	R	•	•	•
16	1.0	R	R	•	•
20	1.0	R	R	4.0	•
25		R	3.0	3.5	4.5
30	–	1.0	2.5	2.5	4.0
35	–	–	1.5	2.0	3.5
40	–	–	–	1.5	3.0
45	–	–	–	1.0	2.5
50	–	–	–	–	1.5

• Not used   – No preset   R Hot rectification required

2.3 Typical requirements for preset (in mm) of the flange of a plate girder prior to welding.[3]

prevent this it may be necessary to grip the sheet in jigs strong enough to cause the contraction to be taken up as plastic strain in the weld metal. Alternatively it may be necessary to resort to alternative means of fastening such as spot welding or riveting.

In the case of a weld between plates of intermediate thickness (say 10–50 mm), buckling is not normally a problem but in making a weld from one side between two initially flat plates there may be distortion such that the final section is that of a broad V. This type of problem is avoided either by making a double sided weld, by jigging or by setting the plates initially at an angle that is just compensated by shrinkage. Figure 2.3 shows a typical worksheet for presetting the flange of a T beam.[3] The type of problem is endemic in the fabrication of welded structural sections, where the finished product must usually meet stringent dimensional tolerances. In particular, the items must be straight, and various techniques are used to ensure this; for example, welding simultaneously on either side of the joist, or back-step welding, where the weld is made in short runs coming back towards the weld already deposited.

Residual stress
In the as-welded condition a fusion welded structure contains residual stress. The simplest case is that of two equal plates welded using a balanced double sided procedure. Shrinkage of the weld metal causes a tensile strain in the weld and its immediate surroundings; this is balanced by compressive strains in the plate remote from the weld. By custom and practice these strains are interpreted as being due to a stress equal to strain × elastic modulus. Figure 2.4 shows the results of

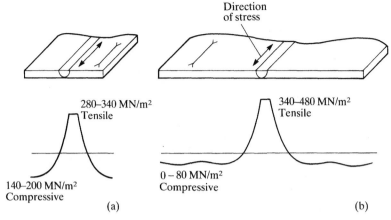

2.4 Typical residual stress fields in flat carbon steel plate welded with coated electrodes:[4] a) Narrow plate (100–200 mm); b) Wide plate (over 300 mm).

measurements on such welds in a pair of narrow and a pair of wide plates respectively.[4]

In arc welding of thick plate, or in deep penetration electron beam or laser welding it is possible for triaxial stresses to appear, and there may be a residual stress distribution in the through-plate direction as well as transverse to and along the weld. In a double sided multipass weld the progressive shrinkage of the outer passes causes the central regions to be compressed. Figure 2.5 shows the results of a survey of a submerged-arc weld in 165 mm thick Mn–Mo steel.[5] Tensile stresses are positive, compression stresses negative. The stress distribution shown in Fig. 2.5 is, on the whole, beneficial and the largest stress persisting after stress relief is compressive. Transverse cracks in the central region of thick welds become invisible to ultrasonic testing in the as-welded condition because of the compressive state but are usually relaxed and detectable after stress relief.

Stress corrosion cracking
Residual stress is damaging under conditions where stress corrosion cracking is possible; for example austenitic chromium-nickel steel in chloride solutions or carbon steel in contact with wet H_2S or cyanides. It may cause cracking in welds that have suffered embrittlement or in plate material where the restraint is high and where an initiating crack is present. The fracture toughness of a weldment is also reduced by the presence of residual stress but not normally to the extent of causing a brittle fracture. In ductile metals not subject to stress corrosion cracking, like pure aluminium or copper, residual stresses have little or no effect on the behaviour of the welded joint.

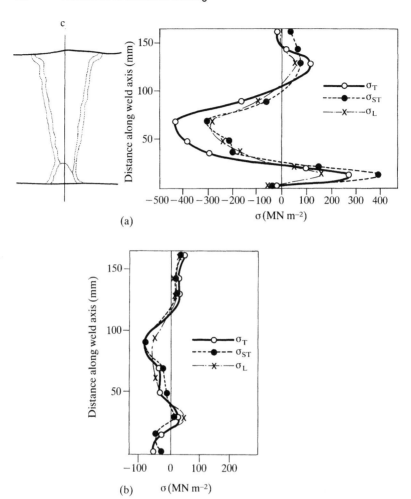

2.5 The distribution of residual stress along the centreline, c, of a submerged-arc weld in 165 mm thick Mn-Mo steel. Post-weld treatment; a) 15 min at 600 °C; b) 40 hours at 600 °C.[5]

Lamellar tearing

Steel plate is manufactured by rolling from a relatively thick ingot or slab to relatively thin plate. In the course of this process inclusions of oxide or sulphur are also rolled out to form, in the worst case, planes of lamellar weakness. If the plate is then subject to a stress at right angles to the plate surface, it may fail through these relatively weak planar areas. In welding an attachment to the surface of the plate the contraction of the weld sets up through-thickness strains and if the plate material is susceptible, it may fail by lamellar tearing. Figure 2.6 illustrates such

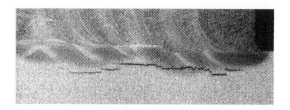

2.6 Lamellar tear under a T butt weld in carbon-manganese steel (×5). Photograph courtesy of TWI.

a defect. The cracks are located close to the fusion boundary. The probability of this type of failure is greatest where the design requires T and corner welds, and it is a potential problem in the manufacture of frames and bed plates for machine tools and in the fabrication of oil rigs. One way to avoid lamellar tearing is to design joints in such a way as to minimise residual strain in the through-thickness direction. Another is to remove the surface layers of the plate to a depth of say 5 mm and fill with a layer of weld metal, stress relieving if necessary.

Alternatively, or additionally, the plate itself is subject to some form of quality control. One method is to measure the through-thickness ductility, which should be greater than say 25%. The reduction in area in a normal (longitudinal) tensile test also provides some measure of susceptibility. The Engineering Equipment and Materials Users Association Publication No. 150 specifies steels for use in offshore structures, and includes a grade, designated Z, with enhanced through-thickness properties. This specification requires the steel mill to supply evidence that adequate through-thickness ductility is routinely maintained, and does not call for through-thickness tests.

Lamellar weakness is associated with the presence of rolled-out inclusions, particularly sulphides, and clean steel has a greatly improved resistance to lamellar tearing. Such steel, produced for example by calcium treatment combined with vacuum degassing, has been used for highly stressed parts of offshore structures.

The weld pool

In arc welding processes, there is an area of the weld pool that is exposed to the vapours of the arc column. This is the arc root, and whether it is anodic or cathodic the area must be substantially free from oxide or slag to allow the passage of the arc current. Except where only a pure inert gas is present there will be a reaction between the liquid metal and the arc atmosphere at this point, including the absorption of hydrogen, oxygen and nitrogen. Figure 2.7 shows the equilibrium solubility of hydrogen in various metals above and below the melting point.

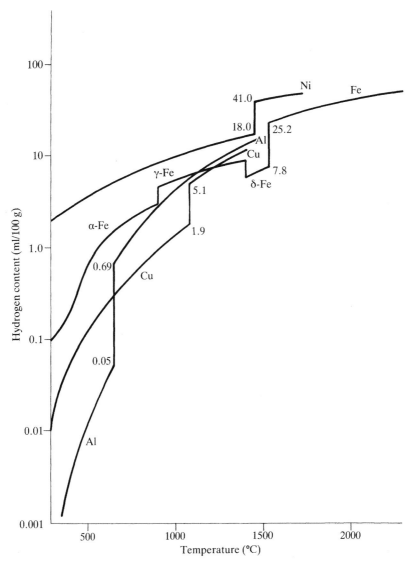

2.7 The equilibrium solubility of hydrogen in aluminium, copper, iron and nickel at 1 atm pressure.[9]

At the rear of the weld pool there is a fall in temperature and gas may be evolved, whilst the metal starts to solidify under conditions of tensile stress. Under unfavourable conditions porosity or cracking may result, and the final solidification structure may or may not have good mechanical properties.

Absorption of gases

For some time, welding technicians have been aware that the nitrogen content of steel weld metal is often higher than would be expected from measurements of the equilibrium solubility. This fact, together with the need to develop open arc welding systems, provided an incentive to make a systematic investigation of nitrogen absorption in liquid metal. A notable contributor was M Uda, of the Japanese National Research Institute for Metals.[6] He melted buttons of iron by means of an arc burning in mixtures of argon and nitrogen and measured the nitrogen content so obtained. The results are shown in the top curve of Fig. 2.8. Up to a nitrogen partial pressure of 0.036 atm the amount of gas absorbed is proportional to the square root of partial pressure, indicating that it is in diatomic form. The apparent solubility, however, is about twenty times that at equilibrium. Above a partial pressure of 0.036% the nitrogen content is constant and close to the equilibrium value for 1 atm nitrogen. Gas is still absorbed at a level above normal equilibrium, but excess nitrogen bubbles out of the melt, forming a spray of small metal droplets. The lower curves show that similar effects are observed in GTA and GMA welding.[7,8]

The conclusion to be drawn from this and other tests is that, in arc welding, nitrogen may indeed be absorbed to a greater extent than under non-arcing conditions, and that the final nitrogen content is the difference between the amount absorbed and that which is evolved before solidification.

Other tests have shown that oxygen behaves in a similar manner, except that whereas with nitrogen the excess gas is ejected, oxygen combines to form FeO which separates out as a slag. In the saturation region the oxygen content is then determined by its equilibrium with the slag.

The absorption/partial pressure relationship for welds and arc melts in argon-hydrogen mixtures is different. At low partial pressures the apparent solubility is the same as would be expected under non-arcing conditions. However in GT arc melting there is a limiting partial pressure of H_2 above which the hydrogen absorption starts to go down (Fig. 2.9).[10] In these tests a metal spray was observed at partial pressure over 0.1 atm and it would seem likely that, as in the case of nitrogen the amount of gas that dissolves at the arc root is above the equilibrium value, and the surplus bubbles out in the immediate vicinity. When the hydrogen partial pressure reaches about 0.36 the cloud of metal drops is dense enough to interfere with absorption by reducing the effective partial pressure. This self-limiting effect is unlikely to operate in welding, however, because the arc root is moving away from the bubbling area.

Thus there is good evidence to suggest that under arcing conditions hydrogen, nitrogen and oxygen may all be absorbed to an abnormally

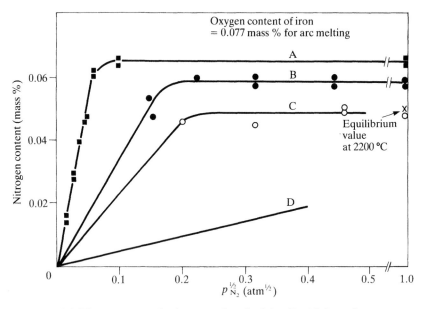

2.8 The amount of nitrogen absorbed by liquid iron from argon-nitrogen mixtures as a function of the square root of nitrogen partial pressure.[6-8] Total pressure = 1 atm: A – Arc melting; B – GTA welding; C – GMA welding, electrode positive; D – Non-arc levitation melting.

high level. Oxygen and hydrogen desorb rapidly to equilibrium values but this may not happen with nitrogen.

Reactions may generate gas in the liquid metal. In particular, the reaction between carbon and oxygen to form CO is quite common in steel. When it occurs in the melting drop at the electrode tip in MMA welding this may be beneficial in projecting small droplets into the weld pool. If CO forms in the weld pool, however, this may cause spatter and porosity. CO formation is particularly undesirable in GMA welding of steel and for this process it is necessary to use deoxidised wire.

In welding of undeoxidised copper, hydrogen may react with copper oxide to form steam, generating porosity. In GTA welding this problem is avoided by using a deoxidised filler rod.

Gas-metal reactions for nickel follow the same pattern as for iron, but the solubility of nitrogen is very much lower. Under non-arcing conditions the solubility at 1 atm nitrogen and 2000 °C is 18 ppm by mass, whilst under the arc the absorption reaches 40 ppm at a partial pressure of 0.0005 atm. This indicates a hundredfold increase in the apparent solubility. Oxygen combines with nickel to form a partially soluble oxide and the oxygen content of the weld metal is determined by the equilibrium constant of this reaction. Quantitative data on hydrogen absorption is lacking, but as with iron a spray of fine metal

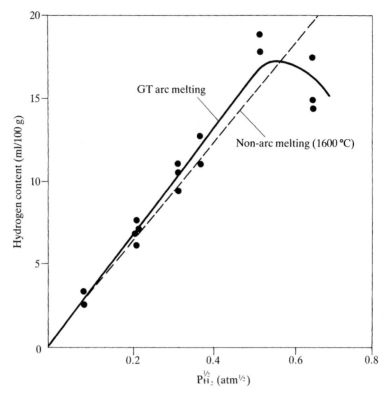

2.9 The amount of hydrogen absorbed in GTA arc melting as a function of the square root of hydrogen partial pressure in an argon-hydrogen mixture. Total pressure is one atmosphere. The fall in absorption at a partial pressure of about 0.25 atm is thought to be due to bubbling, which causes a mist of metal drops and effectively reduces the hydrogen partial pressure.[10]

droplets is observed at low partial pressures of H_2 and it is likely that excess hydrogen taken up at the arc root bubbles out behind the arc. Nitrogen porosity in GTA welding of nickel may be prevented by adding a few per cent hydrogen to the argon gas shield; the hydrogen probably degasses the metal as it bubbles out.

Both oxygen and nitrogen react with aluminium to form insoluble compounds and the amount of oxygen or nitrogen in solution in liquid aluminium is exceedingly small. The solubility of hydrogen on the other hand is, at elevated temperature, similar to that in iron and nickel. However it falls from about 20 ml/100g at 1500 °C to 0.7 ml/100g at the melting point, and then to 0.05 ml/100g in the solid metal. These features may be responsible for the very low tolerance of aluminium for hydrogen in welding operations.

Typical gas contents of steel welds made by various processes using properly dried consumables are listed in Table 2.1. In self-shielded welding the bulk of the nitrogen is in the form of aluminium nitride and the weld metal is less susceptible to strain-age embrittlement than that produced by coated electrodes or submerged-arc welding. However in self-shielded welding the impact properties of the weld may be reduced by using too long a stick-out (30 mm instead of 15 mm). Figures for hydrogen are those obtained in standard tests where the weld is quenched immediately on completing a short run. Hydrogen contents of real welds are lower and diminish with time (see later in this chapter).

Porosity

The simplest cause of porosity is that the weld metal has absorbed too much gas and is still bubbling when it solidifies. Table 2.2 summarises some of the available data on maximum percentage of gas in the arc vapour that will permit sound welds. Some of these figures relate to argon-gas mixtures and are strictly relevant only to GTA welding but on the whole they are in accord with experience. In MMA welding, for example, sound welds are made in steel with an arc atmosphere of up to 50% H_2. In marked contrast both iron and nickel develop porosity with a relatively small contamination by nitrogen. The limiting concentrations of gas for sound casts are also listed in Table 2.2, and here there are further anomalies; cast metal is insensitive to nitrogen but requires a lower concentration of hydrogen for porosity than in arc welding.

Nitrogen porosity is not uncommon in fusion welding of steel. A root pass in pipe made with a basic coated electrode is usually porous and this is thought to be nitrogen porosity. When rutile or cellulosic electrodes are used for the same purpose the weld run is sound, either because of the greater volume of gas evolved from the electrode coating or because the hydrogen, on bubbling out of the weld pool, also removes excess nitrogen. GMA welds other than the root pass in a pipe may also suffer nitrogen porosity if the gas shield is defective due, for example, to a cross wind. Tests using shielding gas containing 3% nitrogen have shown that GMA welds made with flux-cored wire are less subject to such porosity than those made with solid wire[11] (Fig. 2.10).

CO porosity is possible in both iron and nickel. In practice the carbon content of pure nickel is too low for CO formation but in iron, CO bubbles will form[17] if

$$[C] \times [O] \geqslant 0.002$$

where [C] and [O] are respectively the carbon and oxygen contents in mass %. Thus iron containing 0.1% carbon would bubble when the oxygen content exceeded 200 ppm. In practice steel is usually well

Table 2.1 Typical gas content of fusion welds in steel ppm[11,12,32]

Process	H*	N		O
		Free	Combined	
MMA basic coated	3-9	120	-	250
MMA rutile	30-45			
MMA cellulosic	50-65			
Submerged-arc	3-13	60	-	200-700
Submerged-arc (basicity BI2)				600
Submerged-arc (basicity BI3)				250
Self-shielded	3-9	50	170	130

*1 ppm = 1.12 ml/100g

Table 2.2 Critical gas content in shield above which weld or cast is porous[7,8,14-16]

Metal	Gas	Arc weld, %	Cast (no arc), %
Al	H_2	0.01	1
Fe	H_2	40	7.5
	N_2	1-5	100
Ni	H_2	50	5
	N_2	0.025	100

enough deoxidised to prevent CO formation but this is not always the case and CO formation accompanied by spatter has been observed in weld pools and in electrode wire.

Porosity may also result from unsteady conditions which chill the weld metal suddenly, for example at stops and starts in manual welds. Such porosity can be avoided by correct technique. It may also take the form of an elongated blowhole, known in steel welds as tunnel porosity. This elongated bubble is thought to be due to rejection of hydrogen across the solid/liquid interface, due to the lower solubility of hydrogen in the solid than in the liquid (see Fig. 2.7). A bubble nucleates near the root of the weld and is then fed by hydrogen so that it grows horizontally. Similar blowholes occur in castings and ingots.

Gas evolution is not the only cause of porosity. For example, if when making a root run in a pipe joint which has previously been tacked together the tack welds are too heavy, it is possible to get localised lack of fusion in the region of the tack and this shows up on a radiograph as an irregular pore. In the GMA welding of aluminium, using too high a current for the electrode diameter may cause turbulent flow in the weld pool and give rise to a form of tunnel porosity.

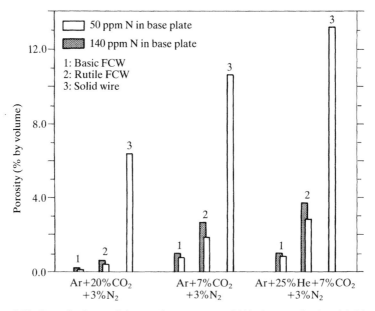

2.10 Porosity in steel due to the presence of 3% nitrogen in the shielding gas. It is lower in welds made with flux-cored wire than with solid wire, possibly because of a different mode of metal transfer.[11]

Slag metal reactions

The composition and properties of any metal that forms a partially soluble liquid oxide or slag may be affected by the character of the slag, but in welding this factor is important only in the case of steel, mainly in submerged-arc welding and welding with coated electrodes. For submerged-arc welding the chemical action of the flux is largely governed by the basicity. The commonly used index for basicity is that adopted by the International Institute of Welding (IIW):

$$BI = \frac{CaO + CaF_2 + MgO + K_2O + Na_2O + \frac{1}{2}(MnO + FeO)}{SiO_2 + \frac{1}{2}(Al_2O_3 + TiO_2 + ZrO_2)}$$

where CaO, etc, represents the mass fraction of the component in question. The flux is acid when BI is 1 or less, neutral in the range 1.0 to 1.5, semibasic in the range of 1.5 to 2.5, and basic when 2.5 or more. Acid components of the slag, and in particular SiO_2, are chemically less stable than the basic components, and at any given temperature dissociate more easily. Acid slags therefore generally produce a weld metal of higher oxygen content than basic slags, as indicated in Table 2.1. Higher basicity also tends to reduce the sulphur content of the weld slightly, and this is normally beneficial. Silicon and manganese are also affected; silicon decreases as BI increases, whilst manganese is higher with a more basic slag.

As a general rule the notch-ductility of the weld metal is improved by lower oxygen, higher manganese and lower silicon, so that basic submerged-arc fluxes are used (particularly in Europe and Japan) for applications where specifications require high impact properties. There are, however, other means of accomplishing this end. The Lincoln Innershield NR 203 Ni-C wire, for example, much used in offshore fabrication produces a weld deposit containing a relatively low free nitrogen content, ½% Ni and typically 130 ppm oxygen. The presence of nickel, combined with low oxygen and nitrogen, promotes notch-ductility.

The rules that govern for submerged-arc fluxes also apply to electrode coatings and cores, so that basic coated electrodes produce carbon steel deposits that usually have better impact properties than those deposited from rutile or cellulosic rods. Likewise basic flux-cored wires are preferred to rutile flux-cored types for low temperature applications.

Dilution and the boundary layer

There is always some fluid motion in the weld pool and in submerged-arc welding, for example, this is very rapid. Some of the plate material is melted and mixed with the filler metal, and because of the weld pool flow the weld metal is uniform in composition, even when the alloy content of the filler differs from that of the plate material. The final weld metal composition in such a case depends on the degree of dilution, defined as

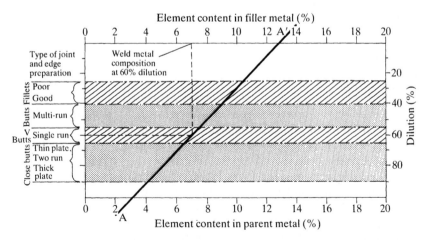

2.11 The dilution nomogram for fusion welds in aluminium.[18] To find the weld composition at any dilution, a rule is laid between the appropriate compositions on the filler and plate scales (A–A′); the weld metal composition is then read off on either scale. In the example shown, the weld metal composition at 60% dilution is 7%.

$$D = \frac{\text{Mass of parent metal melted}}{\text{Total mass of fused metal}}$$

This quantity varies according to the type of joint. Figure 2.11 is a nomograph due to Houldcroft[18] that illustrates the degree of dilution and composition change expected with GMA welding of aluminium alloys. In the case of steel, the dilution is less and is dependent on the process and welding procedure.

At the fusion boundary of the weld pool there is a velocity boundary layer associated with the flow and in dissimilar metal welds the composition changes across this layer from that of the fused zone to that of the parent metal. This transition zone may be observed in cross-sections of austenitic stainless steel overlays on ferritic steel, and is typically about 50 μm (5×10^{-5} m) in thickness.

Solidification and liquation cracking

In some alloys there is a long temperature interval between the start of solidification and its completion, and these metals may present welding problems. One such material is cast iron, which is subject to partial melting at the edge of a fusion weld made, for example, with a nickel-iron electrode. This and other undesirable effects are minimised by making short weld runs and keeping a low interpass temperature. In other cases the weld metal itself may have an unfavourably long solidification range and be subject to solidification cracking. In yet another variant, some constitutents of the parent metal may liquefy, giving rise to liquation cracking.

Solidification cracking

Alloys subject to this defect fall into two main classes. In the first, the extended solidification range is an intrinsic feature of the alloy system. In the second, an impurity element is responsible for producing a low melting compound and causing a hot-short condition.

The first category is typified by the binary aluminium-silicon alloys. The weldability of these and other aluminium alloys was investigated by Pumphrey and Jennings[19] in 1948/9 and the conclusions of this work remain valid and relevant to the present day.

Figure 2.12 is the constitution diagram for the aluminium-silicon system, showing the equilibrium constitution (solid lines) and the constitution under steady cooling (broken lines). For a 1% silicon alloy there is a difference in temperature between the start of solidification and its completion of about 80 °C. Within this region the partially solidified metal has low strength and zero ductility, as shown in Fig. 2.13.

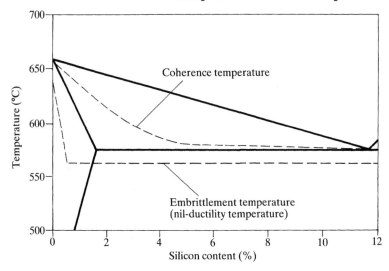

2.12 An equilibrium diagram for aluminium-silicon alloys (full lines) showing also coherence and nil-ductility temperatures on cooling (broken lines). The lower broken line represents the supposed solidus under cooling conditions.[19]

The wider the brittle temperature range, the greater the strain to which the partially solidified metal at the rear of the weld pool is subject, and the greater the risk of cracking.

It will be evident from Fig. 2.12 that the brittle temperature range is reduced as the alloy content increases and this is the main weapon against the constitutional hot-cracking tendency. Thus, the Al¾Mg1Si heat treatable alloy, a composition which is very sensitive to hot-cracking in welding, is joined using a 5Mg or 5Si filler, preferably the former. The 3% and 5% Mg alloys may safely be welded with a matching filler, whilst the AlZnMg heat treatable alloy is welded with a 5Mg filler.

The second hot-cracking mechanism is the formation of an intergranular liquid film of an impurity, notably sulphur and phosphorus in steel. Both these elements combine with iron to form low melting point compounds. They are also surface-active and migrate preferentially to grain boundaries. Thus they may act by either forming liquid intergranular films or by reducing intergranular cohesion. Sulphur is the more troublesome element mainly because of its lower solubility at elevated temperature. It is more soluble in ferrite than in austenite; in equilibrium at 1365 °C, ferrite dissolves up to 0.18%S whereas the solubility of S in austenite is 0.5%. Under continuous cooling the solubility is reduced further.

Much therefore depends on whether the steel solidifies as austenite or ferrite. An iron-carbon alloy with a carbon content of less than 0.1%

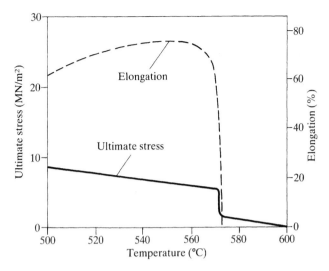

2.13 The mechanical properties of aluminium – 1% silicon alloy on heating to temperatures close to the solidus.[19]

solidifies in equilibrium as ferrite, with C between 0.1% and 0.17% as a mixture of austenite and ferrite, and above 0.17C as austenite. The brittle temperature range for any given P or S content would be expected to increase with carbon content and, as shown in Fig. 2.14, this is indeed the case.[20] Thus, high carbon high tensile steels such as AISI 4130 are especially prone to hot-cracking. Austenitic stainless steels may also solidify either as ferrite or austenite, depending on the balance between ferrite-forming elements (Cr equivalent) and austenite formers (Ni equivalent). Those that solidify as austenite are particularly crack sensitive on welding.

The effect of elements other than carbon on the susceptibility of ferritic steel to hot-cracking is shown in Fig. 2.15.[21] Nickel acts as an austenite stabiliser in the same way as carbon. Welding with coated electrodes is not normally possible with a nickel content over 4%. GMA welding with a 9% Ni wire is however possible.

The effect of various elements on hot-cracking of carbon-manganese steels in submerged-arc welding may be assessed by a cracking index

$$\text{Ucs} = 230\ \text{C} + 190\ \text{S} + 75\ \text{P} + 45\ \text{Nb} - 12.3\ \text{Si} - 5.4\ \text{Mn} - 1$$

where C, etc, represent mass per cent. Cracking of fillet welds is likely if Ucs $>$ 20 and, of butt welds if Ucs $>$ 35.

Hot-cracking in welding of ferritic carbon and alloy steels is avoided by keeping the carbon content of the weld metal down and the manganese content up, and by minimising the sulphur and phosphorus

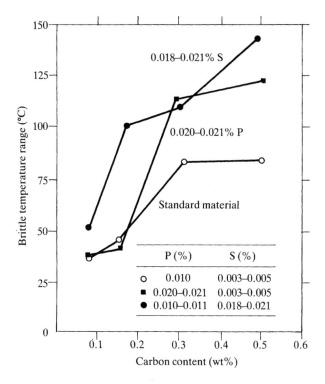

2.14 The effect of carbon content on the brittle temperature range of carbon steel having various sulphur and phosphorus contents.[20]

contents. In practice the hot-cracking of carbon steel welds is rare. It can occur, however, and where there is a cracking problem this possibility must not be overlooked.

In austenitic stainless steels solidification cracking is avoidable by controlling the composition of the weld deposit so that it solidifies as ferrite. After solidification the structure transforms mainly to austenite, but at room temperature a small amount of ferrite remains. By keeping the room temperature ferrite to a range of 3–10% in the weld deposit, the required control over the solidification structure is obtained. For special purposes where a fully austenitic weld metal is required, the S and P content must be maintained at a low level, preferably below 0.002%, with increased manganese content and possibly addition of rare earth elements to combine with sulphur.

The room temperature ferrite content of weld metal may be related to composition by means of a constitution diagram in which the amount of each phase present is plotted as a function of chromium equivalent and nickel equivalent respectively. There are two versions of this plot, the Schaeffler diagram, shown in Fig. 2.16, and the Delong

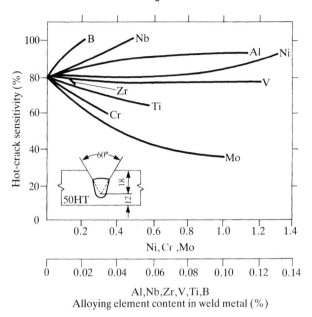

2.15 The relationships between alloying element content and hot-cracking sensitivity, measured as a percentage of weld run cracked, with groove preparation as illustrated.[21]

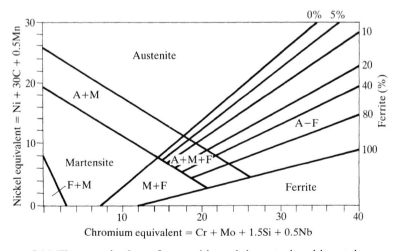

2.16 The constitution of austenitic stainless steel weld metal as a function of composition. The Schaeffler diagram.

diagram, shown in Fig. 2.17. The Delong diagram was developed by the US Welding Research Council (WRC) in order to take account of the nitrogen content of the weld. Nitrogen is a strong austenite stabiliser, so the WRC plot should give greater accuracy. Where the nitrogen content has not been determined by analysis it is assumed to be as follows:

Self-shielded arc welds: 0.12%
GMA welds: 0.08%
Other processes: 0.06%

The Delong diagram also gives a ferrite number, and this is sometimes quoted in specifications. Both diagrams correlate broadly but not exactly with measurements made using a magnetic gauge or by metallography.

A second means of avoiding solidification cracking in austenitic steels is by control of interpass temperature. This must be held at a maximum of say 250 °C so that the time spent in the brittle temperature range, where this exists, is minimised. Control of the room temperature ferrite content works well for the 18Cr8Ni type austenitic stainless steel when welding with coated electrodes or using GMA or GTA processes. It must be remembered however that this is an indirect control measure and that it may not be satisfactory under other conditions.

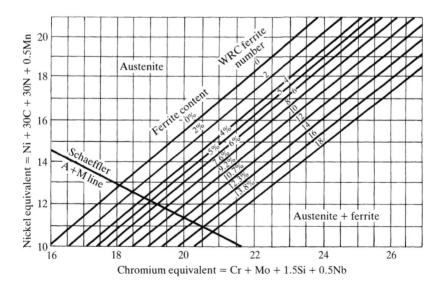

2.17 The ferrite content and ferrite number of austenite stainless steel with metal as a function of alloy and nitrogen content: the Delong (US Welding Research Council) diagram.

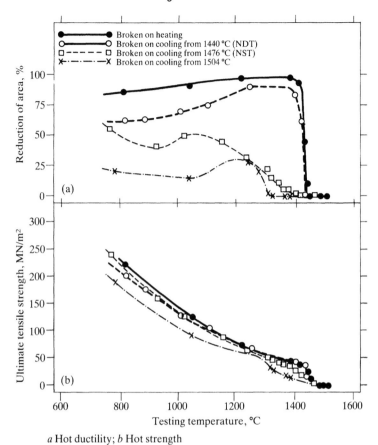

a Hot ductility; *b* Hot strength

2.18 Tensile strength and ductility of SAE 4130 steel (0.35 C 1Cr¼Mo with 0.032 S and 0.032 P) as a function of temperature, showing a wide nil-ductility temperature range near the solidus (reproduced by courtesy of The Institute of Materials).[22]

Liquation cracking

Liquation cracking may occur when a susceptible alloy is welded using a crack-resistant filler alloy. Low melting constituents form liquid films or regions in the parent metal close to the weld boundary and cracks are initiated. The essential features of such cracking are usually similar to those of solidification cracking, namely that there is a temperature range close to the solidus where the metal has coherence but is completely brittle. The medium carbon, high tensile steels such as AISI 4130 are one group that suffers this defect and Fig. 2.18 shows the results of hot tensile and ductility tests during a simulated weld thermal cycle.[22] These alloys are welded in sheet form for aerospace applications, and the cracking takes the form of microcracks 0.01 to 0.1 mm in length, which can propagate in service. The risk of crack formation increases with carbon, sulphur and phosphorus contents. High strength heat treatable aluminium alloys may also be subject to liquation cracking.

Austenitic stainless steel is occasionally subject to cracking in the parent metal adjacent to the fusion boundary, and such cracks appear to be initiated by liquation. The cracks may however propagate into other regions, by some mechanism that is not well understood. Single phase aluminium bronze (95Cu5Al) may suffer an even more severe condition where cracks initiating close to the weld boundary can propagate several centimetres into the surrounding plate.

The solidified weld metal

Tensile strength and toughness

As a general rule, weld metal produced by fusion welding has a relatively coarse grain. Solidification takes place from the fusion boundary inwards, and grains are nucleated from the edge of the high temperature heat affected zone, where the structure has been coarsened to a greater or lesser degree by the weld thermal cycle. The resulting structure is columnar and exceptionally contains regions of equiaxed primary grains. Nevertheless the tensile properties of weld metal are usually equal and sometimes superior to those of the parent metal.

For metals having a face-centred cubic structure such as austenitic stainless steel, aluminium, copper and nickel the general mechanical properties, including toughness, are not greatly affected by grain size. Matters are otherwise for ferritic steel, where coarse grain is associated with lower notch-ductility and reduced yield strength. However it is not the primary grain size that governs in this case, but the fineness of the structure that results from the austenite-ferrite transition.

Figure 2.19 illustrates a typical fine grained microstructure of carbon-manganese steel weld metal.[23] The transformation starts at the austenite grain boundary, then needles of ferrite develop in the interior of the grain. Carbon is rejected during the growth of these needles, so there is a residuum of carbides, martensite and retained austenite in the interstices. This structure is known as acicular ferrite, and the effective grain size is the mean diameter of the needles. Characteristically, such weld metal has good notch-ductility and a yield and tensile strength higher than that of the parent metal. The tensile properties are not affected by stress relief heat treatment in the 550–650 °C range because grain growth is inhibited by the interstitial carbides.

Less favourable structures may also develop; for example ferrite

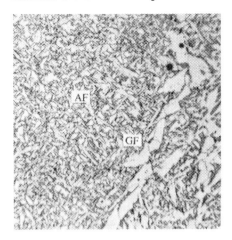

2.19 A typical fine grained structure in carbon-manganese weld metal. AF = acicular ferrite: GF = grain boundary ferrite.[23]

plates or needles may grow across the grain to form a lamellar structure such as that shown in Fig. 2.20. Acicular ferrite is favoured by the presence of fine particles which act as nuclei; these may be oxides or nitrides. Thus, in submerged-arc welds an oxygen content of 200 ppm (typical of a basic flux) is optimum for impact strength. The presence of titanium which forms titanium nitride, may also be beneficial.

Fine transformation structures are the result of the relatively high cooling rate associated with fusion welding. In low alloy steel such cooling rates may develop high levels of hardness in the weld metal, necessitating a post-weld heat treatment. The hardness of carbon-manganese steel welds may also need to be controlled, particularly for environments where stress corrosion cracking is possible. In multi-run welds the underlying passes are to some degree tempered by succeeding runs, and it is not uncommon to make a final run, known as the temper pass, which is then ground off. Quenched and tempered steels are a special case. The weld thermal cycle in a multipass weld may simulate quenching and tempering and therefore it is possible to maintain the original properties of the steel in the weld and heat affected zone provided that the cooling rate is sufficiently high. For such steels the heat input rate must be restricted in accordance with the steel manufacturer's requirements, and it may be necessary to place an upper limit on the preheat temperature.

A special advantage of the augmented properties of weld metal is that it is possible to use a low carbon content filler and still obtain a strength that matches the parent metal. Thus SAE4130, typically 0.35C0.7Mn1Cr 0.25Mo may be welded using a 0.9C1.6Mn1Cr 0.25Mo filler. Maintaining a low carbon content is essential to minimise the risk of hot-cracking.

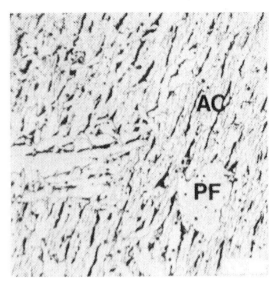

2.20 Lamellar structure of carbon-manganese weld metal. AC = ferrite with aligned martensite, austenite and carbides. PF = polygonal ferrite. Such structures are associated with lower notch-ductility.[23]

Notch-ductility

The impact properties of carbon-manganese steel weld metal are commonly specified for structures exposed to arduous or subzero temperature conditions, and the requirements of such specifications have become more stringent in recent years. Therefore it is not always possible to obtain a sufficient margin of safety using a carbon-manganese filler, and alloying elements may be added. Figure 2.21 shows impact transition curves for weld metal in multipass welding of thick carbon steel.[4] Welds made with coated electrodes are unalloyed and would meet a typical offshore fabrication requirement of 45 J at −40 °C in the cap but not at the root. For GMA welding the addition of oxidising gases to the shield reduces impact properties so a 1.5%Mo wire is used; even so the specification minimum is only just met. Submerged-arc welding with a Ni-Cr-Mo wire is satisfactory, as is open arc welding with the nickel-bearing Innershield NR203Ni wire.

The lower impact properties of the root pass for coated electrode and GMA welds are probably due to strain-age embrittlement. Table 2.1 shows that basic coated weld deposits have a relatively high free nitrogen content. The combination of straining due to subsequent weld runs with heating in the range of 100–300 °C would be expected to produce a strain-ageing in steel having this level of nitrogen. The self-shielded deposit shows the opposite tendency. Because of the high aluminium content such deposits suffer only slightly from strain-age embrittlement,

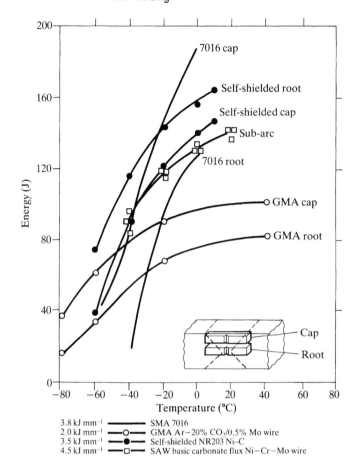

2.21 Charpy V notch impact strength as a function of temperature for several types of weld metal.[4,13]

and the better performance of the root pass is probably due to the better shielding obtained in this location.

Elevated temperature properties

In contrast to the room temperature conditions, there is a tendency for matching welds in some heat-resisting materials to have a lower creep strength than the parent metal, and creep cracks sometimes develop in weldments. This question will be considered in Chapter 3 and in sections dealing with plant operations at elevated temperature.

The heat affected zone (HAZ)

This is the unmelted portion of the parent material that has been subject to thermal cycling during the welding operation and where the original properties have been altered, usually for the worse. The extent of this zone depends on the heat input rate. In an electroslag weld in carbon steel it is 8–12 mm in width, whereas for manual welds it usually extends no more than 3–4 mm beyond the fusion boundary. Even when it is so small, the properties of the heat affected zone are important. In several cases, a crack formed in a brittle part of this zone has propagated, causing a brittle failure.

The high temperature region of the HAZ

When a metal is held for a period of time at a temperature close to the melting point grain growth occurs. Moreover, since grains in the solidified weld metal nucleate on the existing parent metal grains at the fusion boundary, such grain growth affects the grain size of the fused zone. As pointed out earlier, grain size does not greatly affect the properties of metals having a face-centred cubic lattice structure; that is to say, austenitic chromium-nickel steel, aluminium copper, nickel and other non-ferrous metals.

Ferritic steel may be affected in two ways. Firstly there may be embrittlement due to passing through the overheating temperature range. Secondly, the transformation products associated with a coarse-grained austenite may have inferior properties.

Overheating and burning are phenomena associated mainly with the manufacture of steel forgings. Overheating may occur if a steel is heated to a temperature above 1200 °C and then cooled at a rate between 10 and 200 °C per minute. It is manifested by matt facets in a fracture surface and a reduction in ductility, impact strength and fracture toughness. Burning is associated with heating at temperatures of 1400 °C and over, and this results in a severe loss of toughness, and a completely intergranular fracture. In both cases sulphur and phosphorus are taken into solution whilst heating up to the peak temperature and then precipitated at the austenitic grain boundaries during cooling. If the cooling rate is below 10 °C/min, sulphur redissolves and there is no precipitate; if it is higher than about 200 °C/min, the sulphur is held in supersaturated solution. In welding processes other than electroslag welding mean cooling rates down to 1000 °C are of the order 10^3 °C/min so that embrittlement due to intergranular precipitation in the coarse-grained region is unlikely. In electroslag welding however cooling rates can be within the critical range and indeed the impact properties of the coarse-grained zone are low in the as-welded condition.

The time spent in the overheating range is shorter than that in forging operations and this is reflected in the smaller austenite grain size, shown in Fig. 2.22 for various heat input rates when welding low alloy steel using coated electrodes.[24] Forging steel held for 30 minutes at temperatures between 1250 °C and 1350 °C, on the other hand, had a grain size ranging from 0.6 to 2.7 mm.[25]

For any given thermal cycle the austenite grain size may be affected by the presence of particles such as carbides or nitrides. Such particles may inhibit grain growth up to the temperature at which they go into solution and thereby reduce the time at the grain-coarsening temperature range. Figure 2.23 shows this effect for steel subject to a simulated weld thermal cycle and containing various nitrides and carbides.

The character of the transformation in the coarse-grained steel depends on the alloy content and cooling rate, but in all cases the ferritic grain size reflects to some degree the prior austenite grain size. In particular, a coarse structure favours the growth of ferrite blocks and tends to inhibit the formation of fine grained structures. This in turn can affect the notch-ductility of this region unfavourably. Titanium

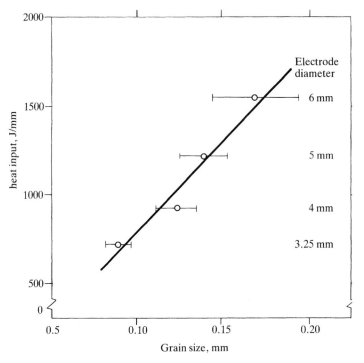

2.22 Grain size in the coarse-grained region of the heat affected zone of a fusion weld in ½Cr¼Mo¼V steel as a function of heat input rate.[24]

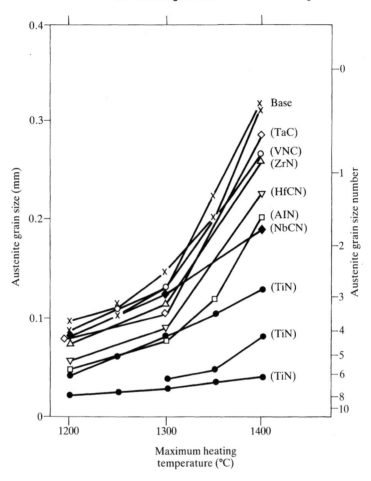

2.23 Austenitic grain size as a function of maximum heating temperature in carbon-manganese steel containing a variety of carbides, nitrides and carbonitrides.[23]

nitride particles, present in certain microalloyed steels, help to reduce the prior austenite grain size and thereby improve the properties of the welded joints. However titanium carbides may be partially dissolved in the coarse-grained region. Reheating during a subsequent pass in a multipass weld may reprecipitate these carbides, causing some degree of embrittlement and countering the beneficial effect of lower grain size.

Austenitic stainless steel may be affected by reactions in the high temperature part of the HAZ. Titanium is added to 18Cr8Ni types to combine with carbon and prevent chromium carbide precipitation at about 650 °C, which can render the steel susceptible to intergranular corrosion by acid solutions. However at temperatures above 1100 °C titanium carbide starts to go into solution and on subsequent cooling chromium carbides precipitate. Consequently a region close to the fusion boundary is sensitised and may be corroded by hot mineral acids. This type of corrosion is known as 'knife-line attack'. The risk of such attack is reduced in niobium-stabilised stainless steel because niobium carbide goes into solution at a higher temperature than titanium carbide.

The medium temperature region of the HAZ

This is the region down to say 300 °C where many of the more significant metallurgical changes take place. Heat treatable aluminium alloy and work-hardened non-ferrous metals are softened. Austenitic chromium-nickel steels and nickel based corrosion resistant alloys may be subject to carbide precipitation leading to a susceptibility to intergranular corrosion, but with little effect on mechanical properties. Carbon and low alloy steels commonly harden due to rapid cooling through the austenite-ferrite transition, and such hardening may lead to cracking.

Aluminium and alloys
Aluminium and the non-heat treatable Al-Mg and Al-Mn alloys may be strengthened by cold reduction. Such work-hardening is almost entirely lost in the heat affected zone of a fusion weld and there is no way in which the properties can be regained. Heat treatable alloys are also softened in the heat affected zone (including the high temperature region) but in such cases a partial recovery of the initial properties may be possible. Figure 2.24 shows the softening due to a GTA weld in the AlMgSi alloy 6063 in the fully heat treated condition. There is almost complete softening for a distance of about 10 mm from the fusion line, and no recovery of properties after 30 days ageing at room tempera-ture.[26] Ageing for eight hours at 180 °C results in a substantial recovery as shown in Fig. 2.25. The AlMgZn alloy 7005 behaves differently in that ageing does occur at room temperature and the properties of a welded joint may be recovered almost completely without post-weld treatment. Post-weld ageing may be possible in pressure vessels and may be worthwhile for aerospace applications; for land based vessels and structures the non-heat treatable Al-Mg alloys are normally used. Alternatively it may be possible to locate the weld in a region of low stress and still take advantage of the properties of a 6063 type alloy.

Intergranular carbide precipitation: austenitic stainless steel
If an austenitic chromium-nickel steel of normal composition 18Cr10Ni and containing carbon is heated within the temperature range 425–800 °C carbides precipitate at the grain boundaries. The amount of precipitate depends on time, temperature and the amount

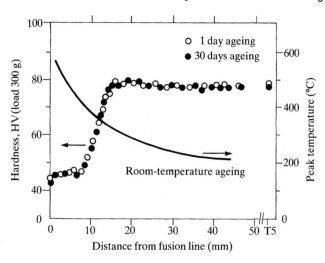

2.24 Hardness traverse across the heat affected zone of a GTA weld in 6060-T5 aluminium alloy (Al 0.4 Si 0.6 Mg, heat treated) after ageing for 1 and 30 days at room temperature.[26]

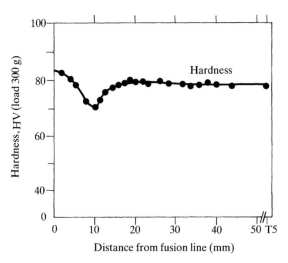

2.25 Hardness traverse across the weld shown in Fig. 2.24 after ageing for eight hours at 180 °C.[26]

of free carbon in solution; the highest precipitation rate is at about 650 °C. The effect of such precipitation is to make the steel susceptible to intergranular corrosion in acid solution. In fusion welding the time spent within the precipitation range is sufficient to sensitise a steel if enough uncombined carbon is present.

Susceptibility to intergranular corrosion is assessed after sensitisation for 30 min at 650 °C by one of two tests; the Huey test, which employs boiling nitric acid, and the Strauss test, with a boiling copper sulphate/sulphuric acid solution. Both these tests represent severe conditions that are rarely met in practice. They nevertheless provide a guarantee that the steel will not suffer intergranular corrosion in the medium temperature HAZ (a condition known as weld decay) under a wide range of operating conditions and over long periods of time.

There are two ways to ensure that a steel will pass a standard intergranular corrosion test. The first is to reduce the carbon content, ideally to 0.02% or below or, more commonly to a maximum of 0.03%. Such extra low carbon steels are now readily available and are often the best solution to the problem. Alternatively, steels may be stabilised by the addition of titanium (Ti4 × carbon content) or niobium (Nb8 × carbon content). These elements form carbides which reduce the free carbon to below 0.02%. Resistance to weld decay is also improved by adding molybdenum and type 316 stainless steel (18Cr12Ni2½Mo) may be used for most applications without stabilising additions. The low carbon variant (316ELC) is used for more severe duties.

As already indicated, there is a wide range of corrosive media in which weld decay is improbable, and in which it is perfectly satisfactory to use type 304 stainless steel, which has a maximum carbon content of 0.08%. All operating conditions must be taken into account however. For example, austenitic stainless steel is used for resistance to H_2S at elevated temperature where conditions are non-aqueous and intergranular attack is not possible. However, when the plant is not operating, the surface layer of sulphide may be converted by condensed water to polythionic acid, a substance that does cause integranular corrosion even in mildly sensitised material.

Intergranular carbide precipitation: nickel base alloys
The susceptibility to integranular corrosion as the result of carbide precipitation increases with the nickel content of any given alloy. Figure 2.26 gives the results of a modified Huey test on Ni17CrFe alloys of varying Ni and Fe content, showing an increase in weight loss up to 60Ni, and a large increase about this level.[27] The Hastelloy alloys, used for resistance to severe chemical environments, contain 50–60Ni and are indeed subject to weld decay problems. For use in the as-welded condition the carbon content is kept below 0.02% (Hastelloy B2) and in the case of Hastelloy C276, 0.35% vanadium is added as a stabiliser. In the case of pure nickel a 0.02% max C type is employed for exposure to severe conditions.

Carbon and low alloy steel
The medium temperature range of the HAZ is the region where the austenite-ferrite transformation takes place but where there has been

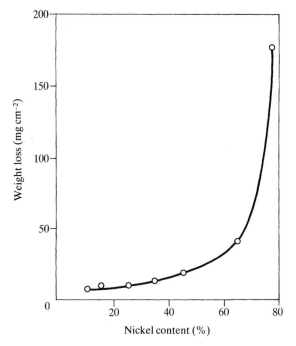

2.26 The influence of nickel content in a Ni17CrFe alloy on the weight loss in a boiling nitric acid solution.[27]

little or no austenitic grain growth. The grain size of the austenite is then similar to that of the original ferrite and, when this transforms, a finer structure results. This region is known as the grain refined zone, and in carbon-manganese steels the notch-ductility is often higher than that of the parent material. At the same time the cooling rate is normally high enough for some hardening to occur. The degree of hardening depends upon the alloy content and the cooling rate. For carbon-manganese steels the effect of composition is assessed by means of the carbon equivalent (CE) and the most commonly used formula for this quantity is that adopted by the International Institute of Welding

$$CE = C + \frac{Mn}{6} + \frac{Cu + Ni}{15} + \frac{Cr + Mo + V}{5}$$

where C, Mn, etc, are per cent by mass. This formula is used in BS 5135 for the calculation of preheat requirements. The higher the CE value, the greater the expected hardness and the higher the preheat. High hardness values, say over 350 VPN, are associated with susceptibility to hydrogen cracking and to embrittlement, but the degree of embrittlement for any given hardness depends on the steel composition. It increases with carbon content and is minimised by the presence of nickel; thus a nickel alloy steel may be adequately tough with a hardness of 400 VPN.

The peak hardness occurs a short distance from the fusion boundary. Figure 2.27 gives the results of microhardness traverses outwards from the fusion boundary for welds made with 4 mm diameter coated electrodes in $\frac{1}{2}$Cr$\frac{1}{2}$Mo$\frac{1}{4}$V and 2$\frac{1}{4}$Cr1Mo steel plate.[28] Although the hardness of the 2$\frac{1}{2}$Cr1Mo material is relatively high this steel is fairly tough even in the as-welded condition. Some degree of hardening may also occur in this region due to the precipitation of microalloying elements in the form of carbides. Such elements may be present in solid solution in the parent plate, or they may have been taken into solution in the high temperature HAZ of a previous pass.

The cooling rate is often defined by the time taken to cool from 800 °C to 500 °C. There are two competing effects here. Welding processes or procedures giving a high heat input rate generally have a long thermal cycle, and this results in relatively coarse grain but a smaller hardening effect. The reverse is true for small heat input rates and small weld pools. As a general rule the second alternative gives the best results because alloy welds are given a preheat and post-weld heat treatment and as-welded hardness is irrelevant whereas reduced toughness due to coarse grain is not improved by heat treatment. Carbon-manganese steels are not usually subject to heat treatment other than in thick sections and here the higher cooling rate is advantageous except when there is a hardness limitation. For this material the carbon equivalent is relevant; for a CE of 0.3 the medium temperature region will have a better impact strength than the low temperature region, which may be subject to strain-age embrittlement. With a CE of 0.45, however, the opposite will be the case. CE = 0.45 is about the upper permissible limit for weldable carbon-manganese steel without special precautions.

One of the constraints in the fabrication of offshore structures for the North Sea is the need to meet a hardness limitation in the HAZ. This is frequently required to minimise the risk of stress corrosion cracking in service. Specifications call for a hardness limit of between 300 and 350 Vickers at a point 1–2 mm below the plate surface. Such a limitation was easier to meet with earlier generations of carbon-manganese steel plate, but the introduction of clean (low sulphur and oxygen) steels, which are desirable for their better resistance to lamellar tearing, caused problems. Sulphide and oxide inclusions act as nuclei for the austenite-ferrite transformation, and when they are reduced in number the transformation starts at a lower temperature and results in a harder structure. Tests are carried out on individual batches of steel using a bead-on-plate weld run to provide a simulated HAZ, and also on procedure test samples. In some instances it has been necessary to

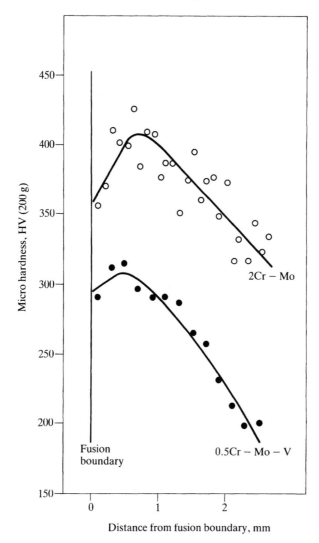

2.27 Hardness distribution across the heat affected zone of a fusion weld in low alloy steel (reproduced by courtesy of The Institute of Materials).[28]

modify the steel composition, for example by reducing the carbon content and increasing the manganese-carbon ratio to meet hardness requirements over the required range of welding procedures.

It was at first thought that because of the higher HAZ hardness, the clean steels would be more susceptible to hydrogen cracking than an equivalent composition with higher oxygen and sulphur. This has not proved to be the case and BS 5135 may be used for establishing preheat in either case (see Chapter 5).

The low temperature region of the HAZ

General
This is the region where the peak temperature is around 300 °C or less, and which may be subject to strain-age embrittlement, or it is any part of the weld where the temperature has fallen below 200 °C and is therefore vulnerable to hydrogen embrittlement and cracking.

Strain-age embrittlement
When a steel containing nitrogen is strained and heated, either simultaneously or later, to a temperature of about 200 °C, incipient precipitates of nitrides or carbonitrides form, and the impact transition temperature shifts upwards by up to 50 °C. There were several brittle failures of welded bridges in Belgium and Germany during the 1930s. The steel used was made by the basic Bessemer process and had a relatively high nitrogen content. The failures had an important effect on steel making development, since they led to the production of the aluminium-treated steel St 50, which has augmented yield strength and good notch-ductility, and was the first of the group of steels that were subsequently designated 'microalloyed'. That apart, strain-age embrittlement remains a welding problem as may be seen from Fig. 2.21. Apart from reducing the impact strength of the root runs in a multipass weld in carbon-manganese steel, strain-ageing may also have an embrittling effect on the tip of any small crack that forms in the HAZ. Such a mechanism is thought to be responsible for the success of the Wells wide plate test in reproducing brittle fractures. In the original form of this test a large welded plate is subject to tension in a direction axial to the weld. Prior to welding fine saw-cuts are made transverse to the weld preparation, and it has been observed that cracks form at the tip of these cuts during welding. Given sufficient strain-age embrittlement, such a crack may, when exposed to tensile stress, propagate into the parent metal.

Hydrogen embrittlement and cracking
There are three factors that affect the risk of hydrogen cracking during welding; first, the amount of supersaturated hydrogen present, secondly the levels of tensile stress, and thirdly the fracture toughness of the metal in the welded joint. The effect of hydrogen on the notched tensile strength of a quenched and tempered steel is shown in Fig. 2.28. There is a reduction of strength at temperatures between −100 °C and 200 °C, and the most severe reduction is at about 40 °C. The cracking in this test, and in a weld, is a time-dependent process. Hydrogen diffuses preferentially to the areas of highest strain, and when the concentration reaches a critical level, a crack is initiated. In welding, the cracks usually appear immediately after completing the joint, but they may be delayed for a period of up to 48 hours, or in exceptional circumstances, even longer.

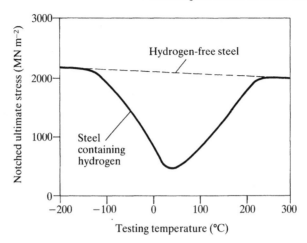

2.28 The tensile strength of a quenched low alloy steel charged with hydrogen. The testpiece comprised a cylinder with a V notch.[4]

The location of hydrogen cracks depends on the metallurgy of the joint. In welding a medium carbon high tensile steel with a low carbon filler, the most susceptible microstructure is in the HAZ close to the fusion boundary (as indicated by Fig. 2.27) and this is where hydrogen cracks appear. In steels welded with matching electrodes, on the other hand, the maximum cooling rate is in the weld metal and this is the usual location for cracks. In cases where there is longitudinal restraint the cracks form transverse to the weld axis. In other cases where bending stresses predominate the cracks may be parallel to the weld axis (Fig. 2.29).

Hydrogen control
The amount of hydrogen in a weld may be measured in one of several ways. In the first, the weld is quenched immediately after completion and then plunged into glycerine or mercury at about room temperature, and the amount of hydrogen that bubbles out is measured. This is the diffusible hydrogen. Alternatively a sample of the quenched weld metal is heated at 650 °C and the amount of hydrogen in the gases evolved is determined. This is the total hydrogen content. Intermediate heating temperatures may be used to speed up the determination of diffusible hydrogen. Such measurements are one way of controlling the hydrogen potential of an electrode coating, a submerged-arc flux, or a flux-cored wire. For electrode coatings and submerged-arc flux an alternative method is to control the moisture content. There is a correlation between moisture content and diffusible hydrogen, as shown for basic types in Fig. 2.30.[29] Such correlations hold good for individual flux formulations, but they may be affected by the raw materials that are used. Consequently the relationships shown in Fig. 2.30 do not

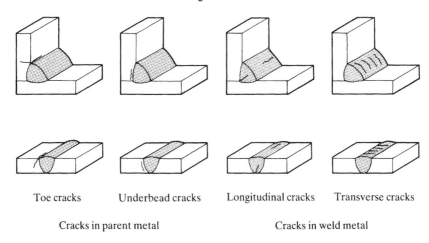

| Toe cracks | Underbead cracks | Longitudinal cracks | Transverse cracks |

Cracks in parent metal   Cracks in weld metal

2.29 Typical forms of hydrogen cracking.

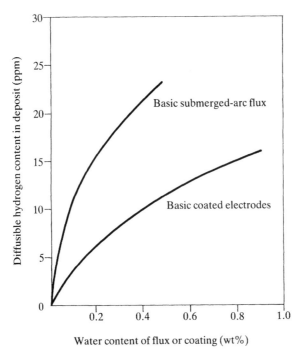

Water content of flux or coating (wt%)

2.30 The hydrogen content of a weld deposit as a function of the water content of an electrode coating or submerged-arc flux.[29]

have universal application. In the USA moisture content is preferred as a measure of control, and Table 2.3 gives the AWS limits for moisture in hydrogen controlled electrodes. The IIW recommendations for hydrogen controlled rods propose the categories listed in Table 2.4. These

Table 2.3 AWS limits for moisture in hydrogen controlled electrodes

Specified minimum ultimate strength of weld deposit (lb/in^2)	(N/mm^2)	Maximum moisture content (mass %)
70 000	482.6	0.6
80 000	551.6	0.4
90 000	620.5	0.4
100 000	689.5	0.2
110 000	758.4	0.2
120 000	827.4	0.2

Table 2.4 IIW designations for the hydrogen potential of welding consumables

Very low	Up to 5 ml/100g deposit
Low	5–10 ml/100g deposit
Medium	10–15 ml/100g deposit
High	Over 15 ml/100g deposit

categories are used in BS 639, also in BS 5135 as one of the factors in determining preheat requirements.

Standard methods of determining the diffusible hydrogen content of the weld deposit from a coated electrode are set out in BS 6693 and ISO 3690, the latter being the IIW method. A 100 mm weld run is laid down on the surface of a strip, typically 15 mm wide and 10 mm thick. The strip is held in a copper clamp. Immediately after welding samples are taken from the centre 30 mm of weld and are plunged into mercury at 25 °C. The amount of hydrogen evolved over a period of 72 hours is measured using a eudiometer. The results may be expressed in terms of ml hydrogen per 100 g of deposited metal (amount of electrode melted) or 100 g of fused metal (total amount of metal melted). As noted above, the quantity used in categorising electrodes and for establishing preheat to BS 5135 is the diffusible hydrogen content of *deposited metal*.

The reproductility of this technique is not particularly good. For the 72 hour test the 95% confidence limits as between different laboratories have been measured as ± 36%. Increasing the diffusion time to 14 days reduces this variability to ± 18%. Variation is largely due to differences in methods of cleaning and of welding. Good cleaning is important because a heavy oxide film acts as a barrier to hydrogen evolution. Such problems have led a number of writers to propose that hot extraction should be used and that the standard technique should be measurement of total hydrogen at 650 °C. These proposals are unlikely to be adopted in the near future.

Where hydrogen control is required it is essential that welding consumables, including coated electrodes, wire for GMA welding and wire and flux for submerged-arc welding be stored in their original containers on racks or pallets at a temperature of 20 °C and relative humidity 60 ° or lower. For quality work, basic coated electrodes are dried immediately before use by spreading the rods on trays in a drying oven at the temperature and for the time recommended by the manufacturer. They are then transferred to the workplace in electrically heated quivers maintained at a temperature of 70 °C, from which they are drawn by the welder. Basic submerged-arc flux is dried in a similar way on trays, typically at a temperature of 300–350 °C, or in a shaft heater or in a rocking shelf drier.[30]

Where specifications permit it may be preferable to employ basic coated electrodes that are formulated for use without redrying. Such rods should be usable for a period of one shift without significant moisture pick-up, and should be subject to a periodic check. The American Welding Society Structural Welding Code AWSD1.1 permits the exposure of hydrogen controlled electrodes to atmosphere for periods of up to 10 hours provided that the user has verified that after such exposure the moisture content does not exceed the limits listed in Table 2.3.

One method of hydrogen control is to use austenitic chromium-nickel electrodes. The diffusivity of hydrogen in austenite at the critical temperature range below 200 °C is exceedingly low, such that the hydrogen is trapped in the fused zone and hardened metal in the HAZ is not exposed to an intolerable influx of the gas. In consequence it is possible to weld hardenable alloy steel with austenitic electrodes without preheat, and at one time it was not uncommon to use this method for welding chromium-molybdenum process pipe in petroleum refineries. Cost, and doubts about the long term integrity of such a dissimilar metal joint, have led to the abandonment of this practice but the property of low diffusivity remains an important factor in making it possible to clad alloy steel with stainless steel by arc welding processes.

Preheat and interpass temperature control

Metallurgical effects

Preheat is one of the more important means of preventing the formation of hydrogen cracks in welds. One effect of preheat is to reduce the cooling rate and thereby to minimise the degree of hardening close to the fusion line. Figure 2.31 illustrates this effect in carbon-manganese steel weld metal. Samples were taken from one of the underlying (heat affected) weld runs in a multipass weld in 12.7 mm plate. Increasing the 'preheat' from −50 °C to 200 °C resulted in an increase in the cooling time from 800 °C to 500 °C and a modest but significant reduction in hardness.[31] However the observed effects are hardly sufficient to account for the efficacy of lower preheats such as 100 °C to prevent

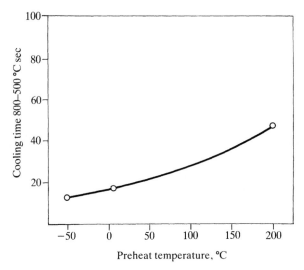

2.31 Cooling time as a function of preheat. 12.7 mm plate on 12.7 mm backing bar E7016 electrode at 3.0 kJ/mm.[31]

hydrogen cracking in appropriate circumstances. In all probability, the most important factors are the lower degree of hydrogen embrittlement at elevated temperature, as illustrated in Fig. 2.26 and the greatly reduced cooling rates below 200 °C, as indicated earlier ([2.7]).

The object of controlling the interpass temperature for most carbon-manganese and ferritic alloy steels is to minimise the risk of hydrogen cracking. This is done by ensuring that the environs of the weld do not fall below a safe level during the whole of the welding operation. In such cases the interpass temperature is specified as a *minimum* value.

Where there is a risk of solidification or liquation cracking, as for example with austenitic chromium-nickel steels, the interpass temperature is controlled so as to minimise the time spent in the brittle temperature range, and in this case the specified interpass temperature is a *maximum*. A maximum interpass temperature may also be specified for preheated carbon and low alloy steel if it is judged that there is a risk of solidification cracking.

There is a third case, that of quenched and tempered steels. To maintain the unwelded tensile properties in a fusion weld, it is necessary to avoid too low a cooling rate. Therefore both the heat input rate and the preheat must be maintained below a recommended value. At the same time there is a need to prevent hydrogen cracking. In such cases both a minimum and maximum interpass temperature must be maintained.

In all cases the position at which the interpass temperature is measured must be determined. This is typically required to be 25 mm on either side of the weld centre-line.

Means of determining preheat and interpass temperatures

Preheating is a greater problem in fabrication of structures than it is for pressure vessels and piping, firstly because it is more difficult to apply, and secondly because there is a greater economic penalty. The penalties that result from hydrogen cracking due to inadequate preheat are however more severe, and codes for structural steel are specific in their requirements.

So far as preheat and interpass temperatures are concerned, the American Welding Society's Structural Welding Code AWS D1.1 divides the processes into two groups: MMA welding with electrodes other than low hydrogen types on the one hand, and low hydrogen electrodes, submerged-arc, gas-metal arc and flux-cored welding on the other. The first category is restricted to steels of modest tensile strength, and has no preheat requirement for thicknesses up to 19 mm. Specified preheat temperatures increase as thicknesses increase above this level. For the other processes and steels the preheat temperatures are specified as a function of tensile strength and thickness from nil to a maximum of 150 °C. The preheat for quenched and tempered steel is lower than that for normalised types, and a maximum level is specified. Interpass and preheat temperatures are the same. Where no preheat is required, it is specified that should the metal temperature be less than 0 °C, it must be heated to 21 °C and maintained at that temperature during welding. In the AWS system a low hydrogen electrode is defined by the moisture content of the coating, as listed in Table 2.3.

BS 5135, on the other hand, relies on the diffusible hydrogen content of the joint, regardless of the process, together with the carbon equivalent of the steel to define the category of the weld. For various combinations of these two variables there are charts showing the preheat and interpass temperatures as a function of heat input rate and combined thickness of the plates being welded. BS 5135 was developed by a combination of laboratory testing and practical experience, and has now been validated by several years' experience in industry, notably in fabrication of oil rigs in the North Sea.

In process piping a typical requirement for carbon steel is that the preheat and interpass temperature shall be 100 °C when the joint thickness exceeds 19 mm. The preheat temperatures of ferritic alloy steels used for corrosion or creep resistance are normally determined by the alloy content, and typical values are listed with post-weld heat treatment temperatures in Table 2.5. Austenitic chromium-nickel steel should never be preheated prior to welding because this increases the risk of solidification cracking. The same rule applies to nickel-base

Table 2.5 Preheat and post-welding heat treatment temperatures

Steel type	Preheat, °C	Post-weld heating, °C
Carbon steel < 19 mm	None	None
Carbon steel > 19 mm	100	580–650
C½Mo	100	650–690
1Cr½Mo	150	650–700
2¼Cr1Mo	200	690–740
5Cr½Mo)		
9Cr1Mo)	200	700–760
12CrMoV)		
3½Ni	None	580–620
9Ni	None	None
Austenitic Cr-Ni Steel	None	Normally none (see text)

alloys. Cast iron is sometimes preheated before repair welding and an optimum temperature is about 300 °C. Thick sections of aluminium and copper may need to be preheated to avoid lack of fusion associated with high thermal conductivity, and preheats up to 600 °C have been used for copper in extreme cases.

The extent and means of preheating

There is little theoretical guidance to be found as to the distance on either side of the weld that must be uniformly heated to simulate uniform heating of a complete structure or vessel. It has been established that to obtain cooling rates equal to that in welding an infinite plate, the plate width on either side of the weld should be ten times its thickness, and it is likely that a similar relationship would apply to preheat. Code requirements are more modest.

The AWS Code DI.1 and BS 5135 require that the minimum specified preheat be maintained for a distance equal to the plate thickness but not less than three inches (75 mm), both laterally in advance of welding. This provision allows the use of a torch or other heater running along ahead of the weld. Such a technique is sometimes used for example in submerged-arc welding of circumferential welds in cylindrical storage tanks. A torch is mounted on the welding carriage so as to heat the joint immediately in front of the welding head. There are however inherent uncertainties about the effectiveness of a mobile preheat, and unless this is part of an established welding procedure, careful testing should be carried out to demonstrate its reliability.

Shop-fabricated pressure vessels are often preheated using multiple open gas flames. This rather crude device works quite well as a rule but there is always a danger of flame impingement and uneven heating. A more sophisticated method is to use infra-red radiant panel heaters, which may be fuelled by gas or may be electric resistance elements (Fig. 2.32). Longitudinal seams may also be preheated by resistance heaters mounted in stainless steel channels.

Process piping was at one time preheated either by a gas torch or by the wire and beads method. This employed a nichrome resistance wire threaded through ceramic insulating beads, the elements being covered with insulation. Power was supplied by a welding generator. Subsequently finger elements, consisting of nichrome wire heaters enclosed

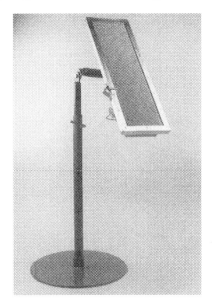

2.32 Infra-red combustion unit for preheating up to 300 °C, using natural gas, propane or butane (reproduced by courtesy of Cooperheat).

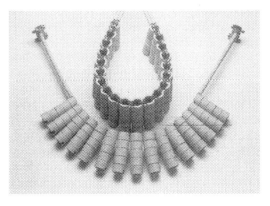

2.33 Finger elements for preheat and post-welding heat treatment of pipe welds (reproduced by courtesy of Cooperheat).

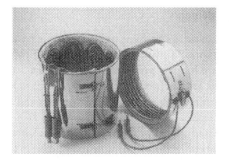

2.34 A wraparound heater for preheat and post-welding heat treatment of pipe welds (reproduced by courtesy of Cooperheat).

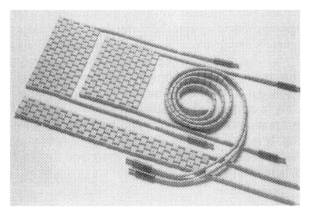

2.35 Ceramic pad elements for pre and post-weld heat treatment of a variety of shapes (reproduced by courtesy of Cooperheat).

in sintered alumina sleeves (Fig. 2.33), were introduced. These are mounted on either side of the joint and may be left in place while the weld is made and then used for post-weld heat treatment. Wraparound heaters designed for a fixed pipe size are easier to set up but less flexible (Fig. 2.34). Other wraparound heaters are available for pipe including one which is fixed by magnetic clamps.

Flat surfaces may be preheated by means of ceramic pads (Fig. 2.35) whilst for irregular surfaces a braided heater, consisting of nichrome wire contained in a flexible ceramic fibre insulator may be used. Special purpose power sources are available for all electric heaters together with regulators for temperature control. Heaters are insulated for power economy and for safety by ceramic fibre mats or mineral wool held in place by stainless steel mesh or straps.

Preheat temperature measurements

The simplest method of measuring preheat temperature is by means of temperature indicating crayons (Tempilstiks). These are composed of a mixture of waxes that melt at a specified temperature. They are drawn across the surface and when the required temperature is reached the streak of wax melts. Tempilstiks are often used for pipe welding.

Thermocouples are an alternative or additional means of measurement. Chromel-alumel couples are most commonly employed and there are various means of contacting the surface. A roller thermocouple may be used for a rotating pressure vessel. Alternatively, the thermocouple may be held in contact with a leaf spring and skidded across the surface. For stationary objects a thermocouple with a spring loaded tip may be pressed against the surface or held in place by a permanent magnet. A much more positive arrangement is to weld the thermocouple to the surface using a capacitor discharge machine. Another technique is to weld a threaded hollow stud to the surface. This stud is slotted to allow the insertion of the thermocouple, which is then held firmly in contact with the surface by means of a nut. The latter two methods are mainly used in post-weld heat treatment rather than for preheating. It is essential that after removal of welded-on attachments the surface be ground smooth and examined for cracks by the magnetic particle method.

Post-weld heat treatment

Metallurgical effects

Post-weld heat treatment at 580–650 °C was originally conceived as a means of relieving residual stress, and was indeed known as stress relieving. This remains an important objective. Where restraint is severe, contraction stresses can build up to a level where they can initiate a brittle crack. High levels of stress also increase the risk of hydrogen cracking. For these reasons it is not uncommon when welding thick sections to carry out an intermediate heat treatment when the weld is, say, half completed.

Residual tensile stress in the weld metal and HAZ would be expected to reduce the resistance to brittle fracture by augmenting any applied stress.

The other beneficial effect of post-weld heat treatment is that it tempers and softens hard transformation products in the HAZ, and eliminates strain-age embrittlement.

Post-weld heat treatment is required for thicker sections of carbon and carbon-manganese steel in structures and pressure vessels; for example, BS 5500 calls for such treatment for thicknesses in excess of 30 mm. It may also be required for critical structural parts and for caustic service. It is commonly specified for alloy steel piping, boiler drums and pressure vessels.

Whilst the metallurgical effects of post-weld heat treatment are generally beneficial, there are some negative features. In the manufacture of heavy-wall pressure vessels there may be a multiplicity of heat treatments. Individual strakes, for example, may be heat treated after welding, then again after joining together in pairs, then after welding to the heads and so forth, and the total heating time may well be in the region of 40–60 hours. Such extended heating may reduce the tensile properties, so that it is always necessary in such cases to conduct simulated heat treatments on procedure test samples to verify that the final properties will be adequate.

There may also be an embrittling effect. Table 2.6 shows the results of fracture toughness tests made on samples from the heat affected zone of a 165 mm thick weld in Mn-Mo steel.[5] The full section sample showed an improvement after long term treatment, possibly because of stress relief, but the two smaller samples were embrittled to a significant degree. Experience suggests that similar effects can occur in carbon and carbon-manganese steels, both in the weld metal and in the parent plate. For example in a large site-fabricated carbon-manganese steel pressure vessel the testplates were cut into two halves; one half was given a simulated heat treatment and tested in the laboratory, whilst the other half went through heat treatment with the vessel itself. This second testpiece, which was subject to the same long slow cool as the vessel, failed its impact test requirements whereas the laboratory test, which was not subject to such slow cooling, passed. It is evident that procedure testing should simulate the heat treatment cycle of the pressure vessel in all respects.

There is also a second problem that is peculiar to certain alloy steels, namely stress relief cracking, a defect that is also known as stress relaxation cracking and reheat cracking. The ½Cr½Mo¼V steel used for steam piping in the UK is particularly susceptible to such cracking, which is promoted by the presence of vanadium and boron. The steel in the coarse-grained region of the heat affected zone relaxes during heat treatment and is at the same time embrittled by carbon precipitation and if the embrittlement is severe enough intergranular cracks occur. Similar cracking can take place in service at elevated temperature due to imposed stresses. Quenched and tempered steel listed in ASTM A709 may also be subject to stress relief embrittlement and cracking in the coarse-grained heat affected zone. AWS D1.1 recommends that post-weld heat treatment should not be applied to these steels, and where it is necessary, say for dimensional stability, that the holding temperature should not exceed 590 °C.

Table 2.6 Fracture toughness ($Mn/m^{3/2}$) of weld HAZ in ASTM A533 (Mn-Mo) steel at 196 °C[5]

	15 min PWHT	40 hr PWHT
Full section	32.7	40.2
25 mm × 25 mm	57.8	44.2
10 mm × 10 mm	75.4	50.3

Post-weld heat treatment temperatures

Stress relief

Final post-weld heat treatments are intended to reduce the welding residual stress down to a harmless level, and to reduce hardness in the weld and HAZ. Typical figures for preheat and post-welding heat treatment for process piping and pressure vessels are listed in Table 2.5. BS 5500 (unfired pressure vessels) gives more detailed requirements with relaxations for thin-walled vessels and variations for optimum creep strength and other properties.

It is customary to specify that the items be held at temperature for one hour per 25 mm thickness with a minimum period of one hour. Most codes and specifications give a maximum heating and cooling rate, and state the maximum temperature at which the weldment can be placed in the furnace. Similar restrictions on cooling rate may be placed on welds that are heat treated locally. Some codes permit a decrease of the maximum temperature if the holding time is increased. In some instances it is necessary to observe an upper limit temperature. One such case is 3½Ni steel. Addition of nickel lowers the austenitising temperature so too high a value could result in unwanted partial transformation.

Intermediate heat treatments for heavy-wall pressure vessels are normally carried out at a temperature 50 °C or so lower than the final value. In this way any reduction in tensile properties may be minimised. Such intermediate heat treatments must also form part of the welding procedure test. They are often lumped together to reduce the time spent on testing, but as indicated earlier the best practice is to reproduce the real heat treatment cycle as closely as practicable.

Hydrogen diffusion

Whereas the requirements for stress relief heat treatments are laid down in considerable detail in documents such as the ASME Boiler and Pressure Vessel Code, BS 5500 and AWS DI.I, there is little to be found in the literature about hydrogen diffusion treatments, although they are not uncommonly used. A hydrogen release treatment is specified prior to the tensile testing of all weld metal specimens in certain specifications, for example AWS A5. The object of this treatment is to eliminate 'fish eyes'. These are bright circular spots on the fracture surface of the tensile specimen, and they are caused by hydrogen-induced cracking around non-metallic inclusions. Their effect is to reduce the ductility as normally measured in tensile testing, and it is considered that the hydrogen release treatment gives a better measure of the long term properties of weld metal.

In actual welds, the object of hydrogen diffusion or hydrogen release treatment is to reduce the hydrogen content of the weld to a level where there is no further risk of hydrogen cracking. One way to do this is to leave an interval of time between completion of the weld and starting non-destructive examination. For example, in some welds on North Sea oil rigs, the minimum such time interval has been specified to be 48 hours. In pipeline work there is necessarily an interval between welding and radiography of more than an hour and in practice this is normally sufficient to ensure long term integrity. However hydrogen cracking has been observed after a period of several days under conditions of severe restraint and such circumstances need special consideration.

The process of diffusion is speeded up by raising the temperature. Figure 2.36 shows that the diffusity of hydrogen may increase by one or more orders of magnitude if the weldment is heated to say 200 °C.[32] The variables that govern the diffusion of hydrogen out of a metal slab are illustrated in Fig. 2.37. This shows the proportion of hydrogen remaining in a half thickness that initially had a uniform hydrogen content and was then exposed to air at constant temperature. The proportion of hydrogen lost is a function of the quantity Dt/w^2 where D is diffusion coefficient, t is time and w is the half thickness. 95% of the hydrogen is lost when $Dt/w^2 = 1.2$, and the time required to accomplish this is proportional to the square of the plate thickness and inversely proportional to the diffusivity.

In a real weld the hydrogen diffusivity falls sharply as the temperature falls. Thus it would be expected that the hydrogen content would fall rapidly during the initial period of cooling and then slowly as the weld reaches room temperature. Measurements show that this is indeed the case and that a weld will initially lose 50–90% of its hydrogen.

Figure 2.38 shows diffusible hydrogen content of a 10 mm thick weld made with cellulosic and low hydrogen rods as a function of time after completion of the weld.[33] The initial hydrogen contents were 50–70 ml/100g and 4–5 ml/100g respectively, and half an hour after welding they were about 5 ml/100g and 1 ml/100g. Suppose that, in the interest of long term integrity, it is required to reduce the hydrogen content by a further 95% using a diffusion treatment at 150 °C.

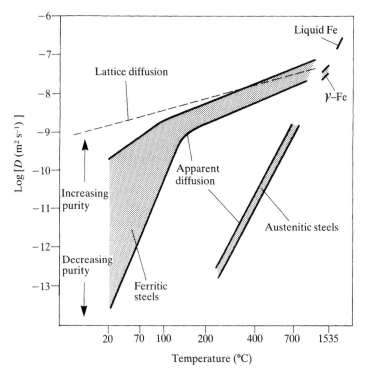

2.36 The diffusivity D of hydrogen in steel as a function of temperature. At lower temperatures the gas is partly trapped in discontinuities
and this reduces the apparent diffusivity.[32]

From Fig. 2.36 the diffusion coefficient is, conservatively, 1×10^{-9}
m²/s. Then the required heat treatment time for a 10 mm thickness
weld is

$$t = 1.2 \times \frac{(5 \times 10^{-3})^2}{1 \times 10^{-9}} = 3 \times 10^4 s = 8.4 \text{ hr}$$

Temperatures ranging from 150 °C to 400 °C have been used for diffusion treatment. Most of the weld-out joints used in mating the deck to
the hull of the Hutton tension leg platform were given a hydrogen
release treatment at 150 °C for 12 hours.

In underwater welding the increased ambient pressure causes a
higher uptake of hydrogen than for an equal process at atmospheric
pressure. Therefore higher preheat temperatures are used and a hydrogen diffusion treatment at 200–250 °C may be specified.

Normalising and quenching
The term normalising is used for heat treatment in the austenitising
temperature range, say at 850 °C followed by cooling in still air down

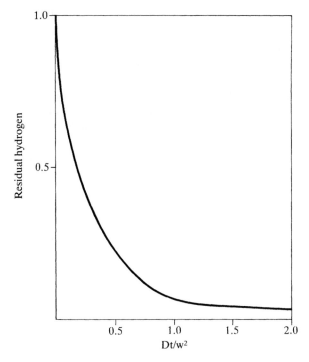

2.37 Proportion of hydrogen remaining in a slab of thickness 2w after exposure to air for time t at constant temperature.

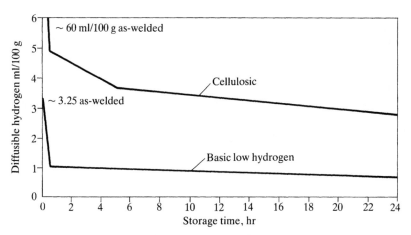

2.38 Diffusible hydrogen content of weld in 10 mm V joint as a function of time after welding.[33]

to room temperature. The effect is, in thin sections, to produce a reasonably fine ferritic grain size and correspondingly good notch-ductility. Such treatment may be applied to longitudinal welds in tube or pipe made by the electric resistance welding process. Downstream of the welding station there is in some cases an induction coil which raises the weld and the heat affected zone to the normalising temperature. The intention is to improve the properties of the joint but how far this is accomplished is difficult to tell because such welds are not often tested, other than indirectly by a flattening test on the pipe.

Normalising or quenching and tempering is required to restore the original steel properties in an electroslag weld. For pressure vessels such treatments are possible for cylindrical strakes that can be put in the furnace with the axis vertical. Horizontal cylinders lose dimensional stability at the normalising temperature and though in principle a local treatment is possible it is rarely attempted. Cylindrical strakes in the German steel 24CrMo10 have been quenched and tempered after electroslag welding but this is unusual; by using the correct combination of consumables and welding procedure it is normally possible to obtain an acceptable match to the properties of a quenched and tempered steel in the as-welded or stress-relieved condition.

Post-weld heat treatment of austenitic stainless steel
It is relatively rare to specify any post-weld heat treatment of austenitic stainless steel because the temperatures required are high and because local heat treatment may do more harm than good. Welded parts that are small enough may be given a furnace treatment provided that there are no problems of distortion. Occasionally it is required that welds in piping be heat treated against intergranular corrosion by polythionic acids. This type of corrosion can occur during downtime in hydro-desulphurising process units as described earlier in this chapter. One way to prevent polythionic acid attack is to give the joint a stabilisation heat treatment at between 850 °C and 950 °C. The purpose of such treatment is to stress-relieve, to complete carbide precipitation and to remove microstresses associated with the precipitates. Heating is for 1 hour per inch of thickness followed by cooling in still air. It is necessary to support the pipework so as to avoid any stress on the joint during heating, otherwise cracking may result.

Heat treatment techniques

Furnaces

Most shop fabrication is given its post-weld heat treatment in a furnace. On large construction sites it may also be economic to erect mobile furnaces for heat treating pressure vessels and piping spools. (Furnace treatment is usually less costly and more reliable than other methods.)

Furnaces are usually gas or oil fired, with the burners set behind walls or otherwise arranged to avoid any possibility of direct flame impingement. There is a bogie or trolley running on rails which carries the parts to be treated. Uniformity of temperature throughout the furnace is important and codes specify a maximum temperature difference during the heating and soaking periods. Normally a furnace will be fitted with a number of fixed thermocouples but for major pressure vessels and other critical items thermocouples may also be attached to the vessel itself. To minimise thermal stress there is usually a maximum furnace temperature at which a part may be inserted, and a maximum cooling rate is specified. For fixed tubesheet heat exchangers it is necessary to use low rates of heating and cooling to minimise temperature differences between the centre of the tube bundle and the shell; otherwise the tube/tubesheet joint will be damaged. When piping sub-assemblies are furnace treated they must be properly spaced and supported to avoid distortion. Paint marking must be supplemented by hard stamping for identification.

In the case of large site-erected vessels such as coke drums and storage spheres the vessel itself may, in effect, become the furnace. A chimney with a butterfly valve is fitted to the top manhole and a high velocity burner (Fig. 2.39) is fired through a bottom manhole. Thermocouples are attached to the outside and the whole vessel is insulated. The outlet valve and burner are adjusted in order to obtain the required temperature cycle. A very uniform temperature distribution may be obtained in this way.

Local heat treatment

Field joints in pipework and structures, and the closing seams of large pressure vessels are frequently given a local treatment. Finger elements, wraparound heaters and ceramic pads used for preheating (Figs. 2.33, 2.34 and 2.35) are also applicable to post-weld heat treatment of joints in piping and structural joints. Large pressure vessels may require heaters placed on internal scaffolding, and a typical set up is shown in Fig. 2.40. Pressure vessel and pipe welds are also sometimes stress-relieved by medium frequency induction heating, using a few turns of heavy cable. This system is used in Continental Europe but not so much in the UK or USA. Insulation is required over the heated area and also on either side, so as to maintain the temperature gradient within the limits set by the relevant code. Thermocouples are attached by capacitor discharge welding or by means of a threaded stud and must be thermally insulated so as not to receive direct radiation from the heaters.

2.39 High velocity gas burner for post-weld heat treatment of large vessels and structures (reproduced by courtesy of Cooperheat).

Vibratory stress relief

In some applications, of which a few fall within the scope of this book, post-welding stress relief is required to eliminate the distortion which may occur after machining, during transport or after a period of time in service. Vibratory stress relief may in such cases be used in preference to thermal stress relief.

The method employed is to vibrate the whole assembly at a resonant frequency, supporting it during this operation on rubber pads placed at the node points. The vibrator is an electric motor driving a rotary mass and producing a frequency in a range of 10–220 Hz. It is clamped to part of the structure and the vibration frequency (monitored by sensors) is increased slowly to the resonant level. The treatment time can be as short as 1000 cycles but is generally a few minutes. Machined items are treated in the rough-machined condition.[34]

Vibratory stress relief has been applied to vehicle chassis parts,

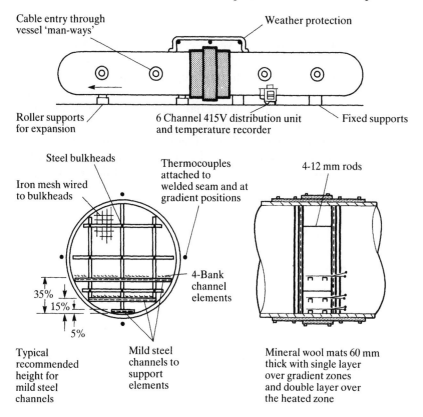

Cable entry through vessel 'man-ways'

Weather protection

Roller supports for expansion

6 Channel 415V distribution unit and temperature recorder

Fixed supports

Steel bulkheads

Iron mesh wired to bulkheads

Thermocouples attached to welded seam and at gradient positions

4-12 mm rods

35%
15%
5%

4-Bank channel elements

Typical recommended height for mild steel channels

Mild steel channels to support elements

Mineral wool mats 60 mm thick with single layer over gradient zones and double layer over the heated zone

2.40 Typical arrangement of internal electric resistance heating elements for local post-weld heat treatment of a circumferential weld in a large pressure vessel (reproduced by courtesy of Cooperheat).

fabricated I beams, rolls used for a variety of purposes and machinery bed plates. It is also useful for austenitic chromium-steel structural parts such as gearboxes and items used in the paper industry.

There has always been a fair amount of controversy about the effectiveness of vibratory stress relief, about the best type of equipment to use (AC versus DC motors, for example), and whether treatment should be at or below resonance level. However it is currently used by large organisations such as British Steel and has a good list of successful applications, so that it can reasonably be regarded as an established technique which is of special value for cases (austenitic stainless steel, for example) where thermal stress relief has positive disadvantages. It does not, of course, provide the metallurgical benefits (softening of hardened regions) of thermal stress relief and is not applicable to pressure vessels, piping and critical structural members.

Records

It is a requirement of pressure vessel codes that a record be kept of thermocouple temperature readings throughout the post-weld heat treatment cycle. It is good practice to keep similar records for the furnace treatment of piping sub-assemblies and the local heat treatment of piping and structural joints. Multichannel recorders are available so that it is possible to make records of pipe temperatures at, say, the twelve o'clock, three o'clock and six o'clock positions for a number of welds heat treated simultaneously. The charts are then identified by weld number and drawing number, and form part of the job of documentation.

REFERENCES

1 Roberts D K and Wells A A: *Brit Weld J* 1954 Vol 1 553–560.
2 Christensen N, Davies V de L and Gjermundsen K: 'Distribution of temperature in arc welding' *Brit Weld J* 1965 Vol 12 54–73.
3 Dewsnap H H: 'Submerged arc welding of plate girders' *Metal Construction* 1987 Vol 19 576–579.
4 Lancaster J F: 'Metallurgy of Welding' 4th Ed Allen & Unwin, London, 1987.
5 Sukuzi M, Kuwana I and Takahashi H: *Int J Pressure Vessels and Piping* 1978 Vol 6 87–112.
6 Uda M and Wada T: *Trans Nat Res Inst Metals* 1968 Vol 10 21–23.
7 Kuwana T and Kokawa H: *Quart J Jpn Weld Soc* 1983 Vol 1 392–398.
8 Kobayashi T, Kuwana T and Kikuchi Y: *J Jpn Weld Soc* 1971 Vol 40 221–231.
9 Brandes E A (Ed): 'Smithells metals reference book' 16th Ed, Butterworth, London, 1983.
10 Uda M: *Trans Nat Res Inst Metals* 1982 Vol 24 218–225.
11 Runnerstam O and Stenbacka N: 'Aspects of pore formation in GMAW' *Weld Met Fab* 1990 Vol 58 (10) 553–554.
12 Terashima H and Tsuboi J: *Metal Construction* 1982 Vol 14 648–654.
13 Yeo R B G et al: 'Welding with self-shielded flux cored wire' *Metal Construction* 1986 Vol 18 491–494.
14 Uda M and Ohno S 'Porosity formation in pure aluminium in non-arc welding' *Trans Nat Res Inst Metals* 1974 Vol 16 67–74.
15 Zaperstein Z P, Prescott G R and Monroe E W: 'Porosity in aluminium welds' *Weld J* 1964 Vol 43 443-s to 453-s.
16 Ohno S and Uda M: 'Effects of hydrogen and nitrogen on blowhole formation in pure nickel at arc welding and non-arc melting' *Trans Nat Res Inst Metals* 1981 Vol 23 243–248.
17 Kuwana T and Sato Y: *Quart J Jpn Weld Soc* 1983 Vol 1 16–21.
18 Houldcroft P T: 'Dilution and uniformity in weld deposits' *Brit Weld J* 1954 Vol 1 470.
19 Pumphrey W I and Jennings P H: *J Inst Metals* 1948/9 Vol 75 203–233.
20 Matsuda F, Nakagawa H, Nakata K, Kohmoto H and Honda Y: 1983 *Trans Jap Weld Res Inst* Vol 12 25–33.

21 Morigaki O, Matsumoto T, Yoshida T and Makita M: 1976 IIW Document No. XII-630-76.

22 Phillips R H and Jordan M F: 'Weld heat affected zone liquation cracking and hot ductility in high strength ferritic steels' *Metals Technology* 1977 Vol 4 396–405.

23 Dolby R E: *Metals Technology* 1983 Vol 10 349–362.

24 Alberry P J, Chew B and Jones W K C: 'Prior austenite grain growth in heat affected zone of a 0.5 Cr-Mo-V steel' *Metals Technology* 1977 Vol 4 317–325.

25 Gardiner R W: 'Effect of overheating on the properties of S82 steel' *Metals Technology* 1977 Vol 4 536–547.

26 Enjo T and Kuroda T: *Trans Jap Weld Res Inst* 1982 Vol 11 61–66.

27 Heathorne M: 'Intergranular corrosion' in Localised corrosion – cause of metal failure, ASTM Special Technical Publication STP 516.

28 Alberry P J and Jones W K C: 'Structure and hardness of 0.5 CrMoV and 2 CrMo simulated heat affected zones' *Metals Technology* 1977 Vol 4 557–566.

29 Evans G M and Baach H: in Metals Technology Conference Sydney, 1976, Paper 4-2.

30 Hauck G: 'Moisture out – but is it dry?' *Joining and Materials* 1988 Vol 1 232–233.

31 Tweed J H and Knott J F: 'The effect of preheat temperature on the microstructure and toughness of a C-Mn steel weld metal' *Metal Construction* 1987 Vol 19 153R-158R.

32 Coe F R: *Welding in the World* 1976 Vol 14 1–7.

33 Rabensteiner C: 'Development of electrodes for line pipe welds' in Metals Technology Conference, Sydney, 1976.

34 Claxton, R A and Lupton A: 'Vibratory stress relieving of welded fabrications' *Weld Met Fab* 1991 Vol 59 541–544.

3 The behaviour of welds in service

General

Prior to the general use of fusion welding for fabrication of structures and pressure equipment, the primary joining method was riveting. Parts made by riveting are discontinuous, such that a fracture which starts in one plate does not necessarily spread into its neighbours. Fusion welding, on the other hand, creates monolithic structures and a running crack can result in catastrophic failure. This possibility became a dramatic reality during and immediately after World War II with the break up and loss of numbers of Liberty ships and T2 oil tankers. Subsequently there has been a less frequent but nevertheless persistent incidence of catastrophic brittle failure, and one of the major objectives of welding research and technology is to contain and, if possible, eliminate this problem. There has been some success in this endeavour, and catastrophic brittle fracture is now a comparatively rare mode of failure, as will be seen in Chapter 6.

Welds also give rise to stress concentrations that may predispose the joint to fatigue failures. The welding residual stress may act as a driving force for various modes of stress corrosion cracking, and differences in chemistry or metallurgical structure may result in selective corrosion of the weld or HAZ. The elevated temperature properties and in particular creep strength and ductility may be lower than that of the parent metal, such that joints may suffer premature failure in high temperature service.

In reviewing these potential problems it is convenient to start by discussing fast crack propagation in plates since the analytical techniques that have been established for this type of fracture may also be applied to various modes of slow crack growth such as fatigue.

Unstable crack growth

General

Fast unstable cracking is the normal mode of failure in brittle materials such as glass. It may also occur in ductile materials when three preconditions are met:

(a) An initiating defect is present in the material. This may be a crack or a stress-raising discontinuity;

(b) There is a tensile stress with a component acting so as to open the crack or other defect;

(c) A source of strain energy is present, and this is released at a greater rate than it is absorbed by plastic deformation along the running crack.

In thin plate, and with very ductile materials, there may be substantial thinning near the failure and the surface may present an angle of about 45° to the plate surface. In thicker plate and less ductile material the central part of the fracture is flat, with an area of ductile 45° failure at the edge. These two failure modes are illustrated in Fig. 3.1. The first type occurs in piping or pressure vessels that have been subjected to over pressure. Running ductile failures have also occurred in pipelines operating at normal pressure. These were associated with a low rupture energy, which is measurable as a low upper shelf energy in a Charpy impact test. The strain energy source in such cases is the compressed gas or liquid.

The second type of profile is typical of what was known initially as a brittle failure but which is now often termed quasi-brittle on the grounds that there is a degree of plastic deformation associated with the failure. In steel the surface shows V shaped irregularities known as chevron markings. These point back towards the origin of the crack and provide a useful clue for failure investigators. Three factors contribute to the formation of a flat fracture surface: restraint, embrittlement and high rates of strain. In thick plate the inner regions are constrained by surrounding materials so that necking and bulk ductile movement may not be possible. Consequently, the failure takes place in a plane at right angles to the stress, and plastic strain is very localised except at the edge of the crack, where a limited amount of flow causes shear lips to form as shown in Fig. 3.1. Embrittlement enhances the degree of restraint and reduces the extent of plastic displacement during the fracture process. High strain rates may also reduce the amount of plastic deformation so that, for example, a metal that fails in a ductile manner in a tensile test may suffer a partially brittle fracture when subject to impact testing.

A flat fracture surface is not unique to running cracks in plate. In the tensile loading of a standard round testpiece, failure is commonly preceded by localised necking of the reduced section. Immediately prior to failure there is a high degree of self-restraint in this region which causes a plane fracture in the centre surrounded by a ductile shear lip: the classical cup-and-cone fracture.

There are various sources of strain energy. The simplest case is that of an infinite plate strained in uniform tension and containing a planar crack, to be discussed in the next section. The strain energy in this case

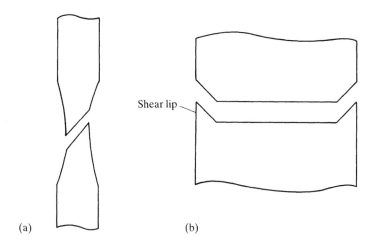

3.1 Failure modes: a) Ductile shear; b) Quasi-brittle flat fracture.

is that stored in the plate as a result of external loading. Most theoretical treatments of fracture mechanics assume this type of strain energy source. An additional contribution may come (locally) from the residual stress associated with fusion welding. All these, and the case of vessels and piping containing pressurised fluids, discussed earlier, are static conditions. There is also the possibility of dynamic loading due, for example, to collisions. A typical case is the impact of a vehicle on a carbon steel pipe branch in process plant. The source of strain energy is the elastically-bent branch. If this is long enough, a brittle fracture initiates at the root of the fillet weld attaching the branch to the main pipe, and propagates almost completely around the circumference, leaving a relatively narrow tongue which bends. Such impact loading increases the probability of brittle failure, as compared with a steady load generating the same level of strain energy.

The development of quantitative fracture mechanics

The problem of brittle strength was considered by Griffith in a classic paper published in 1920.[1] It was proposed that the strength of a brittle solid such as glass was determined by the presence of cracks, and that these cracks would propagate provided that the total energy of the system were reduced thereby. In the case of an infinite plate containing a crack of length 2a and with a tensile stress σ normal to the crack, (Fig. 3.2) there are two contributions to the total energy; the surface energy of the two sides of the crack, $4a\gamma$ per unit width and the strain energy, $\pi a^2 \sigma^2/E$ where γ is the specific surface energy of the solid and E is Young's modulus. The crack will extend when

$$d/da \left(-\frac{\pi a^2 \sigma^2}{E} + 4a\gamma \right) \leq 0 \qquad [3.1]$$

whence the critical stress required to initiate fracture is

$$\sigma_c = (2E\gamma/\pi a)^{\frac{1}{2}} \qquad [3.2]$$

The validity of [3.2] was demonstrated for glass using values of the surface energy obtained independently.

In his paper Griffith stated:

> 'the general conclusion may be drawn that the weakness of isotropic solids, as ordinarily met with, is due to the presence of discontinuities, or flaws, as they may be more correctly called, whose ruling dimensions are large compared with molecular distances. The effective strength of technical materials might increase ten or twenty times at least if these flaws can be eliminated.'

This prediction has only been realised in recent years.

The Griffith relationship was first applied to metals by Zener and Holloman in 1944.[2] From that time onward this concept gradually became established, although there were difficulties because, for metals, the second term in [3.1] included a contribution which took account of the work done in plastic deformation, and which turned out to be several orders of magnitude greater than surface energy. The problem of accounting for plasticity was of much concern to pioneers in this field and remains a matter of concern to this day.

The need for a better understanding of the brittle fracture of metals became acute in the USA during the 1950s due to the repeated failure of early Polaris rocket motor cases at stresses well below the design value. To deal with this problem NASA and ASTM set up a special committee on fracture testing of high strength materials and, in the space of five years, developed a completely new branch of engineering: linear elastic fracture mechanics. The results of their work were presented at a symposium in 1964 and published as ASTM Special Technical Publication No. 381; the classical presentation of fracture mechanics.

At a later stage standard tests for measuring fracture toughness were established. The object of the theoretical work was to provide a means for predicting the conditions under which a pre-existing crack or other defect would propagate in an unstable manner. The fracture toughness tests enabled such an analysis to be applied to a particular material or, for example, to a welded joint. Procedures for assessing the risk of

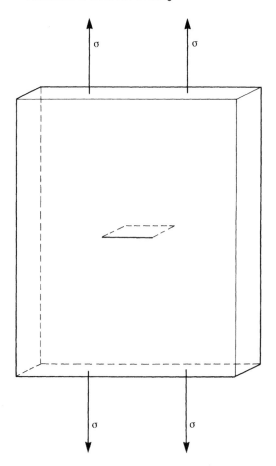

3.2 Centre-cracked plate subject to a uniform stress σ at right angles to the crack surface.

unstable brittle crack propagation have subsequently been set up for a number of types of crack-like defects in relation to the fracture toughness of the material and these are to be found in the British Standard Document PD 6493, in Section XI of the ASME Boiler and Pressure Vessel Code and in the IIW document SST-1157-90.

The stress analysis of cracks

The starting point for computations is to identify three possible fracture modes, as illustrated in Fig. 3.3.[3] These are the opening mode, I, the longitudinal shear mode II and the tearing mode III. In brittle fracture of metals only the opening mode I is operative. Nevertheless it is customary to retain a suffix I for the relevant quantities such as the stress intensity factor, which is designated K_I.

The second step is to obtain the stress distribution in a material containing a crack. This is expressed in two dimensional polar co-ordinates (r, θ), illustrated in Fig. 3.4. The material is supposed to be perfectly elastic and the crack is assumed to be sharp. The tensile and shear stresses are obtained as a set of infinite series in which the first term contains the factor $(a/r)^{3/2}$ whilst the equivalent factor for higher terms is 1, $(r/a)^{1/2}$, $(r/a)^{3/2}$ etc. For small values of r/a the first term is dominant and higher order terms are ignored. The equations are therefore valid for distances from the crack tip where r is small compared with the crack length.

The analysis yields results for tensile and shear stresses in the three principal directions and, for example, those parallel and transverse to the plane of the crack are

$$\sigma_x = \frac{K_I}{(2\pi r)^{1/2}} \cos \theta/2 \left[1 - \sin \theta/2 \sin 3\theta/2 \right] \qquad [3.3]$$

$$\sigma_y = \frac{K_I}{(2\pi r)^{1/2}} \cos \theta/2 \left[1 + \sin \theta/2 \sin 3\theta/2 \right] \qquad [3.4]$$

The quantity K_I is the stress intensity factor and is given by

$$K_I = \text{Constant} \times \sigma a^{1/2}$$

where σ is the applied stress whilst the value of the constant depends on the form and orientation of the crack relative to the applied stress. For example, in the case of the centre-cracked plate shown in Fig. 3.2 the constant is $\pi^{1/2}$ and

$$K_I = (\pi a)^{1/2} \sigma \qquad [3.5]$$

Reference 3 lists values of the stress intensity constant for a number of different forms of test specimen and types of crack.

The third step is to assume that the crack will extend in an unstable manner when the stress field reaches a certain level of intensity. This intensity is directly proportional to K_I and the value of K_I corresponding to unstable crack growth is K_{IC}. K_{IC} is treated as a material property and is known as the fracture toughness, with dimensions (in SI units) of $MN/m^{3/2}$ or $N/mm^{3/2}$.

There are two basic types of stress distribution to be considered. The first applies to a condition in which displacements in the z direction are zero. In other words, there is complete constraint such as to prevent any lateral movement. The transverse stress σ_z is then

$$\sigma_z = \nu (\sigma_x + \sigma_y) \qquad [3.6]$$

where ν = Poisson's ratio

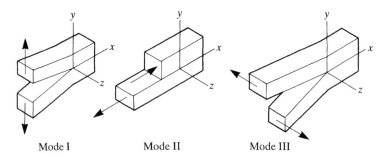

Mode I   Mode II   Mode III

3.3 The basic modes of crack surface displacements (after Paris and Sih[3]).

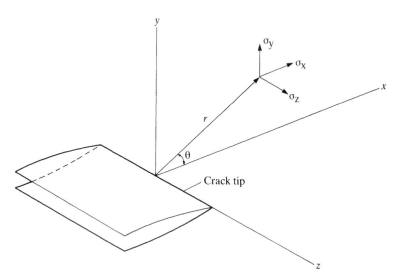

3.4 Two dimensional polar co-ordinates for the stress field associated with a sharp crack.

This condition is known as plane strain. The second possibility is that there is no limitation to transverse displacement, and

$$\sigma_z = 0 \qquad\qquad [3.7]$$

The second condition is that of plane stress.

   In a real crack in plate material the central regions are, as already pointed out, constrained by their surroundings and plane strain conditions exist. At the surface of the plate, on the other hand, the transverse constraint is small and conditions may approximate to plane stress. In a thin plate the two surface regions may merge so that plane stress obtains across the section.

Certain limitations of this analysis must be borne in mind. Firstly, it is an approximation. This is probably unimportant because crack extension must initiate at or close to the pre-existing crack tip, where r/a is small. Secondly, it ignores plastic behaviour. This is a major problem that will be discussed later in this chapter. Thirdly it applies to static loading only and dynamic loads (impact) must be given special consideration. Finally, it defines the fracture toughness in dimensions that have no physical meaning and do not relate directly to any known physical property.

The surface energy of a crack

The strain energy release rate G_I associated with crack extension may be calculated for the perfectly elastic case and it is

$$G_I = \frac{(1 - v^2)K_I^2}{E} \tag{3.8}$$

for plane strain and

$$G_I = \frac{K_I^2}{E} \tag{3.9}$$

for plane stress.

In the case of a centre-cracked infinite plate $K^2 = \pi\sigma^2 a$ and

$$G_I = \frac{(1 - v^2)\pi\sigma^2 a}{E} \tag{3.10}$$

Crack extension occurs at a critical value of the strain energy release rate G_{IC} and comparing [3.10] with [3.2] it is evident that for a brittle, perfectly elastic material:

$$G_{IC}/(1 - v^2) = \gamma \tag{3.11}$$

For a plastic material which is suffering a quasi-brittle fracture the plastic region is small in extent and it may be argued that a similar analysis will apply. γ is then replaced by the plastic work term and if the plastic extension at fracture is δ_C and the mean failure stress is $\bar{\sigma}$

$$G_{IC} = (1 - v^2)\,\delta_C\bar{\sigma} \tag{3.12}$$

The dimension of G_I is energy per unit area; for example, J/m². It is sometimes given the rather misleading title of 'crack extension force'.

Ductile engineering steels have values of K_{IC} ranging typically from 100–300 MN/m$^{3/2}$ (3162 − 9487 N/mm$^{3/2}$) and the corresponding values of G_{IC} are 5×10^4 J/m² to 5×10^5 J/m². This may be compared with the surface energy γ of iron, which is of the order 1 J/m². A carbon-manganese steel having a K_{IC} of 10 MN/m$^{3/2}$ (316 N/mm$^{3/2}$) would be in a notch brittle condition. With such a material, the critical crack length for a centre-notched plate stressed to half the yield stress would be about 1.5 mm.

All these considerations relate to the initiation of an unstable brittle crack in a uniform isotropic material. Once the crack has initiated the strain energy release rate increases with increasing crack length (hence the instability), so that the value of fracture toughness required to arrest the crack is greater than that for initiation, and is not easy to define quantitatively. Control procedures are, for the most part, directed towards the prevention of crack initiation. In doing so the relevant property is that of the most embrittled part of the structure, which is usually in the heat affected zone of the weld. It may be argued that this is an over-conservative technique since a crack in a small volume of notch-brittle material could well be arrested by notch-ductile surroundings. On the other hand, there have been cases where cracks have run extensively along welds.

The effect of plasticity

The stress distribution around the tip of a crack as defined by [3.3], [3.4], [3.6] and [3.7] assumes perfect elasticity. Immediately around the

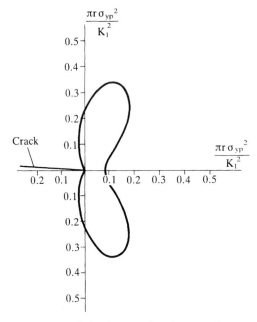

3.5 Extent of plastic zone in plane strain.

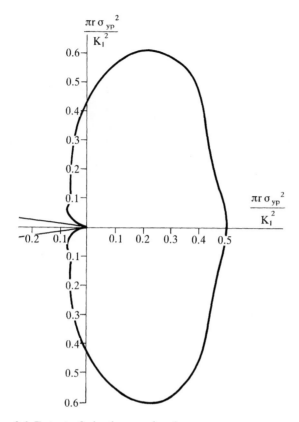

3.6 Extent of plastic zone in plane stress.

crack tip, however, plastic yielding must occur. The extent of the plastic zone may be calculated if it is assumed that the stress distribution in the remainder of the metal is unaffected. Using the von Mises criterion for yielding under multiaxial stress:

$$\sigma_{yp} = \frac{1}{\sqrt{2}} \, [(\sigma_y - \sigma_x)^2 + (\sigma_x - \sigma_z)^2 + (\sigma_z - \sigma_y)^2]^{\frac{1}{2}} \qquad [3.13]$$

together with [3.3], [3.4] and [3.6] (for plane strain) and [3.3], [3.4] and [3.7] (for plane stress) the form of the plastic region for these two cases may be calculated. Poisson's ratio is assumed to be 0.3. The results are shown in Fig. 3.5 (plane strain) and 3.6 (plane stress). The actual form of this region has been determined experimentally by straining and then etching a cracked specimen of silicon iron under conditions of plane strain.[4] The result is shown in Fig. 3.7, and this appears to validate the calculations at least qualitatively.

The supposed distribution of the plastic zone across the section

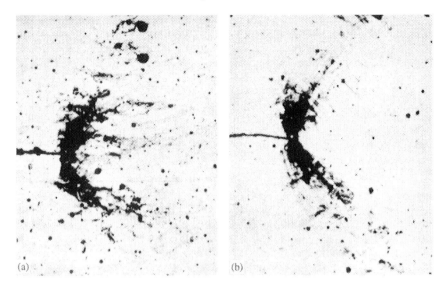

3.7 Plate of silicon steel 0.42 in thick containing a fatigue crack, strained at right angles to the plane of the crack, then etched to show the extent of the plastic zone:[4] a) Plate surface; b) Plate midsection.

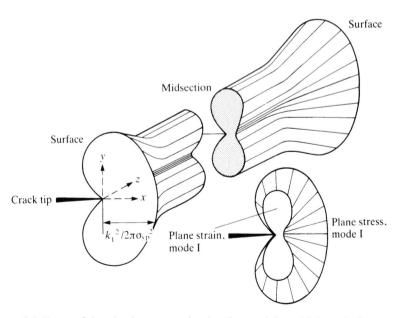

3.8 Form of the plastic zone at the tip of a crack in a thick steel plate (reproduced by courtesy of UKAEA).[4]

thickness is illustrated in Fig. 3.8. At the surface the conditions approximate to plane stress with a relatively large plastic zone. This diminishes in size towards the central regions, where plane strain conditions dominate. If the plate is thin enough, the two regions of plane stress meet and merge and fracture is of the ductile type sketched in Fig. 1a. Otherwise the fracture mode 1b occurs.

The micromechanism of fracture

In a brittle substance, fracture occurs by separation across crystallographic planes (crystals of mica split along their weakly bonded basal planes for example). The fracture of metals, on the other hand, is always accompanied by some degree of plastic deformation. This occurs prior to fracture (in the case of a running crack by the formation of a plastic zone ahead of the crack tip for example). It also occurs during the fracture process due to the formation of local regions where there is plastic movement and tearing.

At temperatures below the ductile/brittle transition, fracture surfaces show facets with secondary cracks and ridges where the metal has been drawn at right angles to the surface before tearing. Figure 3.9 shows such a facet on the surface of a hardened and tempered 12Cr wire broken in tension at room temperature.[5] This is a quasi-cleavage fracture. The mechanism of ridge formation is shown diagrammatically in Fig. 3.10.

Above the transition temperature the fracture is ductile and under the microscope has the dimpled appearance shown in Fig. 3.11. The appearance of such fracture surfaces is said to be due to microvoid coalescence. It is supposed that micro-inclusions nucleate voids, either by cracking or by separation at the interface with the metal. The voids grow and the ligaments between them extend until they finally separate.

It frequently happens, particularly in the transition temperature zone, that a fracture will contain patches of quasi-cleavage and dimpled fracture in the same area.

Crack opening displacement

In order that plane strain conditions should dominate it is necessary when measuring K_{IC} to establish a minimum testpiece thickness. In the ASTM standard for fracture toughness testing E-399 there is an empirical rule that the thickness should be not less than $2.5 \times (K_{IC}/\sigma_{yp})^2$. For the high tensile quenched and tempered steels used in aerospace applications this rule presents few problems, but for the majority of structural and pressure vessel steels at room temperature the specimens may become prohibitively large. Figure 3.12 plots the minimum test thickness for a Mn-Ni-Mo pressure vessel steel A533 grade B class

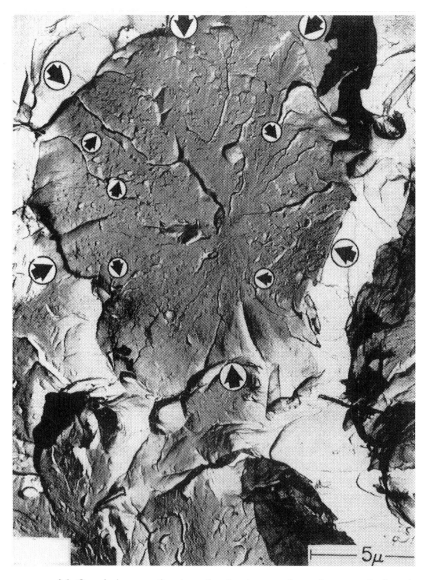

3.9 Quasi-cleavage facet on the fracture surface of a quenched and tempered type 410 stainless steel. Arrows indicate river patterns which lead away from the centre of the facet (source: ASTM).[3]

I, using data from Wessel.[6] At room temperature, even for this relatively high tensile material, the testpiece thickness must be about 250 mm, with other dimensions in proportion. For carbon-manganese steels valid fracture toughness tests would be impracticable.

Such considerations led Wells to propose a technique which would be applicable to steels in the medium and lower temperature range.[7]

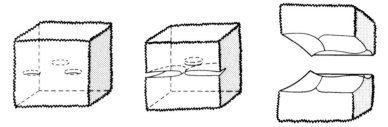

3.10 Growth of quasi-cleavage facets forming tear ridges on the fracture surface (source: ASTM).[3]

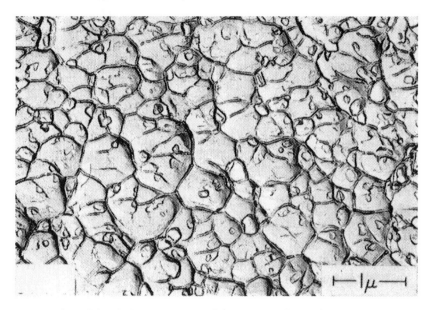

3.11 Dimpled appearance of fracture surface of aluminium alloy 7075-T6, caused by microvoid coalescence (source: ASTM).[3]

The principle of this test is based on a model of crack behaviour generally attributed to Dugdale. It is supposed that when a crack of length 2a in an infinite plate is subject to uniform tension it takes the form shown in Fig. 3.13. Over the length 2a the behaviour is elastic but at the ends plastic zones are formed. The loading is simulated by an imaginary crack of length 2a where the region −a to +a is elastically deformed whilst that from −a to −a, and from +a to +a is subject to a uniform stress equal to the yield point. It may be shown that the displacement at the tip of the original crack is:

$$\delta = 8 \frac{\sigma_{yp}}{\pi E} a \ln \sec \left(\frac{\pi \sigma}{2 \sigma_{yp}} \right)$$ [3.14]

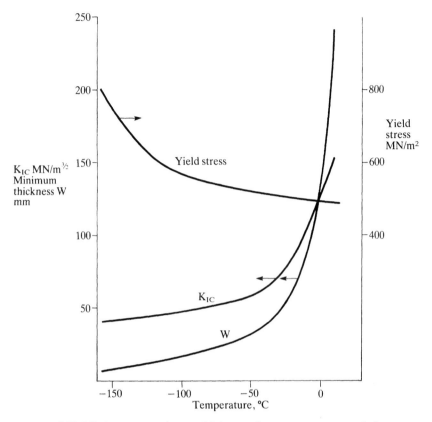

3.12 Minimum specimen thickness for measurement of fracture toughness ASTM A533 Grade B Class 1 (after Wessel).[6]

Expanding the ln sec term and ignoring all but the first term leads to:

$$\delta = \frac{\pi\sigma^2 a}{E\sigma_{yp}} \qquad [3.15]$$

Comparison with [3.10] shows that:

$$G_I = (1 - \nu^2)\, \delta\sigma_{yp} \qquad [3.16]$$

Thus, there is an equivalence between the crack opening displacement δ and G_I and, by implication, K_I. Further, it is possible, using a relatively simple notched bend test (the CTOD test, to be discussed in more detail later) to determine the critical value of crack opening displacement δ_c at which the crack will start to extend in a brittle manner. This quantity is then related to G_{IC} and K_{IC} as indicated by [3.16].

If δ_c is the critical value of δ for unstable crack extension, then applying a safety factor of 2, the maximum allowable crack size is:

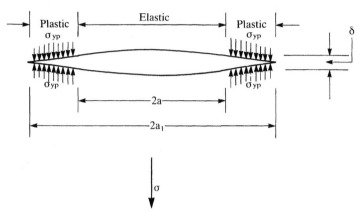

3.13 Wedge opening displacement model of a crack.

$$\overline{a}_m = \frac{\delta_c E \sigma_{yp}}{2\pi\sigma^2} \qquad [3.17]$$

This expression is used for the low stress region of the 'design curve' in PD 6493. The design curve is used to calculate the maximum defect size. At higher design stress there is a correction factor in the case of ferritic steels.

The J integral analysis

The second means of dealing with crack extension under elastic-plastic conditions was developed by Rice.[8] The strain energy density is integrated along a line that surrounds the tip of the crack, and it is shown that the result is independent of the path which the line takes. The integral is designated 'J', hence the somewhat esoteric name for the test procedure.

Performing this integration for the perfectly elastic case, discussed earlier, leads for plane strain to:

$$J = \frac{(1 - v^2)K_I^2}{E} \qquad [3.18]$$

and comparing this with [3.8] it is evident that for small scale yielding J is equal to G_I, the strain energy release rate. Rice demonstrated this equivalence by an independent method. Like G_I, J has the dimensions Joules/m².

In principle, J is measured by obtaining load displacement curves for a series of crack lengths. The area of the curve is the value of J. These values are plotted against crack length and by extrapolating the curve to zero crack length the critical value of J for crack initiation is obtained.

Fracture toughness testing

Measurement of K_{IC}

Methods of K_{IC} measurements are standardised in ASTM E 399, which requires the testing of a flat specimen in three point bending or an edge notched tension specimen. In both cases there is a machined notch which is extended by fatigue loading to develop a sharp crack. As noted earlier, both the specimen thickness B and the crack length a must exceed 2.5 $(K_{IC}/\sigma_{yp})^2$ where σ_{yp} is the 0.2% offset yield strength of the material. Consequently it is used primarily for materials of high yield strength and for embrittled steel. (It has been used for monitoring long term irradiation damage in nuclear reactors for example.)

Figure 3.14 illustrates the two types of specimen for K_{IC} testing. The bend specimen is subject to three point loading with the outer supports spaced at a distance of 4W. The other type is loaded by pins in a tensile machine. Prior to loading the fatigue crack is formed. Fatigue cracks tend to propagate faster in the centre of the section, and to minimise such curvature and maintain the crack sharp and flat the stress intensity range must be maintained within specified limits.

Prior to loading a clip-in displacement gauge of the type shown in Fig. 3.15 is fitted to the specimen. The clip may engage an integrally machined knife edge; alternatively knife edges may be attached to the surface of the testpiece. The four strain gauges shown in the sketch provide a measure of displacement, and the result of the test is a load displacement curve. A critical load P_Q determined from this curve, is the load at which pop-in (see below for an explanation of pop-in) or complete fracture occurs, or when the curve departs from linearity by a specified amount. A provisional figure for $K = K_Q$ is then obtained using standard formulae; for example in bending:

$$K_Q = P_Q S/BW^{3/2} \int (a/W) \qquad [3.19]$$

where S is the span of the bend rig and the other symbols are as in Fig. 3.14. f (a/W) is a polynomial for which values are tabulated. Finally the value of K_Q is tested for conformance with the formula B, a > 2.5 $(K_Q/\sigma_{yp})^2$; if so $K_Q = K_{IC}$. If not, the test is repeated with a larger specimen.

'Pop-in' is a feature of some K_{IC} and also of other fracture toughness tests. A brittle fracture initiates and is then arrested after propagating a short distance. This behaviour may be due to the formation of a

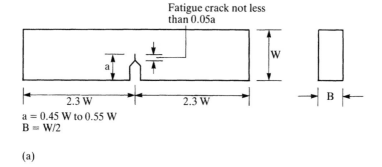

Fatigue crack not less
than 0.05a

a

W

2.3 W   2.3 W   B

a = 0.45 W to 0.55 W
B = W/2

(a)

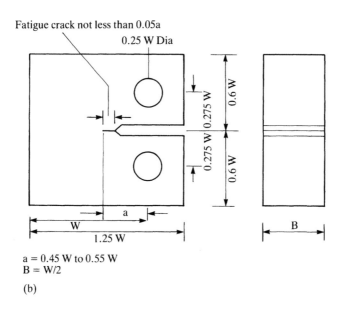

Fatigue crack not less than 0.05a

0.25 W Dia

0.275 W 0.275 W

0.6 W

0.6 W

a

W

1.25 W

B

a = 0.45 W to 0.55 W
B = W/2

(b)

3.14 Specimens for fracture toughness testing: a) Bend; b) Compact tensile.

quasi-brittle crack in the centre followed by the formation of shear lips that are capable of bearing the load. Alternatively it may be due to the testing machine characteristics; the displacement associated with the initial fracture being sufficient to reduce the applied load. Whatever the cause, such a condition is unlikely to prevail in a real structure and pop-in is taken to be equivalent to failure in calculating K_{IC} or (see below) δ_C.

All fracture toughness tests, K_{IC} included, are subject to scatter. It is customary therefore to obtain three results for any one batch of material.

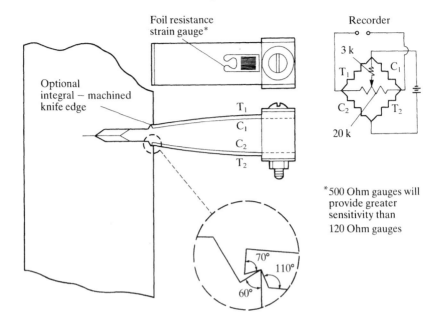

3.15 Clip gauge used for measuring displacements in fracture toughness testing (source: ASTM E 399).

CTOD testing

The crack tip opening displacement (CTOD) test is standardised in BS
5762, the latest edition of which must be consulted for details. It was
originally called the crack opening displacement (COD) test, but the
present title is considered to be more appropriate. The specimen
dimensions are similar to those shown in Fig. 3.14 but the compact
tensile type is rarely used. Almost all CTOD tests are in three point
bending. The clip gauge is also similar to that sketched in Fig. 3.15.
CTOD testing has been used mainly for carbon-manganese and low
alloy steel in the ductile/brittle transition temperature range, and has
found much use in weld procedure tests for work on North Sea offshore
structures. As a rule the notch is at right angles to the plate surface and
the dimension W is the full plate thickness, but other orientations are
possible. For a review of the development of the test and its associated
problems see Ref. 9. Development started in the early 1960s. At that
time the notch was machined and a direct measure of the crack opening was possible using a rotating paddle gauge. After changing to a
fatigue cracked specimen the relationship between the clip gauge displacement and the crack opening was far from obvious and after much
consideration the formula given in BS 5762 was adopted:

$$\delta = \frac{(1 - v^2)K^2}{2\sigma_{yp} E} + \frac{0.4(W - a) V_p}{0.4 W + 0.6a + Z} \qquad [3.20]$$

where the first term represents the elastic component of δ, V_p is the plastic element of the clip gauge displacement and Z is the height of its attachment above the surface of the specimen. The value of K is given by:

$$K = \frac{YP}{BW^2} \qquad [3.21]$$

where P is load and Y is a stress intensity coefficient, and is tabulated as a function of a/w.

Finally a plot of load against displacement is obtained. This may take various forms. The simplest is that in which a brittle fracture or pop-in occurs as illustrated in Fig. 3.16a. The value of CTOD calculated in such a case is designated δ_C. Alternative types of diagram and the corresponding designations of δ are:

(a) CTOD when slow crack growth starts δ_i
(b) CTOD at brittle fracture preceded by slow crack growth δ_u
(c) CTOD at maximum load under conditions of stable crack growth δ_m[11]

The value of δ to be used for specification purposes or for an engineering critical assessment is a matter for agreement between the parties concerned. Where brittle behaviour prevails there is no problem, but for stable crack growth it may be argued that δ_i is too conservative because the structure would remain intact if loaded beyond this point. There may also be some difficulty in determining δ_i; this requires that the load/crack length relationship be determined for a number of specimens and extrapolated back to zero.

The inherent variability of results is such that a number of tests are required if the lower bound value is to be established. Where the steel is relatively tough with a COD of 1 mm, for example, there is a limited amount of scatter but for lower mean values of say 0.2 mm the percentage scatter is greater and a larger number of tests are, in principle, required. The relative importance of the structural element concerned is also a matter that needs to be considered. For general purposes, and where other guidance is lacking, a CTOD value of 0.15 mm is often regarded as a lower limit for acceptability in the case of carbon-manganese and ferritic alloy steel.

J integral tests
The ASTM standard relevant to this test is E 813. The object of the test is to establish a value for J for the initiation of crack growth. J is the

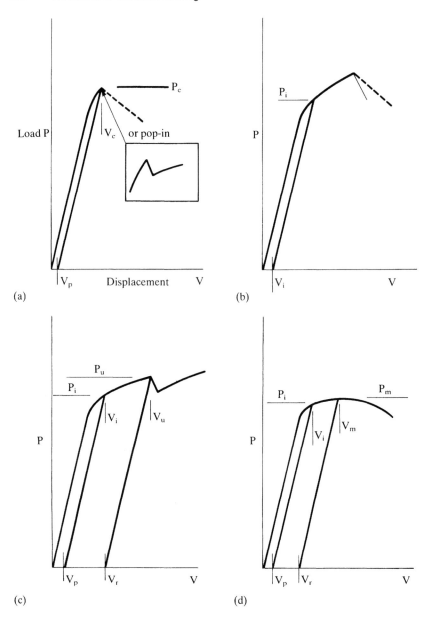

3.16 Typical forms of load/displacement curves in CTOD testing.

area of the load displacement curve for this condition. J integral testing is applicable to elastic-plastic conditions in the same way as CTOD. In other words, it may be used for carbon and low alloy steel in the transition temperature range.

The specimen plus clip required by E 813 are those illustrated in Fig. 3.14 and 3.15, and the bend test is carried out with three point loading and a span of 4W. There are two alternative ways to make the required measurements. The first is to use a number of specimens. Each one is loaded to a different level and the crack extension obtained by heat tinting and breaking the sample. The value of J is obtained for each test by measuring the area A under the load crack length curve. There are elastic and plastic contributions to J:

$$J = J_{el} + J_{pl}$$

$$= \frac{(1 - v^2) K^2}{E} + \frac{2A_{pl}}{Bb_o} \qquad [3.22]$$

where K is a function of P and a/W, as in CTOD testing, and b_o is the width of the remaining ligament $(W - a)$. Log J is then plotted against log a and a least-squares fit obtained. This curve is replotted as J versus a and extrapolated to intercept the 'blunting line' defined by $J = 2\sigma_y$ (a − 0.2 mm) and shown in Fig. 3.17. This gives the value of $J = J_Q$ for zero crack length. If J_Q meets the condition:

$$b_{o,} \ B > \frac{25J_Q}{\sigma_y}$$

where

$$\sigma_y = \frac{J_{yp} + J_{UTS}}{2} \qquad [3.23]$$

then $J_Q = J_{IC}$. From this quantity it is possible, if required, to obtain an equivalent figure for the stress intensity:

$$K_I (J) \quad \text{or } K_J = \left(\frac{J_{IC} E}{1 - v^2} \right)^{\frac{1}{2}} \qquad [3.24]$$

for plane strain.

The other method is to use a single specimen and interrupt the test at intervals to obtain the crack length and the corresponding value of J. The crack length is determined by electrical resistance or by compliance measurements. Compliance is the relationship between displacement and load, and this is a function of (a/W). This relationship may be determined by experiment or calculation. Thus, the crack length can be determined when the test is interrupted by the slope of the unloading line. This method has the advantage that it eliminates the scatter due to using multiple samples. However it is difficult to apply to inhomogeneous materials and welds, or to very ductile metals.

In some tests slow crack growth has been observed at values of J for

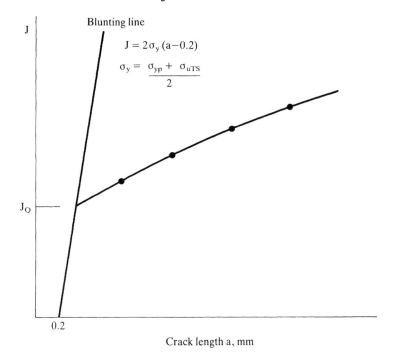

3.17 Evaluating J_Q from J integral tests.

which only elastic deformation should occur. The displacement of the blunting curve by 0.2 mm makes allowance for such initial irregularities.

Empirically it has been determined that valid J_C values are obtained if:

$$B > 0.35 \, (K_J/\sigma_{yp})^2 \qquad\qquad [3.25]$$

Comparison with the requirements of E 399 for K_{IC} testing suggests that J integral tests should permit a sevenfold reduction in sample thickness. Thus, it is of value for materials whose toughness makes the K_{IC} testing impracticable, and also for those cases where inherent variability makes it desirable to test a large number of specimens.

Other methods of measuring fracture toughness
The Wells wide plate test employs a welded plate with the weld parallel to the tensile forces. Before welding, fine saw-cuts are made in the weld preparation at right angles to the line of weld. These cuts propagate during or after welding to provide a sharp crack. The plate is tensioned using hydraulic devices and measurements are made of the stress and strain at fracture.

This test has also been used to validate the predictions of smaller scale tests by inserting a crack-like defect of known size into the weld. For example, a thin plate is edge-welded on to the surface of the weld preparation to provide an area of lack of fusion. The plate is then tested with the weld at right angles to the load.

The equivalent energy method measures the ratio of areas under the load displacement curve at the critical point and at some point in the elastic region of the curve. It employs samples of about the same size as J integral testing. Equivalent energy testing is empirical in character and has not been standardised.

Instrumented Charpy impact tests employ a fatigued cracked specimen and record a load extension diagram for the fracture. This is a dynamic test and the results show much greater scatter than other methods. Also the results lie in a scatter band below those obtained by K_{IC} tests. It has the advantage that the testpieces are small and relatively cheap, and can be used for testing small areas such as the heat affected zone of welds. However the test has not been standardised and the absolute values obtained for fracture toughness need to be checked by other methods. A critical review of the instrumented Charpy and equivalent energy methods will be found in Ref. 9.

R curve determinations have a somewhat different purpose. This test employs either bend, compact tensile or centre-notched specimens and measures the crack growth as a function of stress intensity factor under increasing tension. The curves obtained represent the increase in fracture resistance of the material due to work-hardening. It is possible to use R curves for determining the critical stress intensity factor for a running ductile crack in the region above the transition temperature. This test is standardised in ASTM Standard E-561.

The need to provide a fatigue crack is a major disadvantage of fracture toughness testing. In practice this requires the use of costly equipment and is time consuming. For this and other reasons such testing is not suitable for routine work. It is however employed in establishing welding procedures and for research, and the accumulation of data is such that conservative figures have been established for most engineering materials. Thus it is possible to make an assessment of the potential hazard presented by cracks that are found in a structure or pressure vessel during service. Techniques for making such assessments are discussed in Chapter 6.

The Charpy impact test

The standard Charpy test uses a 10 mm × 10 mm specimen with a 2 mm deep V notch, placed upon an anvil and broken by a pendulum weight. The energy absorbed is measured by the height of the swing of the pendulum after fracture. In the UK and USA this is recorded simply as energy but in Continental Europe it is not uncommon to divide the energy by the cracked area of 800 mm² to give results in J/m². The fracture appearance is also used as a criterion. The proportion of the fracture surface occupied by fibrous or shear fracture may be measured and one means of defining a transition temperature is where the fracture is 50% fibrous. This is the fracture appearance transition temperature FATT 50.

Alternatively the transition temperature is defined in terms of impact energy, the 27 J (equivalent to 20 foot-lb force) or 40 J transition for example. Such figures are of course arbitrary but nevertheless may provide a useful datum for control purposes.

A third measure provided by impact specimens is the lateral contraction at the root of the notch. For certain materials this may provide a more sensitive quality criterion than energy; more often it is used as additional information.

To provide for material thinner than 10 mm there are substandard thicknesses of Charpy specimen, and where specifications require a minimum Charpy energy, lower values are specified for the thinner specimens. It is customary to require three specimens for each test and to specify a minimum value for the average of the three, plus a minimum for any single specimen.

There is a correlation between Charpy V notch energy and K_{IC} for the upper shelf region:

$$K_{IC} \simeq \sigma_{yp} \left[\frac{0.646C_v}{\sigma_{yp}} - 0.00635 \right]^{\frac{1}{2}} \qquad [3.26]$$

where C_V is Charpy energy in joules and σ_{yp} is in MN/m².

In the transition region the situation is more complex and a number of investigators have concluded that there is no relationship between C_V and K_{IC}. However, according to Barsom and Rolfe:[9]

$$\frac{K_{IC}}{E} = 0.221 \, C_v^{1.5} \qquad [3.27]$$

where K_{IC} is in N/mm$^{3/2}$, E is in N/mm² and C_v is in Joules.

Dolby[10] found a correlation between the transition temperatures measured by Charpy V and COD for individual types of weld deposit. Figure 3.18 shows the relationship for a manual metal arc multipass deposit. However lower strength metal has a higher strain rate sensitivity, and this shifts the impact transition to a higher temperature. Therefore the slope and position of the correlation curve depends on the yield strength of the weld metal (Fig. 3.19).

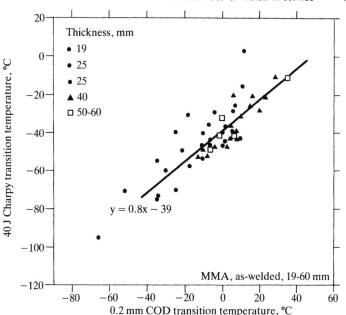

Thickness, mm
• 19
• 25
• 25
▲ 40
□ 50-60

y = 0.8x − 39

MMA, as-welded, 19-60 mm

3.18 The relation between 40 J Charpy transition temperature and 0.2 mm COD transition temperature for as-welded MMA multipass deposits.[10]

The Pellini drop-weight test

The drop-weight test was developed by the US Naval Research Laboratories in 1952. At that time fracture toughness testing had not yet evolved, and the Navy wished to obtain a more precise evaluation of the ductile brittle transition temperature than was possible using Charpy impact specimens. The object of the test is to determine the temperature (the nil-ductility temperature) at which a plate subject to impact will fail in a brittle manner. A typical sample size is five inches by two inches. A brittle weld (made using a hardfacing electrode) 2½ inches long is laid centrally on one side of the sample and then notched at its midlength. This is laid with the brittle weld face down across two supports, and a tup is released so as to drop and strike the plate in its centre. The sample is cooled to its expected nil-ductility temperature and, depending on the result, repeated at 5 °F intervals up or down until the marginal condition is obtained. Here the plate either cracks in two or cracks across a half-width. This test, which is standardised in ASTM E 208, is used, for example, for assessing the embrittlement of steel by neutron irradiation. A variant, the Pellini explosion bulge test (see Chapter 4) is employed for testing quenched and tempered steel used in submarine construction.

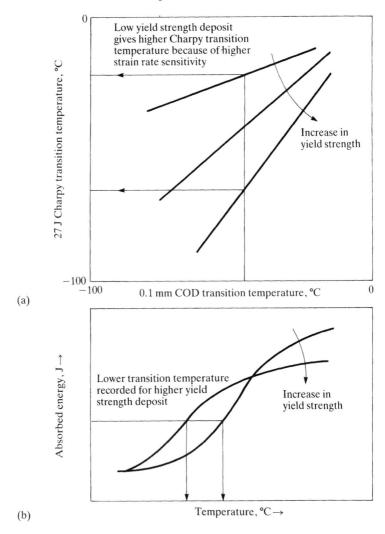

3.19 Schematic diagrams showing: a) Charpy COD correlations and influence of yield strength; b) Effect of yield strength on Charpy V data.[10]

Stable crack growth: slow cracking

There are a number of mechanisms by which a crack may propagate in a steady, non-catastrophic manner in metals. At atmospheric temperature the two most important categories are fatigue cracking and stress corrosion cracking. In both these cases fusion welds may be a weak link through the presence of discontinuities and because of residual tensile stress. At elevated temperature, creep cracking may be a problem and welds can increase the risk of failure due to embrittlement (in creep relaxation cracking) or to lower creep-rupture properties. Austenitic-ferritic joints have a limited life at elevated temperature due to their unavoidable metallurgical discontinuity.

Fatigue

The effect of fusion welding
Cracking by exposure to an alternating stress is the commonest mode of failure in machinery and structures. The problem is especially acute in the case of structural steelwork. For welds made with coated electrodes in particular there is frequently a flux-filled groove at the toe of the weld that acts as a point of initiation for a fatigue crack. Figure 3.20

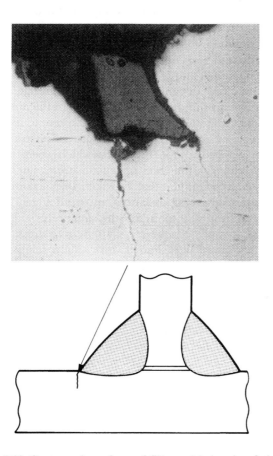

3.20 Cross-section of toe of fillet weld showing fatigue crack propagating from slag filled discontinuity.[12]

illustrates this condition for a fillet weld that has been exposed to fatigue loading. Figure 3.21 compares the fatigue behaviour of such a joint with that of an unwelded plate and one containing a drilled hole. The presence of the weld has greatly reduced the fatigue strength. In part this is due to the weld profile, which intensifies the stress at the toe of the weld. However the stress intensification due to such a profile is not greatly different to that of a drilled hole. The major part of the reduction is associated with the groove, which eliminates the period of crack initiation that is normally required with a smooth surface. Most of the fatigue life of a welded joint is spent in crack propagation.[12]

The effect of stress
The pattern of alternating stress may take many forms. A rotating rod loaded at one end is subject to alternating bending at a steady frequency. Structures are commonly subject to an irregular combination of axial loading with bending, caused for example in the case of an oil rig by wave action, or in a bridge by traffic. Process equipment is less subject to this type of stress but reciprocating pumps can give rise to pressure fluctuations, and it sometimes happens that flow in heat exchangers can set up vibration.

The severity of the effect depends upon the stress range and the mean stress. In a uniform solid only the tensile part of the alternation is damaging and the effective stress range is that from zero to maximum tensile. An as-welded joint however contains residual stress at yield point level, so that an applied compressive stress which would normally be harmless, merely changes the level of tensile stress. Therefore in such a case the total stress range must be taken into account. Even after stress relief there is some residual stress, and particularly for large structures it may be desirable to maintain the same rule.

In assessing the effect of any given level of alternating stress, account must be taken of the type of weld and its orientation relative to the direction of loading. In the UK joints have been classified according to their severity in reducing the fatigue strength, as illustrated in Fig. 3.22. For each combination of joint and stress orientation a fatigue design curve has been established (see later in this chapter) and these rules have been incorporated, for example, in Part 10 of BS 5400 for steel bridges.[13] This procedure differs from that normally applied for design against fatigue loading. In the case of rotating equipment, for example, a fatigue limit is established based on tests made with a smooth bar, and this limit is reduced by a stress concentration factor which is specific to a particular design feature, e.g. a keyway.

Fatigue strength
The results of fatigue tests are commonly plotted on an S–N diagram, which shows the logarithm of the stress range as a function of the

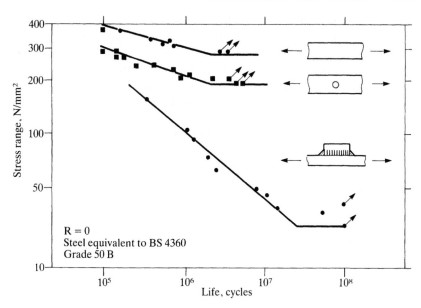

3.21 Comparison between the fatigue strengths of plain steel plate, plate with a central hole and plate with a fillet welded attachment.[12]

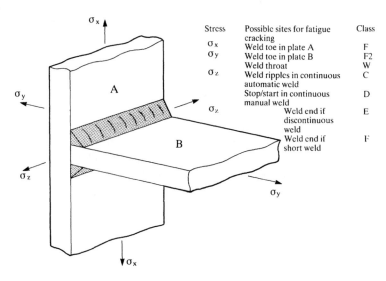

Stress	Possible sites for fatigue cracking	Class
σ_x	Weld toe in plate A	F
σ_y	Weld toe in plate B	F2
	Weld throat	W
σ_z	Weld ripples in continuous automatic weld	C
	Stop/start in continuous manual weld	D
σ_z	Weld end if discontinuous weld	E
	Weld end if short weld	F

3.22 Welded joint classification for design against fatigue.[13]

logarithm of the number of cycles. As shown in Fig. 3.21 there is, for unwelded steel, a fatigue limit, that is a stress range below which fatigue cracking does not occur. The same figure indicates a much lower fatigue limit for welded material but such a limit may not always be found. Likewise aluminium and aluminium alloys, welded or unwelded, do not have a fatigue limit, and for design it is necessary to specify the required fatigue life.

Where the stress range is not constant the number of cycles to failure is usually calculated using Miner's rule. This states that:

$$\Sigma \frac{n}{N} = \left(\frac{n_1}{N_1} \quad \frac{n_2}{N_2} \quad \frac{n_3}{N_3} \cdots + \frac{n_i}{N_i} \right)$$

where n_i is the number of cycles at stress range S_i and N_i is the number of cycles to failure at this stress range, and so forth. In general for safe design $\Sigma \frac{n}{N_i} < 1$ but there may be circumstances where a number less than one is required.

In other cases an equivalent constant stress range may be calculated. For example in PD 6493 it is proposed that if the number of cycles at stress range S_i is n_i, the corresponding constant range S for 10^5 cycles is:

$$S = \left(\frac{\Sigma n_i S_i^4}{10^5} \right)^{\frac{1}{4}} \qquad [3.28]$$

The section thickness may affect the fatigue strength. Thicker sections have lower fatigue properties and it has been proposed that for thicknesses up to 32 mm the allowable stress S_a range should be calculated from:

$$S_a = S \left(\frac{32}{t} \right)^{\frac{1}{4}} \qquad [3.29]$$

where S is the allowable stress range for the joint in question and t is thickness in millimetres.

A second type of fatigue test employs the type of specimen used for fracture toughness testing. The rate of crack growth is measured as a function of the stress intensity factor range ΔK. There is no size effect in fatigue testing so that test pieces of reasonable dimensions can be used. It is found that:

$$da/dN = C (\Delta K)^m \qquad [3.30]$$

Figure 3.23 shows a typical curve; in this instance for 2¼Cr1Mo and 2½Ni weld metals.[14] The plot is logarithmic with a slope m. Putting $\Delta K = ka^{\frac{1}{2}} \Delta\sigma$ and integrating leads to:

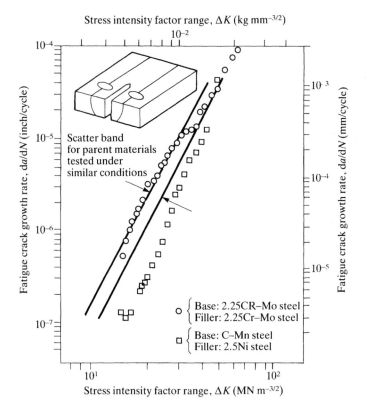

3.23 Fatigue crack growth in butt welds tested in air at 24 °C.[14]

$$(\Delta\sigma)^n N = \text{Constant} \qquad\qquad\qquad [3.31]$$

This is the S–N relationship and n = m.

The straight line relationship between log (da/dN) and log (ΔK) indicated by Fig. 3.23 is not necessarily true for low or high values of the stress intensity range. At both low and high values the slope of the curve is reduced, giving it an overall S shape.

The crack growth rate is little affected by tensile strength or microstructure. This is significant in the case of a welded joint. With a smooth testpiece crack initiation influences the result and the fatigue strength increases with the ultimate strength. In fusion welds, on the other hand, the propagation rate determines the fatigue life and the fatigue properties are not affected by tensile strength. Likewise the path of a fatigue crack is not influenced by material properties, and a crack initiated at the weld toe will travel straight through the variety of microstructures that lie below it.

Design data

The stress concentration factor associated with the weld profile has, therefore, an important effect on the fatigue life. As will be evident from Fig. 3.22 this factor will vary according to the direction of the alternating stress. S–N curves have been established for each of the classes indicated in this figure and are plotted in Fig. 3.24. These curves were set two standard deviations below the mean values for a large number of results collected internationally by The Welding Institute, Cambridge. The mean slope n of fatigue crack growth rate curves for welds is 3, and correspondingly the slope of the curves for classes D–W in Fig. 3.24 is $-\frac{1}{3}$.

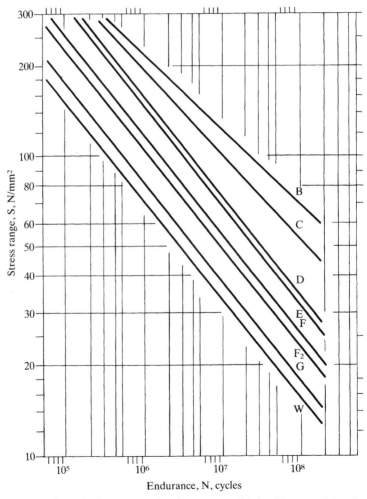

3.24 Fatigue design curves for the classes of joint illustrated in Fig. 3.22.[13]

No fatigue limit is shown. It is considered that welded joints exposed to a stress range not greater than that corresponding to an endurance of 10^7 cycles will not suffer fatigue cracking. However, if joints are exposed to a higher stress range for a period and then to a range below the 10^7 limit, cracking may continue even at the lower stress.

Figure 3.24 represents welds of average quality. Where the profile is unfavourable, which may be the case with manual welds made in the overhead position, the fatigue properties could be lower than those indicated. In general, the fatigue strength is reduced as the reinforcement angle (the angle between the edge of the reinforcement in a butt weld and the plate) is reduced.

Means of improving fatigue performance
It will be self-evident that removal of the notch at the toe of the weld could greatly improve the fatigue performance of joints. There are two ways of doing this: by remelting using the tungsten arc process or mechanically, by machining or grinding. Alternatively the surface may be given a compressive stress by shot blasting or peening.

The largest improvement in properties is achieved by GTA remelting. A run is made along each weld toe so as to remelt the surface layers and eliminate the groove. Figure 3.25 shows the fatigue strength at 2×10^6 cycles for butt and fillet welds in three different steels.[15] These are grades E 355 and E 460 to DIN 17102 and a proprietary high tensile type E 690; the number indicates the yield strength in N/mm^2. As noted earlier, increasing the tensile strength does not increase the fatigue properties. GTA remelting however has a substantial effect; the mean increase in fatigue strength being 60% for butt welds and 80% for fillet welds. Unfortunately the process is costly and not easy to apply under site conditions.

Peening may be carried out using a pneumatic tool fitted with a solid head. It is necessary to make several passes along the joint to ensure consistent results. Four passes, with an indentation 0.6 mm deep have been recommended. Shot peening can give similar results but it is more difficult to control because there is no measurable indentation. Both techniques produce a very thin surface layer and could therefore be vulnerable to corrosion; also, the notch-ductility could be reduced.

The least effective, but most practical technique is grinding. There are two methods: disc grinding or burr grinding. These are illustrated diagrammatically in Fig. 3.26. Figure 3.27 shows median curves for hammer peened and ground welds.[16] Grinding gives an increase of about 30% in the fatigue strength at 2×10^6 cycles as compared with the as-welded condition. Grinding is the most generally favoured technique because the tools are readily available on job sites, the results are visible and inspectable, and costs are not excessive. Burr grinding

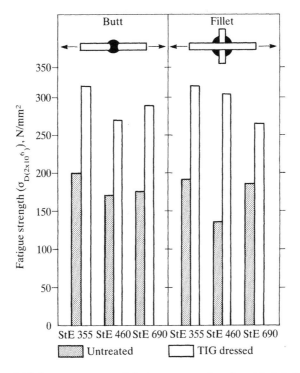

3.25 Improvement of fatigue strength of welded joints by GMA remelting R=0.[15]

is less subject to error than disc grinding but either method is acceptable subject to proper control.

Corrosion fatigue

Corrosion fatigue is the term used for the damage suffered by metals subject to alternating loads in a corrosive environment. Corrosion results in pitting of the metal surface, and the pits act as stress raisers. Cracking is initiated at the root of one or more pits, thus reducing the fatigue life. In this respect the condition is similar to that of a fusion weld in that the initiation phase of fatigue cracking is reduced or eliminated. Also in corrosion fatigue the fatigue limit may not be observed, and where it exists, may change in value according to the nature of the corroding medium. Likewise the corrosion fatigue strength is independent of the ultimate strength in the case of steel.

The effect of corrosion on unwelded metals may be expressed as a damage ratio, which is the corrosion fatigue strength at some specified number of cycles divided by the endurance limit in air. This ratio

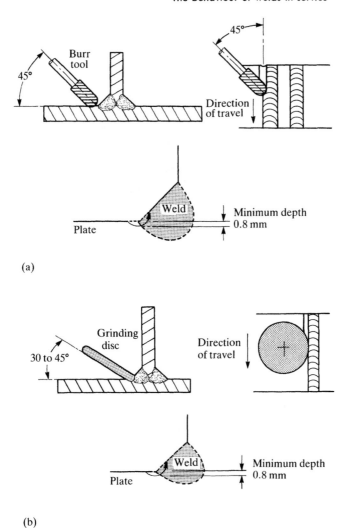

3.26 Methods of grinding the toe of a weld: a) Burr grinding; b) Disc grinding. [14]

increases with the corrosion resistance of the metal. In tap water the damage ratio for carbon steel is typically in the range 0.4 to 0.75, for aluminium it is 0.5 whilst for austenitic chromium-nickel steel and copper it is 1.0.

Figure 3.28 shows the results of tests carried out in sea water at 5 °C in connection with the UK Offshore Steels Research Project.[17] These curves are for cruciform welds in carbon steel cycled at 0.16 Hz, and there is little difference between the S–N curve in air and that in sea

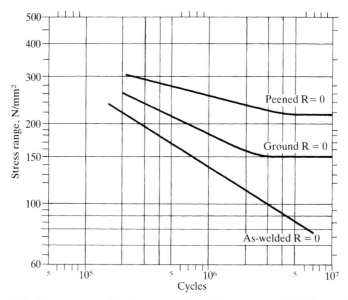

3.27 Comparison of fatigue strengths of specimens with fillet welded attachments in the peened, ground and as-welded conditions (from G S Booth). [16]

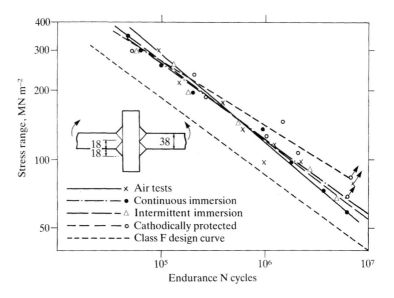

3.28 S–N results for transverse welded joints for exposure in air or sea water, compared with the BS 5400 design curve (reproduced by courtesy of The Institute of Materials). [17]

water. The relevant design curve from BS 5400 is also shown; this evidently gives a reasonable factor of safety. The figures for cathodically protected welds, in particular, lie well above the design curve.

Improvement techniques such as peening, grinding and TIG dressing may not be effective in sea water exposure.

The crack growth rate of carbon and low alloy steels in boiler quality water is higher than that in air. It is also sensitive to the stress ratio R, where:

$$R = minimum \ K/maximum \ K \qquad\qquad [3.32]$$

The crack growth rate is increased for higher values of R. It also increases as the loading frequency decreases, although this effect is less significant than that of R. Figure 3.29 shows these effects diagrammatically.[18] Section XI of the ASME Code contains a chart showing crack growth rates as a function of stress intensity range for carbon and low alloy steels in air and water, and the curves for water exposure make allowance for the effect of R (Fig. 6.19).

Effective protection against corrosion fatigue requires that the danger of pitting must be eliminated. Painting may give some protection under mild conditions if a zinc based primer is used, but for the more severe exposures such as a North Sea oil rig full cathodic protection using sacrificial anodes or impressed current must be employed. UK practice in the design of such structures against fatigue assumes such protection, but in some countries a corrosion fatigue limit equal to half the fatigue limit in air is permitted for unprotected structures exposed to sea water.

Hydrogen-induced cracking

Cracking associated with the presence of hydrogen has already been discussed in relation to welding in Chapter 2. It may also occur during service in process plant that contains hydrogen, and in equipment exposed to corrosion processes that generate hydrogen. The corrosion problem will be considered separately. Hydrogen cracking in process plant is rare, but it has for example caused the catastrophic failure of an ammonia converter.

The necessary preconditions for hydrogen crack growth are the same as in welding: the presence of hydrogen in the metal combined with a susceptible microstructure and temperature in the range -100 to $+200$ °C. Generally the problem is only likely to arise with higher tensile strength steels, in particular those in the quenched and tempered condition. It is conjectured that hydrogen dissolved in the metal will diffuse preferentially to regions of triaxial strain such as those around the tip of a crack or stress raiser. When the hydrogen concentration reaches a

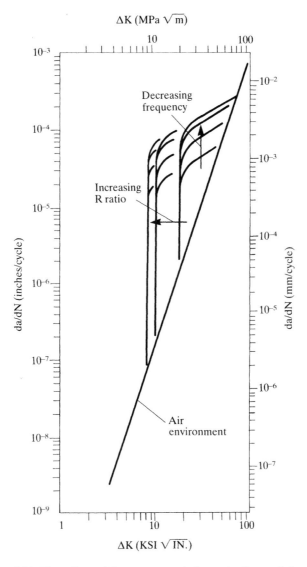

3.29 The effect of frequency and the ratio R = minimum K/maxi-
mum K on the rate of growth of corrosion fatigue cracks in test simulating conditions in a pressurised water reactor (source: ASME).[18]

critical level the crack will extend a short distance, fresh metal will be
strained and the process is repeated. There is a minimum value of the
stress intensity factor required to initiate and maintain such crack
growth. This quantity, designated K_H or K_{IH}, has been measured using
a wide variety of techniques. The results of such tests are plotted in Fig.

3.30 as a function of yield strength. The line is a lower bound curve for unwelded steel; the points below the line are for weld metal.[19]

If the steel has been embrittled it becomes more sensitive to hydrogen cracking. Quenched and tempered 2¼Cr1Mo steel is often used for heavy-wall pressure vessels handling hydrogen, and this steel may suffer temper embrittlement after long term service. Figure 3.31 shows the effect of such embrittlement on K_{IC} and K_{IH}.[20] This figure suggests that in the unembrittled condition K_{IH} for 2¼Cr1Mo steel lies between 50 and 100 MN/m³. The yield strength of this steel is generally 350–450 MN/m², that is on the extreme left of Fig. 3.30, where the lower bound curve shows 50 MN/m for K_{IH}.

Such figures may be required for an assessment of the failure risk in a hydrogen charged steel containing a defect or stress raiser.

Figure 3.32 shows crack propagation rates for a quenched and tempered 13% Ni steel having a 0.2% yield stress of 1415 N/mm². The tests were carried out using a compact tensile specimen in purified hydrogen at three different pressures. The curves have three stages; I, where the rate rises quickly from the lower limit K_{IH} to a plateau in stage II and finally in stage III K approaches K_{IC} for the steel in air and the growth becomes unstable.[21] The growth rate in stage II is in the range 10^{-5} to 10^{-4} m/s, such that a 100 mm section would be penetrated in, at the most, a few hours. K_{IH} is indicated to be 10–20 MN/m³/², which is consistent with Fig. 3.30.

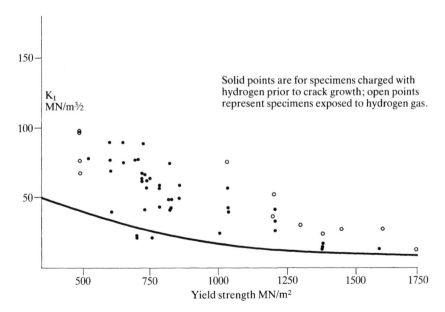

3.30 K_H as a function of yield strength, quenched and tempered steels.[19]

Stress corrosion cracking

General

There are many combinations of corrodent, alloy and stress which can generate cracks, as opposed to a general dissolution of the metal. These combinations fall into two main categories. In the first, atomic hydrogen produced by corrosive action at the surface diffuses inwards and causes the same type of cracking as that described earlier. Alternatively, or additionally, hydrogen may recombine at high pressure along planes of discontinuity, such as laminar inclusions. Such action occurs

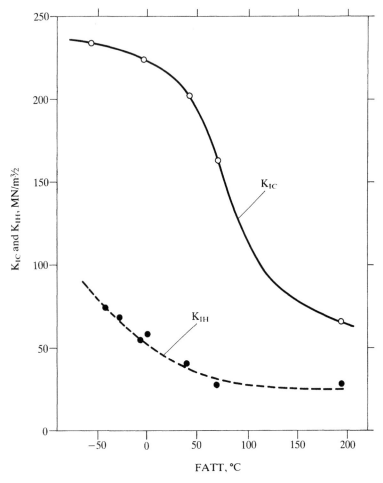

3.31 Room temperature K_{IC} and K_{IH} versus fracture appearance transition temperature (FATT) of 2¼Cr-1Mo steel.[20]

in the presence of compounds that inhibit the recombination of hydrogen atoms to form gas; notably hydrogen sulphide and cyanides.

The other category includes those systems where the metal surface is generally passive due to the presence of an oxide or other film, and where this film may be locally ruptured to expose clean metal, which then dissolves. Initially a pit is formed and this develops into a crack. At the crack tip bare metal is exposed whilst the sides of the crack are repassivated. The crack then propagates either through solution of the bare metal at the tip (anodic dissolution) or because hydrogen is generated by the corrosive action. This hydrogen diffuses into the strained zone ahead of the crack tip. When the hydrogen concentration in the strained region reaches a critical level a fresh length of crack forms and joins up with the original; the same mechanism as for hydrogen cracking discussed earlier. More complex mechanisms are possible.

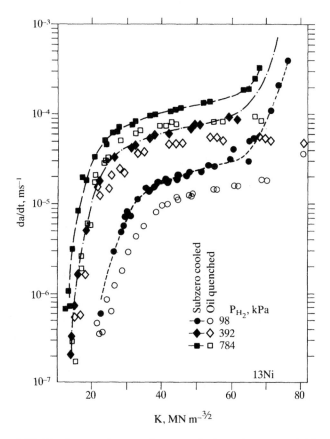

3.32 Crack growth rates in quenched and tempered 13% Ni steel exposed to purified hydrogen at room temperature (reproduced by courtesy of The Institute of Materials).[21]

The presence of hydrogen for example, may promote slip and thereby expose unfilmed metal which is then removed by anodic dissolution.

In stress corrosion, crack growth rates plotted as a function of stress intensity K shows the same general pattern as that of hydrogen cracking (Fig. 3.32). The growth rates in stage II however are in the range 10^{-9} to 10^{-6} m/s. In some systems the growth rate is consistent with an anodic dissolution mechanism; in others the forward diffusion of hydrogen may control the growth rate.

Cracking may also occur when stress is applied to a metal subject to intergranular corrosion, (e.g. austenitic stainless steel that has been sensitised by intergranular carbide precipitation).

By carrying out fracture toughness tests in the relevant corrosive media it may be possible to determine a limiting value of the stress intensity factor below which cracks do not propagate. This value is designated K_{ISCC}. Such data must be treated with caution because behaviour under corrosive conditions may vary unexpectedly.

Welding may predispose the metal to stress corrosion cracking in two ways. In the as-welded condition the residual stress is of yield point magnitude and is quite sufficient to initiate most forms of stress cracking. Moreover, the channel which may be present at the toe of butt and fillet welds acts as a stress concentrator. Stress relief may reduce the susceptibility but is not a certain cure, and in austenitic stainless steel is impractical under normal circumstances. Secondly, metallurgical changes in the heat affected zone may render the steel susceptible (e.g. by creating hard zones or, in the case of austenitic stainless steel, by carbide precipitation).

H₂S corrosion
There are two main sources of hydrogen sulphide. It may occur naturally; for example, in oil and natural gas wells in the Middle East, Canada, and notably in the Lacq gas field in France, where gas contains up to 15% H_2S. Alternatively, H_2S may be generated by sulphate reducing bacteria. Sulphates are abundant in fresh water, sea water and soils, and given the right conditions (namely a temperature between 5 and 50 °C) neutral pH and low oxygen content, such bacteria can flourish.

H_2S may cause general corrosion of steel in aqueous solution at room temperature. At elevated temperatures mercaptans in crude oil break down to form H_2S and may corrode carbon steel whilst hot mixtures of hydrogen and H_2S which form in desulphurisation units can attack carbon, low alloy and stainless steel. At room temperature and in the presence of moisture or in aqueous solution the corrosive action of H_2S on iron and steel results in the formation of atomic hydrogen. In common with other 'poisons' such as cyanide, H_2S inhibits the reassociation of the hydrogen atoms to form hydrogen gas, and the bulk of the hydrogen diffuses into the metal at concentrations well above the normal solubility. If there is a localised tensile stress present in the metal (in the vicinity of a weld, for example), and if the level of stress is high enough, such hydrogen concentrations can result in the formation of cracks that run at right angles to the direction of the stress. This condition is known as sulphide stress corrosion cracking, or sulphide stress cracking. In the absence of stress the hydrogen may desorb at discontinuities such as inclusions. The hydrogen atoms recombine to form gas at high pressure, and this can induce a laminar form of cracking known as hydrogen pressure induced cracking, hydrogen blister cracking, hydrogen pressure cracking, hydrogen-induced stepwise cracking or, more commonly, hydrogen-induced cracking. The term 'hydrogen pressure induced cracking' will be used here to avoid confusion with those forms of hydrogen-induced cracking that are not dependent on the presence of laminar inclusions, such as sulphide stress cracking.

Sulphide stress cracking became a general problem in the petroleum industry in the 1950s, due to the development of deeper, high pressure wells. Parts affected were casing, tubing, wellhead fittings and flow lines. Above ground, failures occured in transmission gas pipelines, and here the cracking was located at the welds, including the longitudinal submerged-arc welds. Two types of crack are observed. Type I is a series of laminar cracks joined by a crack at right angles to the surface, and is really a localised form of hydrogen-induced pressure cracking. Type II is the normal type of transgranular crack which may occur in the weld or heat affected zone (Fig. 3.33).[22]

The susceptibility of a steel to sulphide stress cracking may be assessed by exposing a stressed specimen to an aqueous solution of H_2S and measuring the time to failure. Figure 3.34 is a summary of the results of a large number of tests made with specimens strained to the indicated percentage of the yield stress. These tests were made in 5% sodium chloride solution with an addition of 3000 ppm H_2S. Lower amounts of H_2S give longer failure times, but the lower hardness limit is not much affected.[23]

Based on such information the US National Association of Corrosion Engineers have set up standards for materials in H_2S (sour) service. In general, for ferritic carbon and low alloy steels the maximum permitted hardness is 22 on the Rockwell C scale, which corresponds to a yield strength of about 80 Ksi (550 MN/m^2). Welds in carbon steel are acceptable in the as-welded condition provided that the hardness does not exceed 200 Brinell, testing being carried out on the weld metal. Experience has shown that such testing provides a reliable means of preventing H_2S stress corrosion cracking in carbon steel welds where the carbon equivalent is below 0.45. It is also applied to piping carrying corrosive aqueous fluids containing cyanides.

Gas and crude oil from North Sea wells have a low sulphur content

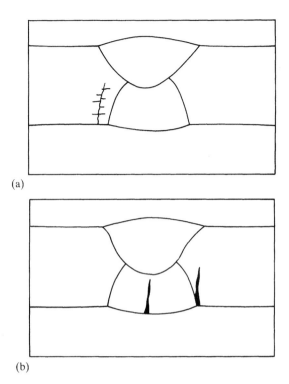

(a)

(b)

3.33 Sulphide stress corrosion cracking, Types I and II: a) Type I
SSCC; b) Type II SSCC.[22]

and are classified as 'sweet'. There has been concern about the possibility
of H_2S formation by bacterial action. To increase the yield from hydrocarbon reservoirs, water is pumped in to maintain pressure and flow,
and there is a risk that infection with sulphate reducing bacteria could
result in an increase in H_2S levels. Biocide treatment may be necessary
in such cases.

It has already been noted that, as well as causing stress cracking,
hydrogen sulphide may also cause a non-localised form of delamination and cracking called hydrogen-induced pressure cracking. This may
or may not be associated with welds, but for completeness will be considered briefly here.

The problem arose in the Persian Gulf during the early 1970s. Several
lines that were carrying sour gas were affected. In one case the line ruptured only a few weeks after commissioning. This was a spiral welded
pipe and the failure occurred adjacent to the welds. The pipe had suffered
hydrogen incursion resulting in lamination and cracking to the point
where it finally ruptured due to the internal pressure.

These failures were attributed to the use of fully killed controlled

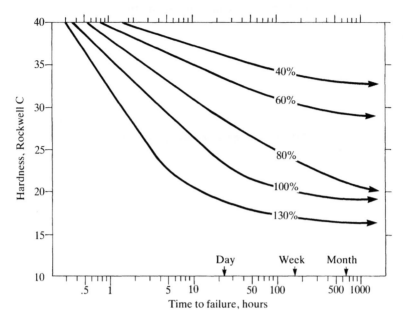

3.34 Approximate relationship between hardness, level of stress and time to failure for steel exposed to 3000 ppm H$_2$S in 5% NaCl solution. Figures on curves are stress in terms of per cent of yield (reproduced by courtesy of NACE).[23]

rolled strip in the manufacture of the pipe. Such material gave better yield strength and impact properties than semi-killed steel, but it contained very elongated Type II manganese sulphide inclusions, which provided sites for hydrogen desorbtion and the formation of laminar cracks. These may take various forms, as sketched in Figs. 3.35 and 3.36. Steelmakers have attacked this problem in two ways; by desulphurisation, so as to reduce sulphur contents down to 0.002% or lower, and by calcium treatment, which results in the formation of more rounded inclusions.

However a second problem has appeared; that of segregation. Most strip is produced from continuously cast slab, and some elements, notably manganese and phosphorus, segregate to the central region of the slab. When rolled out to strip, these segregates may form a hard band which is susceptible to hydrogen cracking even in the absence of laminar inclusions. In consequence some pipe failures occurred opposite the weld. This is again a steelmaking problem, and it has been tackled in various ways: by modifying the design of the casting machine, electromagnetic stirring and modifying the composition. Manganese and carbon contents have been reduced, the tensile properties being

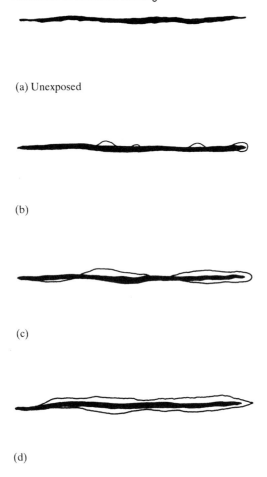

(a) Unexposed

(b)

(c)

(d)

3.35 Schematic representation of HPIC development at an inclusion. With increasing time of exposure b) and d) progressive separation along with inclusion/matrix interface occurs.[22]

maintained by increasing the alloy content. Naturally the problem becomes more severe with the high tensile pipe grades, × 65 and above.

One way of avoiding hydrogen-induced cracking is to dry the gas. Unfortunately this is not always practicable, and when practicable, not always reliable. Bearing in mind the speed of the attack, this means that the main effort has been directed towards avoidance by either improved material, injection of inhibitors, or a combination of the two. Piping for sour service is usually required to meet a test under simulated service conditions. The testpiece, which may include a weld, is immersed in the test solution for a period of 96 hours or more, without any applied stress. It is then removed, the amount of diffusible hydrogen

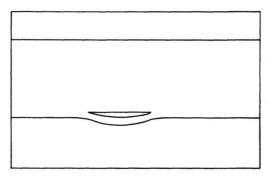

(a) Relatively small surface blistering

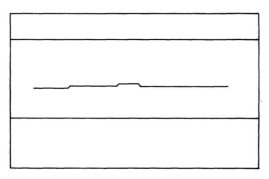

(b) Straight cracking with little or no stepwise
     component

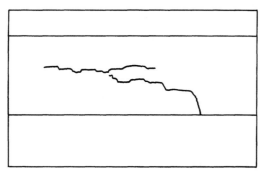

(c) Extensive stepwise cracking with surface emergent
     component.

3.36 Various forms of HPIC damage.[22]

is measured and the degree of cracking is estimated in a microsection. The NACE test solution contains 0.5% acetic acid and 5% sodium chloride, and is saturated with H_2S; this represents acid conditions. A milder reagent is that developed by Cotton; synthetic sea water saturated with H_2S. Many other testing procedures have been used, including full scale rigs.

The severity of the test is important, because it can accept or reject a steel. For example, the addition of 0.3% copper gives a protection in solutions with a pH higher than 5 but not in acid solutions. In these circumstances individual operating companies set up their own specifications and test procedures.

Water

From time to time there have been reports of the cracking of carbon and low alloy carbon steel in contact with high purity (deionised) boiler feed water or wet steam in the temperature range 75–120 °C. In 1960 cracks were found around tubeholes in a number of boiler drums in Germany. The cracks were radial but preferably oriented in line with the long axis of the drum. When broken open the fracture surfaces were found to be rippled, which suggested the possibility of corrosion fatigue. However, similarly rippled cracks were found in the girth welds of a carbon steel line carrying deionised, deaerated feed water for a refinery boiler (Fig. 3.37), and this line had not been subject to fatigue loading. The cracks originated from areas of lack of fusion at the root of the weld.[24]

The catastrophic failure of a steam turbine at Hinkley Point 'A' Power Station in 1969 was widely reported. The prime cause of failure was a stress corrosion crack originating from a keyway in one of the low pressure turbine discs, which was made from a medium strength low alloy steel. As in all such cases water quality was at first thought to be a factor, but exhaustive analyses showed that this was not the case, and it was accepted that stress corrosion cracking in high purity water was possible. An important contributory factor was the absence of flow; crack initiation took place readily when tests were made in stagnant conditions. Such conditions existed in the keyway of the turbine disc and also in the weld shown in Fig. 3.37.

Pitting, and in particular pitting around a sulphide inclusion, can provide stagnant conditions where the water can become acidic and far from pure. Figure 3.38 shows how pits can develop adjacent to a non-metallic inclusion. Given the right conditions such pits can initiate a stress corrosion crack.[25]

The first defence against such cracking is to avoid discontinuities. Steel quality may also be significant. The German experience was with a copper-bearing low alloy steel that showed embrittlement after long term service.

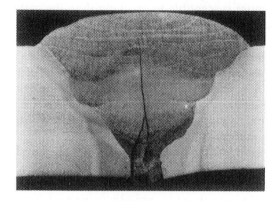

3.37 Crack in girth weld of carbon steel pipe carrying deionised, deaerated boiler feed water (reproduced by courtesy of Elsevier Applied Science Publishers Ltd).[24]

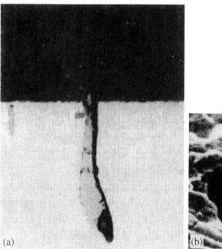

3.38 Local attack adjacent to an inclusion in 3Cr-Mo steel: a) Cross-section ×400; b) Deep etched in Nital. SEM ×1300 (reproduced by courtesy of The Institute of Materials).[25]

Anhydrous ammonia

Pure ammonia is widely used as a fertiliser. After manufacture it is distributed in tankers to agricultural areas where it is stored in various forms of tankage before being injected directly into the ground. Problems have arisen due to cracking both of tankers and storage tanks. The cracks are mainly transgranular. They occur in weld metal and also in cold formed parts such as heads and plates of spherical tanks. Medium tensile steel to ASTM A517 (TI steel) was used for road tankers and this proved somewhat more susceptible to cracking than the carbon steel used for storage tanks.

Investigation showed that the cracking was due to stress corrosion, although the mechanism was not established. However the primary causative agent is oxygen. In commercial ammonia an oxygen content of 5 ppm or greater was sufficient to promote cracking and it was inhibited by the addition of 0.2% water. It is standard practice to make such water additions to commercial anhydrous ammonia but from time to time upsets and cracks are found.

The remedial action is to grind out cracks (which may be quite shallow) and where practicable to stress-relieve. This operation is possible in the case of spheres and bullets but more difficult for cylindrical tanks. Ammonia tanks are subject to periodic inspection for cracking. It is generally recommended that bullets should be stress-relieved or at least be fabricated with hot-formed or stress-relieved heads, and that every precaution is taken to avoid oxygen contamination.

CO-CO_2-H_2O mixtures

Mixtures of these gases occur in a number of chemical processes: in the manufacture of synthetic town gas, and in the formation of a synthesis gas for the conversion of coal to oil, for example. Stress corrosion cracking in such mixtures was observed in Japan and elsewhere and examined systematically by Kowaka and Nagata.[26] The cracking affects carbon and low alloy steels and is transgranular. It occurs in the region of welds and also in metal under tension; for example it occurred in a high pressure gas cylinder. The simultaneous presence of all three components: carbon dioxide, carbon monoxide and water vapour is necessary. The problem is considered to be due to adsorption of CO on the metal surface, such that it acts as a passivating agent. Rupture of the adsorbed layer can then give rise to corrosion cracking as described earlier.

Cracking may occur in a wide range of CO-CO_2 mixtures, as shown in Fig. 3.39. Wetting of the surface is necessary, and it is possible to avoid cracking by keeping the metal surface at a temperature above the dew point of the mixture. The problem is avoided by using austenitic stainless steel as the material of construction wherever condensation can occur, by stress-relieving and possibly, in the case of piping, by steam tracing (heating the pipe).

Other stress corrosion cracking environments that affect ferritic steel

It is frequently required in chemical processing to remove CO_2 from a gas mixture, and this is commonly done by absorption in hot potassium carbonate solution. This solution is corrosive to carbon steel but the attack can be prevented by adding inhibitors. One such inhibitor is arsenic pentoxide, and whilst this is effective in preventing general

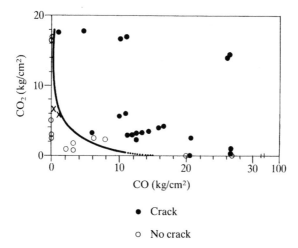

3.39 Effect of partial pressure of CO and CO_2 on stress corrosion crack-ing of carbon steel tested between 18 and 70 °C for 168–1000 hours (reproduced by courtesy of NACE).[26]

corrosion it does promote stress corrosion cracking at welded joints. The problem can be alleviated by stress-relieving, but this is not a cer-tain cure. Where cracking has occurred in CO_2 removal plants of this type, it is common practice to weld a formed strip of steel around the affected joint. Welds repaired in this way have quite a good life, pre-sumably because the stress due to two fillet welds on the outer surface is lower than that due to a butt weld and may indeed be compressive.

Caustic cracking is one of the classic modes of stress corrosion cracking. It is intergranular and the required combinations of concentration and temperature to cause cracking of a butt weld in steel are plotted in Fig. 3.40.

Given the right combination of acidity, concentration and tempera-ture, stress corrosion cracking of carbon steel may occur in carbonate, nitrate and phosphate solutions, whilst high strength steels may crack in water, chlorides and sulphates.

Selective corrosion
The weld and heat affected zone differ in various ways from the parent metal, and this may result in selective corrosion of one or other of these regions. In general, such corrosion takes place in the more highly conductive media, particularly if they are acidic.

Weld metal is segregated in that if there are any impurities present they will be concentrated in the interdendritic regions. Thus, for example, commercial purity aluminium has a good resistance to attack in hot, strong nitirc acid, but under the same conditions weld metal is selectively

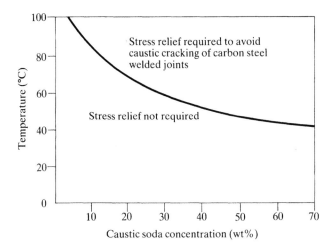

3.40 The conditions under which stress relief of welded joints in carbon steel is required in order to avoid stress corrosion cracking in caustic soda solution.[36]

corroded. Carbon steel weld metal may likewise be selectively dissolved in oxygenated sea water.

The weld metal may also differ in composition from the parent metal. This may be for operational reasons, as in open arc welding, where substantial additions of aluminium are required, or to improve properties, or to improve corrosion resistance. This in turn may set up a galvanic couple so that regions adjacent to the weld are dissolved. Sometimes, however, it may simply protect the weld metal so that the weld remains unaffected when there is general corrosion of the plate (as is the case for open arc welds). Differences in composition may also be due to contamination. Titanium welds made on a steel backing, for example, pick up iron and their corrosion resistance is reduced.

Differences may also be due to transformations. In ferritic alloy steel the corrosion resistance tends to fall as the hardness increases. Where such hardening occurs in the heat affected zone there may be preferential attack in this region. Another classic case is that of weld decay in austenitic stainless steel, where the attack is parallel to but separate from the weld. In practice, weld decay is now a rare phenomenon. Since the advent of argon-oxygen decarburisation, excessive carbon content in unstabilised stainless steel is unlikely. However a form of weld decay may affect nickel base alloys such as Hastelloy, due either to carbide precipitation or to intergranular precipitates of a molybdenum-rich phase.

Selective attack on welds has given problems in North Sea gas and oil fields, as noted in Chapter 4.

Austenitic chromium-nickel steels

Alloys covered by the AISI 300 series specifications, typically the 18Cr10Ni or 25Cr20Ni types, are very susceptible to stress corrosion cracking in certain media. At lower levels of nickel content, where the alloy constitution is basically ferritic with a discontinuous austenite content (e.g. 26Cr4Ni) the risk of chloride stress corrosion is small; likewise as the nickel content increases the risk diminishes, as shown in Fig. 3.41.

Stress corrosion cracking of austenitic stainless steels has been observed in sea water, hydrogen sulphide solutions, acid chloride solutions and strong solutions of caustic alkalis. The most frequent incidence of failure is in chloride solutions, and the standard test for susceptibility is the exposure of a strained specimen in boiling 42% magnesium chloride solution. The cracking is transgranular and is typically branched. Welds are self-evidently a hazard since the residual stress is high and there is no practicable means of removing it. Stainless steel subject to cold work also has a high level of residual stress and is likewise susceptible.

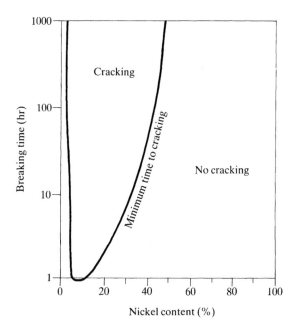

3.41 Effect of nickel content on the stress corrosion cracking susceptibility of nickel-chromium-iron alloys.[36]

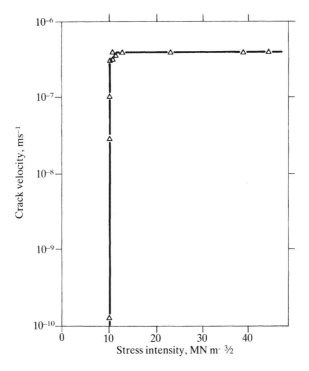

3.42 Crack velocity in the stress corrosion cracking of austenitic chromium-nickel steel in boiling magnesium chloride solution.[27]

There is no general agreement about the mechanism of stress corrosion cracking in the case of austenitic stainless steel. The specific effect of the chloride ion must be noted. This is sometimes ascribed to its small size. However the iodide ion, which is only about 20% larger, does not act in the same way. Hydrogen may contribute in some way to the cracking, but it has not been possible to demonstrate steady crack growth in a cathodically charged precracked specimen of austenitic stainless steel. Crack growth rates for a pre-cracked compact tensile specimen in boiling (140 °C) 42% magnesium chloride solution are shown in Fig. 3.42. Under these conditions $K_{ISCC} = 10$ MN/m$^{3/2}$ and there is a steady stage II growth rate of about 4×10^{-7} m/s.[27]

Growth rate does not define the time to failure. The incubation period prior to crack initiation dominates, and may indeed be thousands of hours. The time for onset of cracking increases as the temperature and chloride concentration falls. Some of the available data are plotted in Fig. 3.43. Cracking at temperatures below 100 °C is rare but it can happen with long exposure times (the author has come across one case where stainless steel in contact with moist NaCl cracked at

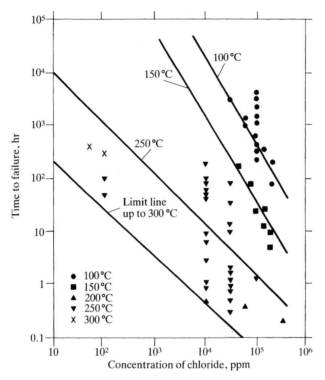

3.43 Relationship between time to failure, chloride concentration of solution and test temperature for 18Cr10Ni austenitic stainless steels. (Originally published in 'Hydrocarbon processing' Gulf Publishing Company, Houston.)

room temperature after three years). At the other extreme, at temperatures over 250 °C cracking can occur with chloride concentrations of less than 100 ppm. In some instances the chloride content of the water may not be relevant. A classic case is that of the heat exchanger with high purity boiler feed water on the shell side and a high inlet temperature on the tube side. The liquid in the annular space around the tube inlet is evaporated to leave a film containing chloride on the tube surface, and failure can be very rapid, even with water containing only a few ppm of chloride.

Little is known about the mechanism of crack initiation, except that pitting is not essential.

The stress corrosion cracking of stainless steel is traditionally associated with acid chloride conditions but tests have shown little effect of pH; if anything lower pH is associated with longer life, and cracking can occur with a pH as high as 12.

There are two main routes for avoiding this problem. The first, which is applicable where conditions are known to be unfavourable, is through material selection. In mildly corrosive conditions where no welding is required a straight chromium steel (e.g. an 18Cr steel) may be satisfactory. For welded components, or for more severe duties, a ferritic-austenitic steel may be the better choice. There are numbers of proprietary steels of this type.

The second approach is to eliminate sources of chloride. For example, hydraulic testing of austenitic stainless steel equipment should, for simple items, be done using good quality potable water, and for critical items, using deionised water. Provision should be made for complete drainage and drying after test. Wetting of insulation must be avoided and piping should be painted prior to insulation. Hard-stamping should be minimised and where essential done with low stress stamps.

Intergranular corrosion of austenitic stainless steel

This problem has already been mentioned in Chapter 2 and is considered in more detail below.

Early versions of austenitic chromium-nickel steel in the UK contained typically 0.12C, 18Cr and 8Ni. This material was welded to support brackets for use in a pickling vat containing sulphuric acid. The brackets failed by intergranular corrosion along a line in the parent metal parallel to the weld, a process which became known as 'weld decay'. It was associated with an intergranular precipitate of chromium carbide, $Cr_{23}C_6$, and the steel manufacturer concerned solved the problem by stabilising the steel with the addition of titanium, to produce an alloy known as FDP (Firth's Decay Proof). This experience also produced the Hadfield test; a boiling copper sulphate/sulphuric acid solution which was similar to that in the ill fated pickling vat, and which was, and still is, used to test steel for resistance to weld decay. The Hadfield test is known in Germany and elsewhere as the Strauss test, and in the USA is standardised in ASTM A262E.

In an unstabilised stainless steel, carbide precipitation results in depletion of chromium in regions close to the grain boundary. Concentration profiles for an 18Cr12Ni2¼Mo steel containing 0.07% C are shown in Fig. 3.44. The sample had been aged at 600 °C for 1000 hours. Chromium and molybdenum have been absorbed by the precipitate and iron and nickel percentages are correspondingly greater.[28] Selective corrosion of the grain boundaries is generally considered to be due to chromium depletion, but local strain associated with the carbides may also play a part.

Precipitation occurs between 425 and 800 °C, and is most rapid at about 650 °C. In a cross-section the carbides appear as somewhat irregular ellipsoids but viewed normal to the grain surface they are

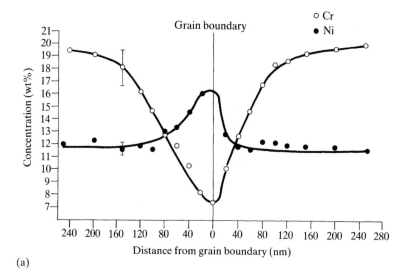

(a)

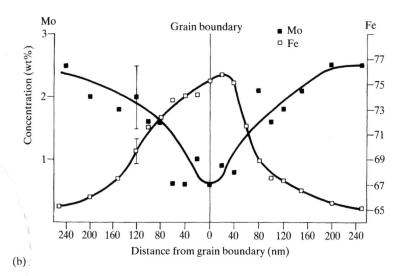

(b)

3.44 Concentration profiles across a grain boundary in type 316 austenitic stainless steel after ageing at 600 °C for 1000 hours (reproduced by courtesy of The Institute of Materials).[28]

fern-like and may spread over a considerable proportion of the grain surface area. Carbides are dissolved by heating for a few minutes at 950–1050 °C. This process is known as solution treatment, and austenitic stainless steel is normally supplied in the solution treated condition. The precipitates may be overaged by heating at 870–900 °C for 2 hours; this removes microstresses, completes the precipitation and eliminates chromium depletion. Such a stabilising anneal has been used as a post-weld heat treatment but is of questionable value because any local heat treatment develops a band of carbide precipitation outside the treated zone.

Carbide precipitation is avoided by the addition of either niobium (AISI type 347) or titanium (AISI type 321). Both these elements form carbides that are stable at the solution treatment temperature, so that the dissolved carbon content is reduced to a very low level. The amount of titanium required for stabilisation is $4 \times$ carbon content minimum, and niobium must be $8 \times$ carbon content minimum. Titanium carbide starts to go into solution at 1100 °C and niobium carbide at 1300 °C. Therefore, depending on the thermal cycle, some of the stabilised carbide may go into solution in the heat affected zone immediately adjacent to the fusion boundary. Chromium carbide may then precipitate during cooling. Stainless steel welds so affected can suffer selective corrosion adjacent to the weld (knife line attack) in strong acid.

An alternative or complementary way of avoiding intergranular corrosion is to reduce the carbon content. The ASTM ELC (extra low carbon) grades require a maximum carbon content of 0.03% (0.04% for pipe), and for all normal exposures these materials are resistant to intergranular corrosion. A compromise composition much used in the USA, but not so greatly favoured in Europe, is AISI type 304, an 18Cr12Ni with 0.08% max carbon. This material is also resistant to weld decay in most environments. However in the 1970s some intergranular cracking was observed in US boiling water reactor systems, where the steel was exposed to oxygenated water at 288 °C. Investigation indicated that carbides precipitated during welding were initially harmless but grew during service at 288 °C until they were sensitive to attack. Nuclear grade type 316 stainless steel, containing 0.02% max C and 0.06–0.12N (for strength) has been developed for this application. A niobium stabilised type 347 with 0.02% max carbon is used in Europe with satisfactory results. Other modified stabilised grades have been used for various applications; for example, a titanium stabilised type 321 with 0.06% carbon maximum for nitric acid exposure.

Intergranular corrosion under low stress conditions is manifest as a band of wastage a few millimetres wide with the inner edge typically 2.5 mm from the fusion boundary. Figure 3.45 is a micrograph of intergranular corrosion showing how wastage occurs through the detachment of grains. If there is tensile stress then multiple cracking will appear and failure may be quite rapid; the condition is termed intergranular stress corrosion cracking but the essential mechanism is the same as for weld decay.

Corrosive media that cause weld decay in a sensitised material include mineral acids, high temperature oxygenated water and

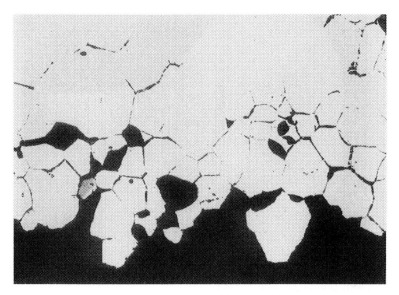

3.45 Intergranular corrosion in an austenitic chromium-nickel steel ×50. (Photo courtesy of TWI.)

polythionic acid, as reported in Chapter 2. Aqueous alkaline solutions at room temperature are harmless in this respect.

Strain relaxation cracking

As already noted in Chapter 2 welds in certain ferritic and austenitic steels have proved to be subject to intergranular cracking when strained and subsequently heated. This condition may obtain during post-weld heat treatment, when residual strain is relaxed, or when the weld is reheated by a subsequent pass, or it may affect piping which has suffered thermal strain and is then maintained at elevated temperature during service. It is also known as reheat cracking or stress relief cracking.

There are a number of factors that predispose an alloy to such cracking. These are:

(a) Alloy content
    Carbon-manganese steels are not affected, but chromium, molybdenum, vanadium and boron (particularly the latter two elements) may be damaging in the case of ferritic steels. In austenitic chromium-nickel steel, niobium and titanium may cause problems. Figure 3.46 shows a relief crack adjacent to a weld in a niobium-bearing stainless steel.
(b) Microstructure

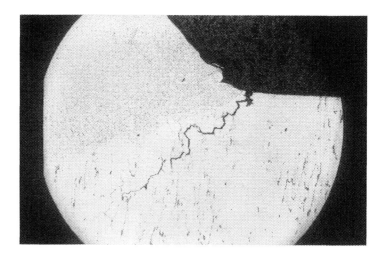

3.46 Reheat cracking of niobium stabilised austenitic stainless steel (×50). (Photo courtesy of TWI.)

The most susceptible structure is that of the coarse-grained region of the heat affected zone of welds, but cracking has been observed in weld metal and sometimes cracks will propagate into the parent metal.

(c) Stress concentration

Cracks are initiated preferentially at the toe of a butt or fillet weld, where there is a combination of stress concentration and susceptible microstructure.

(d) Time and temperature

In the case of ferritic steels the crack growth rate is far below 400 °C; rapid in the range 500–700 °C and low at higher temperatures. Rapid heating through the temperature range 500–750 °C has therefore been recommended for stress-relieving susceptible steels. Likewise rapid heating between 600 and 950 °C has been proposed to avoid cracking of the 18Cr12Ni1Nb austenitic steel.

(e) Thickness

The most severe problems have been encountered with sections thicker than 25 mm, and the risk of failure is much lower for thin material.

Ferritic alloys: effect of composition

There are several formulae relating composition to the risk of strain relaxation cracking. Those due to Nakamura and Ito are, respectively:

$$\Delta G = Cr + 3.3Mo + 8.1V - 2$$
$$P_{SR} = Cr + Cu + 2MO + 10V + 7Nb + 5Ti - 2$$

and in both cases the alloy is considered susceptible if the sum is positive. Neither formula takes account of the strong effect of boron and they do not apply to boron treated steels.

Indeed boron treated steels are notoriously subject to cracking during stress relief. One such was ½MoB pressure vessel steel, which is no longer produced. T1 steel (ASTM A 517 grade F), which is a quenched and tempered structural type also contains boron and is very susceptible. However, a boron treated carbon-manganese steel is safe.

The ½Cr½Mo¼V steel used for main steam piping in UK power stations is another steel notoriously subject to cracking during stress relief and in service, and much work has been done, particularly on the modification of weld procedures, to minimise the problem. In most other countries the alloy used in this service is 2¼Cr1Mo, which is only marginally susceptible to strain relaxation cracking.

Forgings of Mn-Mo-Ni-Cr steel specified in ASTM A508 Class 2 are commonly used in nuclear pressure vessels. This type of alloy is subject to both reheat cracking and underclad cracking (to be discussed below). ASTM A508 Class 1 and 3 are for plate and are generally similar in composition but do not contain chromium and are aluminium treated. These two grades are much less susceptible, possibly because residual AlN particles restrain grain growth in the coarse-grained region of the heat affected zone.[29]

Nichols[30] lists the ΔG rating of various steels together with their observed behaviour. The agreement is not very good. Table 3.1 is derived from Nichols' list and shows the relative susceptibility of some of the commonly used ferritic alloy steels.

Table 3.1 Susceptibility of ferritic alloy steels to strain relaxation cracking

Not susceptible	Borderline	Susceptible	Very susceptible
C-Mn	A517J	A517E	A517F
1Cr½Mo	A508/1	A533B	½MoB
5Cr½Mo	A508/3	HY 80	½Cr½MoV
9Cr1Mo	2¼Cr1Mo	HY 130	
A517 A	A517B	A508/2	
A533 A			
A508/2			

As might be expected, elements that have a detrimental effect on creep ductility such as phosphorus, sulphur, arsenic, antimony and tin also increase the susceptibility to strain relaxation cracking. The relative effect of these impurity elements is a function of alloy composition.

The increased use of clean steels is evidently favourable for avoiding this type of cracking.

Ferritic alloys: cracking mechanism
Two different types of crack have been identified in laboratory specimens cracked under strain relaxation conditions. The first is a brittle intergranular mode characterised by smooth intergranular facets, which occurs at the lower end of the post-weld heat treatment range (450–600 °C). The second shows intergranular microvoid coalesence and occurs at temperatures over 600 °C. These two modes are illustrated in Fig. 3.47.[31]

Crack growth rates for the brittle fracture mode are shown for three different temperatures in Fig. 3.48. These tests were made with a notched pre-cracked bar loaded in four point bending, the material having been given a simulated weld thermal cycle. The crack growth rate increases exponentially with the stress intensity factor and, for any given stress intensity factor, increases exponentially with temperature.

In these and other tests it was found that the sulphur content at, and immediately beyond, the crack tip was substantially increased above the bulk level. One possible reason for this effect is that, like hydrogen, sulphur diffuses preferentially to regions of triaxial strain, and contributes to the cracking tendencies by further weakening the grain boundaries.

Sulphur also appears to play a part in the microvoid coalesence fractures observed at high temperature. Cavities typically average 2 μm in diameter and contain a particle of manganese sulphide with a mean diameter in the region of 0.3 μm. These are particles that precipitate close to the grain boundary when the coarse-grained region cools from the overheating temperature range.[32]

Vanadium plays a part by forming carbide precipitates within the grains. This stiffens the grain so that more of the strain relaxation must be accommodated at the grain boundary.

The sequence of events would therefore seem to be as follows. On heating above say 1200 °C the parent metal close to the fusion boundary suffers grain growth and at the same time sulphur and phosphorus are taken into solution. On cooling a fine precipitation of sulphide forms close to the grain boundary but a significant proportion of the sulphur remains in solution. If a stress raiser such as a crack or unfavourable weld profile is present a zone of hydrostatic tension is formed ahead of the notch. On reheating, sulphur diffuses preferentially into this zone, and when the concentration at the prior austenitic grain boundary reaches a critical level, there is a local failure, the crack moves incrementally forward, and the process is repeated. The intermittent forward motion of the crack has been demonstrated by acoustic emission tests.[33]

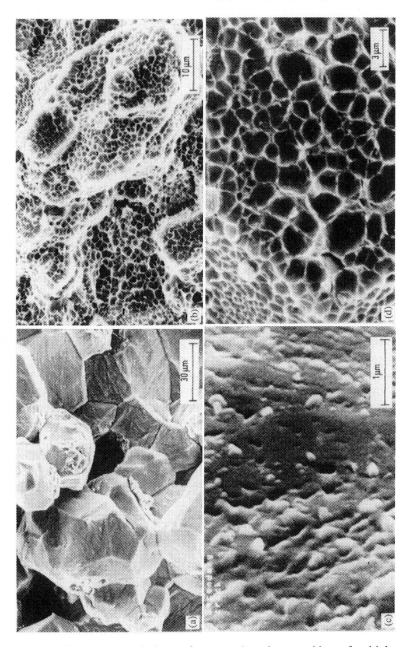

3.47 Fracture morphology of stress relaxation cracking of a high strength structural steel: a) Low ductility intergranular fracture; b) Detail of a); c) Intergranular microvoid coalescence; d) Detail of c) (reproduced by courtesy of The Institute of Materials).

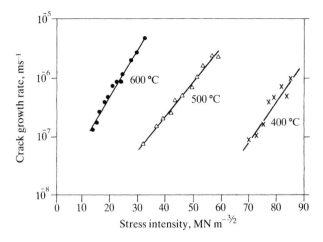

3.48 Crack growth rate in notched band test specimens of 2¼Cr1Mo steel austenised at 1200 °C (after Hippsley).[31]

Such brittle intergranular cracking is promoted by the presence of phosphorus, antimony and tin. These elements segregate to prior austenitic grain boundaries and further weaken them.

At temperatures above 600 °C the crack propagates by the normal creep cracking mechanism of microvoid coalesence, but cracking is made easier by the presence of sulphide precipitates which initiate void formation.

Avoidance measures include selection of a less susceptible alloy, reducing impurity levels, grinding the weld toes and heating rapidly through the temperature range 500–750 °C where crack growth rates are high.

Testing for strain relaxation cracking
Reference 29 lists no less than 26 methods of test, most of which were qualitative in character and sought to reproduce the typical intergranular crack found in real failure so as to explore the effects of, for example, welding variables. More recent work has utilised a fatigue cracked bend specimen for which a stress intensity value K may be calculated. Tests are made either with a constant load or at constant displacement, and crack growth rates are determined by measuring the electrical resistance of the specimen. By testing in inert gas or vacuum it is possible to obtain unoxidised surfaces for fractographic examination. The result has been an accumulation of quantitative results, some of which have been presented in earlier sections.

Austenitic alloys

In early attempts to use supercritical steam for power generation (later abandoned) austenitic stainless steel was specified for the main steam lines, naturally in heavy-wall thickness. The alloy used was 18Cr12Ni1Nb, and this proved to be highly susceptible to strain relaxation cracking both during stress relief and in service. In the coarse-grained region of the parent metal adjacent to the fusion boundary niobium carbide is taken into solution and on reheating is precipitated inside the grains. It thus has the same effect as vanadium in ferritic steels; hardening the grains causes strain relaxation to occur preferentially in the grain boundary regions and intergranular cracking may result. Similar but much less severe effects have been observed experimentally with 18Cr12Ni steels containing titanium and even without any addition. When molybdenum is added however the carbides precipitate at the grain boundaries and this type of steel is not susceptible to strain relaxation cracking.

The various remedial actions used to avoid cracking in the 18Cr12Ni1Nb steel were similar to those for ferritic alloys: grinding the toe of welds, and rapid heating to the stress relief temperature of 1050 °C. In addition much benefit was obtained both in fabrication and repair by the use of a filler alloy of 16Cr8Ni2Mo composition, with which it is possible to produce sound welds with the minimum of delta ferrite in the structure. This type of filler has subsequently been used for welding other austenitic stainless steels for elevated temperature service.

High temperatures and pressures (650 °C and 35 N/mm^2) may be required for future power generation equipment and fully austenitic chromium-nickel steels may be specified for good creep rupture properties combined with stability and corrosion resistance. Much work will be required to develop welding consumables to match such materials.

Underclad cracking

This problem occurred in the stainless cladding of nuclear reactors when the cladding was applied in a single layer using submerged-arc welding with a high heat input rate. Cracking mainly affected ASTM A508 Class 2 or the German equivalent 22NiMoCr3 7. Figure 3.49 shows the location and orientation of the cracks. They appear in that part of the coarse-grained HAZ that is reheated to between 500–720 °C by a neighbouring cladding run. The cracking is intergranular relative to the prior austenite grain boundaries and is oriented generally at right angles to the plate surface. The risk of underclad cracking is greatly reduced by cladding in two layers, the first layer being deposited at a low heat input rate; also by using less susceptible steel.

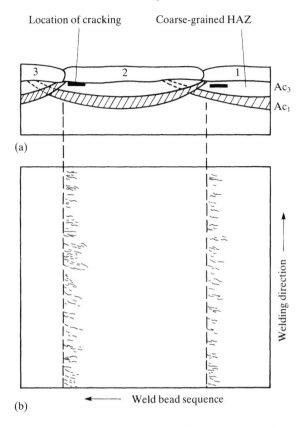

3.49 Location and orientation of underbead cracks in weld deposit clad low alloy steel: a) Transverse section; b) Plan view (reproduced by courtesy of Elsevier Applied Science Publishers Ltd).[29]

Liquation cracks have been found under cladding, and they occur in the same location as those formed by reheating; except of course that they are present before the adjacent cladding layer is deposited. Underclad cracks may be initiated by liquation cracks.

Neither type of crack is considered to present a hazard for the long term integrity of the vessel but they are nevertheless unacceptable to code requirements and must be avoided.

Creep and rupture behaviour of welded joints

Ferritic steels
Ferritic alloy steels that are used for elevated temperature applications where the stress for rupture may govern for design include 2¼Cr1Mo,

9Cr1Mo and 12Cr1Mo, together with a number of variants where strengthening additions such as vanadium and niobium are used.

½Cr½Mo¼V is welded with a 2Cr1Mo filler metal, and almost all the fabrication and service problems with this alloy have been associated with cracking in the coarse-grained region of the heat affected zone, as discussed earlier in this chapter. The creep properties of the weld metal more or less matches that of the parent metal, such that creep tests of cross-weld specimens that do not fail in the coarse-grained region may fail either in the weld metal or the parent metal.

2¼Cr1Mo alloy is welded with an electrode of matching composition. The creep rupture properties of submerged-arc weld metal deposited with a heat input rate of 3.67 kJ/mm are plotted in Fig. 3.50 together with the average curve for plate material of similar composition. In this case the weld metal overmatches the plate for short rupture times; more so in the as-welded condition than when given a normal post-weld heat treatment at 1300 °F. However, in the as-welded condition the rupture ductility, as measured by the reduction of area at fracture, is significantly lower than when stress-relieved, and there is evidence of creep embrittlement.

Creep embrittlement is a phenomenon that has some features in common with stress relaxation cracking. Characteristically the rupture ductility falls with increasing rupture time to a minimum value and then increases again. Carbides precipitate at the grain boundaries during the test, and this reduces the alloy content and causes carbides adjacent to

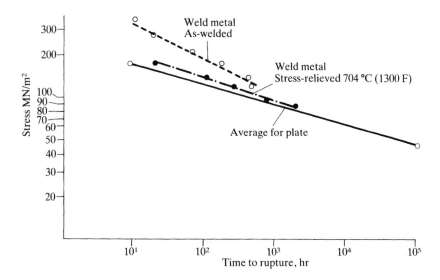

3.50 Creep rupture properties of 2¼Cr1Mo plate and weld metal tested at 593 °C (1100 F).

the boundary to go into solution. At the same time impurity elements, in particular arsenic, antimony and tin are rejected by the matrix volume in which the carbides are growing, causing a local enrichment of these elements and thereby reducing the grain boundary cohesion. Solution of carbides near the grain boundary reduces the local resistance to plastic deformation so that strain is concentrated in the grain boundary regions. This effect, combined with an increased tendency to void formation, results in creep embrittlement. With increasing time to failure the carbide-depleted zone grows wider and is able to accommodate a greater level of strain; in consequence the rupture ductility increases.[35] It may be necessary to take the possibility of creep embrittlement into account when assessing the integrity of equipment operating in the creep temperature range.

Returning to specific alloy systems: there has been considerable interest in the modified 9Cr1Mo alloy developed by Oak Ridge National Laboratory and accepted by ASME with the designation P91. This steel has additions of vanadium and niobium, and at 600 °C its mean stress to rupture at 10^5 hours is just over 100 MN/m² as compared with just under 50 MN/m² for 2¼Cr1Mo. Some investigators however obtained lower rupture values for similar material and yet lower figures for testpieces incorporating a transverse weld. Figure 3.51

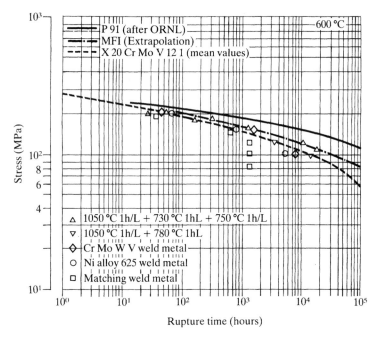

3.51 Results of creep rupture tests on alloy ×10CrMoVNb91, welded and unwelded, compared with results obtained on US alloy P91.[34]

shows the original results obtained by Oak Ridge National Laboratory (ORNL) together with values obtained with the German alloy x10CrMoVNb 9 1. unwelded and welded at 600 °. Also shown in this figure are mean values for the 12Cr1MoV steel X20CrMoV 12 1.[34]

Austenitic alloys
To produce weld metal that will match the creep properties of plate material in the case of austenitic chromium-nickel steels it is necessary to restrict the ferrite content of the deposit, and the 16Cr8Ni2Mo filler metal mentioned earlier has a suitably low proportion of ferrite.

A number of high carbon austenitic chromium-nickel alloys are used for high temperature operations in the petrochemical industry, typical of which is the alloy HK40 (0.4C25Cr20Ni). Such alloys contain a carbide eutectic and weld metal is not subject to hot-cracking, therefore a matching filler alloy is used. Welds made using the gas tungsten arc process have a rupture strength close to that of the parent metal, but those made by basic coated electrodes have significantly lower properties.

REFERENCES

1 Griffith A A: 'The phenomena of rupture and flow in solids' *Phil Trans Royal Soc* London Series A 1920 Vol 221 163–198.
2 Zener C and Holloman J H: 'Plastic flow and rupture of metals' Trans ASM 1944 Vol 33 163–235.
3 Paris P C and Sih G C: 'Stress analysis of cracks' in Fracture Toughness Testing and its Applications, ASTM Special Technical Publication No. 381 ASTM 1965.
4 Hahn G T: in Practical Fracture Mechanics for Structural Steel, UKAEA 1969 Q19.
5 Beachem C D and Pelloux R M N: 'Electron fractography' in ASTM STP No 381 ASTM 1965.
6 Wessel E T: 'Linear elastic fracture mechanics for thick-walled steel pressure vessels' in Practical Fracture Mechanics for Structural Steels, UKAEA 1969 pp H1-H44.
7 Wells A A: IIW Houdrement Lecture *Brit Weld J* 1965 Vol 12 2.
8 Rice J K: 'A path independent integral and the approximate analysis of strain concentration by notches and cracks' *J Appl Mech* 1968 Vol 35 379–386.
9 Barsom J M and Rolfe S T: 'Correlations between K_{IC} and Charpy V notch test results in the transition range' ASTM STP 466 1970 281–302.
10 Dolby R E: 'Charpy V and COD' *Metal Construction* 1981 Vol 13 pp 43–51.
11 Harrison J D: 'The state-of-the-art in crack tip opening displacement (CTOD) testing and analysis' *Metal Construction* 1980 Vol 12 415–422, 524–529 and 600–605.
12 Maddox S J: 'Improving the fatigue strength of welded joints by peening' *Metal Construction* 1985 Vol 17 220–224.

13 Maddox S J: 'Avoiding fatigue failure' *Metal Construction* 1980 Vol 12 531–533.

14 James L A: 'Fatigue-crack propagation behaviour of several pressure vessel steels and weldments' *Weld J* 1977 Vol 56 386–391.

15 Baumgardt H, de Boer H and Musgen B: 'High strength steels for offshore technology' *Metal Construction* 1984 Vol 16 15–19.

16 Booth G S: 'Improving the fatigue strength of welded joints by grinding' *Metal Construction* 1986 Vol 18 432–437.

17 Tomkins B and Scott P M: 'Environment-sensitive fracture: design considerations' *Metals Technology* 1982 Vol 9 240–250.

18 Bamford W H: 'Application of corrosion fatigue crack growth data to integrity analyses of nuclear reactor steels' *Trans ASME* 1979 Vol 101 July pp 182–190.

19 Jarecki A and Lancaster J F: 'Heavy-wall reactors for hydro-treating refinery applications and for coal liquifaction plant' in Fabrication and Construction Aspects of Energy Related Projects, Proceeding of South African Institute of Welding Conference, September 1988.

20 Murakami Y et al: 'Hydrogen embrittlement of Cr-Mo steels temper embrittled during long time service' *J Jpn High Pressure Institute* 1982 Vol 19 132.

21 Nakamura M and Furubayashi E-I: 'Crack propagation of secondary hardened steel in gaseous hydrogen atmosphere' *Materials Science & Technology* 1989 Vol 5 584–589.

22 Golightly F A and Beardsley M E: 'The Nature of damage in plant and pipeline operating in sour conditions' in Sour Service in the Oil, Gas and Petrochemical Industries 1985 Oyez Scientific and Technical Services Ltd.

23 Hudgins C M et al: 'Hydrogen sulphide cracking of carbon and alloy steels' *Corrosion* 1966 Vol 22 238–251.

24 Lancaster J F: 'Failures of boilers and pressure vessels' *Pressure Vessel and Piping* 1973 Vol 1 155–170.

25 Parker J G and Wigmore G: 'Initiation of stress corrosion cracking in metals' *Metals Technology* 1982 Vol 9 216–220.

26 Kowaka M and Nagata S: 'Stress-corrosion cracking of mild and low alloy steels in $CO\text{-}CO_2\text{-}H_2O$ environments' *Corrosion* 1976 Vol 32 395–401.

27 Rieck R M, Atrens A and Smith I O: 'Stress corrosion cracking and hydrogen embrittlement in AISI type 304 stainless steel' *Materials Science & Technology* 1986 Vol 2 1066–1073.

28 James A W and Shepherd C M: 'Some effects of heat treatment on grain boundary chemistry and precipitation in type 316 steel' *Materials Science & Technology* 1989 Vol 5 333–345.

29 Dhooge A et al: 'A review of work related to reheat cracking in nuclear reactor pressure vessel steels' *Int J Pressure Vessels and Piping* 1978 Vol 10 329–409.

30 Nichols R W: 'Reheat cracking in welded structures' *Welding in the World* 1969 Vol 7 (4) 244–261.

31 Hippsley C A: 'Brittle intergranular fracture at elevated temperatures in low alloy steel' *Materials Science and Technology* 1985 Vol 1 475–479.

32 You C P, Hippsley C A and Knott J F: 'Stress relief cracking phenomenon in high strength structural steel' *Metals Science* 1984 Vol 18 387–394.

33 Hippsley C A, Buttle D J and Scraby C B: 'A study of the dynamics of high

temperature brittle intergranular fracture by acoustic emission' *Acta Metall* 1988 Vol 36 441–452.

34 Lundin C D: 'Materials and their weldability for the power generation industry' in Advanced Joining Technologies Ed T H North, Chapman and Hall, London 1990.

35 Swift R A and Rogers H C: 'Study of creep embrittlement of 2¼Cr 1Mo weld metal' *Weld J* 1976 Vol 55 188-s to 198-s.

36 Lancaster J F: 'Metallurgy of Welding'. 4th Ed, Allen and Unwin, London 1987.

4 Structures

General

The welded structures to be considered in this chapter include ships, steel frames for buildings and other purposes such as pipe racks, steel bridges and tubular structures for offshore drilling and oil production rigs. In all these fields there is a continuing need for improvements both in the quality of welded joints and in the properties of steel. Much progress has been made by the steel industry in the direction of higher standards during recent years, and for the most part these developments have lead to better weldability. It will be appropriate therefore to start by giving a brief outline of the way in which steelmaking methods have changed, and how such changes have affected welded structures.

Aluminium alloys are also employed to a significant extent in ship superstructures and for decking and living quarters on offshore oil rigs. These applications will be discussed later in this chapter.

Steelmaking developments

Iron as a structural material

Before about 1770, the materials used for bridges and major buildings in Britain were stone and timber. In the second half of the eighteenth century cast iron was employed to an increasing extent in structures, starting with the bridge that spans the river Severn at Coalbridge (1773). The displacement of charcoal as a means of reducing iron ore in blast furnaces by coke lowered the price of cast iron and increased its availability. Cast iron is brittle and is only safe to use as a structural material provided that any tensile stress is kept low. Nevertheless it found significant use. It was also used in the construction of the Crystal Palace for the Great Exhibition in London in 1853 (a time when cast iron had largely been replaced by iron in other fields of structural work).

Wrought iron first became a major structural material following the invention of the puddling process by Henry Cort in 1784. Development was initially slow. The first iron ship was probably the barge 'Trial', which displaced about 40 tons and was built by John Wilkinson of Coalbrookdale in 1787. The first iron steam ship was the 'Aaron Manby', built in 1821. This was a paddle steamer that made history by sailing from Rotherhithe across the Channel and up the Seine to Paris. Such vessels established the basic techniques for constructing metal ships, and these methods were later employed in steel ships. Figure 4.1 shows how iron and steel displaced timber in ship construction during the nineteenth century. Iron production rose correspondingly from a few hundred thousand tons in 1820 to nearly 8 million tons in 1880.[1]

Cast iron is not a weldable material in the usual sense of the term, since it is subject to hardening and cracking in the heat affected zone and high sulphur contents may cause solidification cracking of the weld metal. However, repair welding is quite often required. This can be accomplished by screwing steel studs into the metal on either side of the crack and making a bridge of weld metal between the studs. Alternately the metal may be welded directly using a 50Ni50Fe electrode, either with high preheat or with low preheat and by making short runs, allowing the metal to cool down between runs. This is skilled and specialist work and such repairs should be considered as a temporary expedient.

Wrought iron is a ductile material consisting essentially of iron interlayered with slag. The through-thickness ductility is therefore low, and it is undesirable to weld attachments to the surface. Iron may be welded edgewise however using rutile or low hydrogen electrodes.

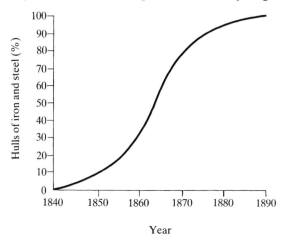

4.1 Ships built of iron and steel during the period 1840 to 1890: percentage of total (reproduced by courtesy of The Institute of Materials).[1]

Steel: early development

In medieval times tools that required a cutting edge were made in various ways. One technique, known as 'steeling', was to forge weld a strip of steel to the iron implement. In England steel bars for this purpose were usually imported from Germany or Scandinavia, where they were made directly from suitable iron ores. Alternatively the iron was case-hardened at the edge; a process that could be carried out locally by blacksmiths. In the seventeenth and eighteenth centuries steel production in England was increased by using the cementation process. Alternate layers of iron bars and carbonaceous matter were laid in a chest of refractory clay and fired in a furnace for three to seven days. The process was complete when blisters appeared on the surface of the bars, and the product was known as 'blister steel'. This material was not uniform in composition, and at a later stage it was improved by 'shearing'. Bars were broken into short lengths, bound together, and then heated to a temperature at which they could be hammer forged into one piece. The 'shear steel' so produced could be made in a range of hardnesses, from a soft grade used for cloth shears to the hardest grade which was suitable for engraving tools. Then in the middle of the eighteenth century Benjamin Huntsman succeeded in melting blister steel in small fireclay crucibles, using a suitable flux. The steel was then cast. Crucible steel, which was much more uniform than shear steel, was made in increasing quantities during the late eighteenth and early nineteenth centuries.[2] Later, the addition of alloying elements made possible the development of high speed steels.

Bulk steelmaking

The manufacture of steel by the cementation of iron was of necessity a small scale process and the product was correspondingly expensive. Thus it was used for tools, instruments, clocks and the like but not for structures. This situation changed radically after the invention in 1856 of the Bessemer process for the conversion of liquid iron directly to steel by blowing air through it. Figure 4.2 shows a section of a typical Bessemer converter. Air is introduced through tuyeres in the bottom of the vessel, and the blow is continued until most of the carbon in the iron has been removed by oxidation to carbon monoxide. The vessel was lined with a silica based refractory which did not react with and remove any phosphorus in the iron. About twenty years after the invention of the acid Bessemer process Sidney Gilchrist Thomas developed a basic lining for the converter and this made possible the refining of iron derived from high phosphorus ores.

The success of the Bessemer process was by no means immediate. Early products were variable in quality and some were brittle. Also the

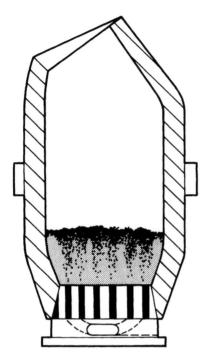

4.2 Diagrammatic cross-section of a Bessemer converter.

price of the steel was higher than that of iron; in 1881, for example, it was 50% higher. However by the last decade of the nineteenth century the price had fallen and steel became the dominant material for ship construction (Fig. 4.3). During this same period the first experiments were being made with electric arc welding using a fusible bare steel electrode. Fusion welding also developed at a slow pace, and it was not until 1941 that the first all-welded cargo ships were produced by the Kaiser Swan Island shipyard in the US.

Bessemer converters did not have a very long life in Britain and the USA, being replaced by open hearth or electric arc furnaces which had a much higher capacity and which did not introduce nitrogen into the steel. In the open hearth furnace carbon was oxidised by additions of iron ore and at a later stage by oxygen injection. In Continental Europe however, particularly in Belgium and Germany, the basic Bessemer or Thomas process was used to a substantial extent for iron derived from domestic phosphorus-bearing ore, even up to and shortly after World War II. Thomas steel contained nitrogen and was subject to strain-age embrittlement, which contributed to the failure of welded bridges mentioned in Chapter 2. Until the brittle failures of welded

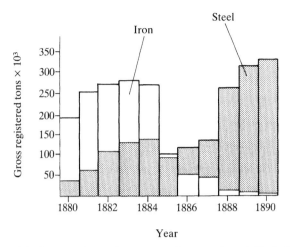

4.3 Gross registered tonnage of iron and steel Clyde-built ships 1880–1890 (reproduced by courtesy of The Institute of Materials).[1]

Liberty ships during World War II, open hearth steel was not thought likely to behave in a brittle fashion. Apart from this problem the bulk steel produced both by the Bessemer and the open hearth process may be classified as weldable, in contrast to cast iron and wrought iron, neither of which are suitable for welded fabrication on a large scale. Typical compositions of cast iron, wrought iron and semi-killed mild steel are shown in Table 4.1.

Table 4.1 Typical carbon and impurity contents of cast iron, wrought iron, semi-killed and killed steel

Material	Comparison mass, %				
	C	Si	Mn	P	S
Cast iron	2.0–6.0	0.2–2.0	0.015	0.05	0.05
Wrought iron	0–0.2	0.04	Trace	0.005	0.10
Semi-killed steel	0.15–0.35	0.02	0.75	0.04	0.04
Killed steel	0.15–0.25	0.30	0.10	0.03	0.03

Recent developments in steelmaking

After World War II Austria needed an increased output of quality steel, and it was decided that the required productivity could best be obtained by means of an oxygen converter process. This possibility had in fact been envisaged in one of Bessemer's 1856 patents, but at that time oxygen was not available in the required quantity, nor would it have been practicable to inject it from below because of erosion of the tuyeres. When tonnage oxygen did become available with the development of the Linde-Frankel process in 1928, the problem of the tuyeres remained, and it was thought that top blowing would not produce sufficient agitation of the bath to distribute oxygen throughout the melt. Experimenters at the Voest works in Linz nevertheless tried top blowing, and it worked. The liquid steel circulated vigorously even though oxygen was blown only on to the top surface, and it continued to circulate until the carbon content fell to a level of 0.05%. Figure 4.4 illustrates the refining mechanism.[3]

Not only was this process technically successful, but the speed of the refining operation resulted in a large increase in productivity. Thus,

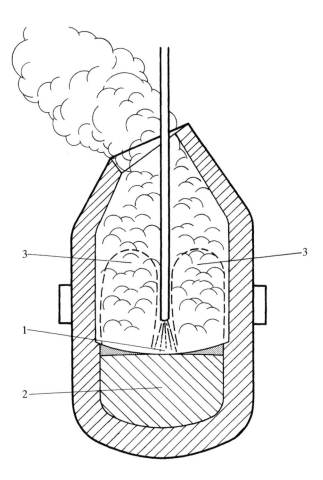

4.4 Refining in the LD vessel takes place in three regions: 1 – oxygen penetrates the slag and reacts with liquid metal producing local temperatures up to 2500 °C; 2 – circulation of the bath from the hot spot; 3 – slag metal reactions.[3]

the amount of steel produced per hour by a 400 ton oxygen converter is fifteen times that produced by an open hearth furnace of the same capacity. Consequently this process, known in Austria as the Linz-Donawitz or LD process, and elsewhere as the basic oxygen process, has, in developed countries, almost completely supplanted the Bessemer converter and the open hearth furnace. The electric arc furnace survives because of its ability to utilise scrap. Basic oxygen steelmaking absorbs less scrap than the open hearth process, so the electric furnace share of total steel production has increased. Figure 4.5 shows the contribution of various processes to world iron and steel output from 1860.[3]

Significant improvements have been made to the original top blown process. Bottom stirring by the injection of argon and/or other gases makes it possible to reduce carbon below the 0.05% limit given above. Depending on the amount of stirring, nitrogen, oxygen, phosphorus and sulphur may also be reduced to low levels. The invention of concentric tuyeres in which a central stream of oxygen is surrounded by hydrocarbon gas made oxygen bottom blowing possible, whilst coal injection increases the possible scrap rate. There is a penalty for hydrocarbon injection in that the hydrogen content of the finished steel is increased.

4.5 Growth of global crude steel production from 1860 and percentage of total made by various processes.[3]

Clean steel

Overall the effect of such development was to make the production of clean steel possible on a tonnage basis, with a corresponding improvement in weldability, through-thickness properties and resistance to hydrogen-induced cracking. Production of such quality material was not practicable with the open hearth furnace because of sulphur pick-up from the fuel.

The term 'clean steel' is applied to at least four different products. The first is the bulk steel which is the subject of this chapter, where the main requirement is to minimise laminar defects that could give rise to lamellar tearing (particularly in offshore fabrication) and hydrogen-induced cracking (particularly in pipelines). A higher level of cleanness is required for automotive bearing alloys and aircraft applications, and this can be met by air melting followed by vacuum treatment. A yet higher degree of cleanness is necessary for military and aerospace applications, and this is achieved by vacuum arc remelting or electroslag remelting.[4] A yet higher degree of cleanness is obtained by means of the cold hearth process. The feed stock (already clean) is electron beam melted in an intensively cooled copper trough. As the liquid metal flows along the trough, heavy impurities sink and are trapped in the skull formed at the surface of the trough, whilst lighter particles float to the surface and are removed. The purified metal is continuously cast in a copper mould.[5] In this chapter we are only concerned with the first category of clean steel.

Laminar defects in steel are associated primarily with the presence of sulphide inclusions. Such inclusions may take one of two forms. The first, designated type I, is spheroidal and has little plasticity at the rolling temperature, therefore it retains its spheroidal shape and is relatively harmless. The second, type II, appears as an intergranular film late in the solidification process, and is plastic at elevated temperature, causing hot shortness and low transverse ductility. Type I sulphides are formed in steels having a relatively high oxygen content, and are present in rimming, semi-killed and silicon killed steels. Type II sulphides are characteristic of low oxygen steels, such as aluminium killed types. To avoid the unfavourable effect of such inclusions there are basically two routes: desulphurisation or inclusion shape control. In some cases these two operations are combined.

Desulphurisation is carried out by reaction with either calcium or magnesium, the sulphides of which are relatively stable. Initial treatment is applied to hot metal from the blast furnace in ladles or torpedo cars upstream of the basic oxygen converter. Magnesium is absorbed in coke or coated with alkali metal salts and is injected into the liquid metal. Alternatively calcium is injected in the form of calcium carbide, lime, soda ash or calcium carbonate. Hot metal can be desulphurised by such methods down to 0.01S or lower, and such levels will be maintained during treatment in the basic oxygen process. Further reductions may be achieved by treatment of the liquid steel in the ladle using a desulphurising slag consisting of lime, fluorspar and deoxidant. A similar slag can be employed for desulphurisation with the argon-oxygen decarburisation process, or in vacuum treatment.

Inclusion shape control is achieved by additions of rare earths or calcium. Zirconium and titanium also globularise sulphides but these elements have other metallurgical effects which may not be desirable. Misch metal is most frequently used, and Fig. 4.6 shows the effect of

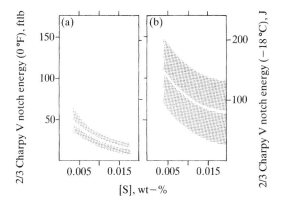

4.6 Effect of sulphur content on Charpy V notch value for steel: a) Without rare earth additions; b) With rare earth additions (reproduced by courtesy of The Institute of Materials).[6]

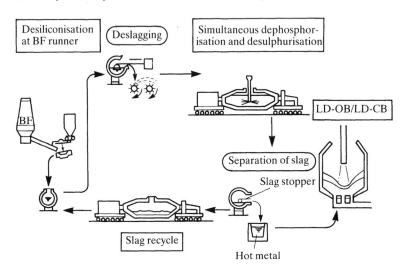

4.7 Hot metal desulphurisation and dephosphorisation (reproduced by courtesy of The Institute of Materials).[7]

such additions on the impact properties of the steel.[6] Like through-thickness ductility, impact energy is adversely affected by laminar inclusions.

Phosphorus may also contribute to planar anisotropy, and dephosphorisation may be carried out at the same time as sulphur removal during the transfer of hot metal from the blast furnace to the converter. One technique is to remove silicon down to 0.02% and then treat with a basic oxidising slag. By this means it becomes possible to produce a hot metal with typically 4.3% carbon, 0.05% silicon, 0.015% phosphorus and 0.005% sulphur. Figure 4.7 illustrates simultaneous dephosphorisation and desulphurisation.[7]

Casting

Steel made in a Bessemer-type converter or open hearth furnace was, with few exceptions, cast into ingot moulds. The nature of the steel so produced depended on the deoxidation practice. Where no deoxidant was added the oxygen combined with carbon to form CO as the liquid steel cooled. This 'rimming' action had two effects. Firstly the solidified metal was porous and completely filled the mould, and secondly there was severe segregation in the ingot, the outer part being almost pure iron, with impurities concentrated in the interior. The second variant was semi-killed or balanced steel, in which enough deoxidant was added to minimise segregation, but there was sufficient porosity to compensate for shrinkage and allow the metal to fill the mould. The third type was fully killed steel, where the deoxidant content was sufficient to kill the rimming action completely, and where a substantial shrinkage cavity formed at the top of the mould. The amount of metal which had to be removed from the top of the ingot was greatest for killed steel and the least for rimming steel, and the cost of the product was in the inverse order. Continental European practice was, as a rule, to make either rimming or fully killed steel, whereas in the UK and USA semi-killed steel was the norm.

Semi-killed steel is characterised by a low silicon content, (typically less than 0.05%) but the current generation of welding consumables is usually well enough deoxidised to compensate for any dilution by the parent material, so it does not give rise to any significant difficulty in fusion welding. Rimming steel may, on the other hand, cause problems. When rolled into plate the impurities, including sulphur, are concentrated in the mid-thickness and this may result in porosity and/or cracking. As a rule however this material may be successfully welded with coated electrodes provided that the procedure is such as to avoid excessive dilution by the parent metal. Killed steel of normal carbon content is weldable without reservation.

In recent years this situation has been radically transformed by the

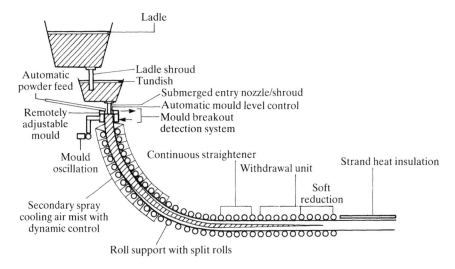

4.8 Layout of a continuous slab casting machine (reproduced by courtesy of The Institute of Materials).[8]

development of continuous casting. Figure 4.8 is a diagram of a continuous slab-casting machine.[8] Liquid metal feeds from the ladle to a tundish and thence to a water cooled mould. The partially solidified metal is conveyed by rolls through a 90° bend to emerge horizontally. It may then be cut into slabs for subsequent re-rolling or go directly to the rolling mill. The widespread adoption of continuous casting has been made possible by the use of the basic oxygen furnace, which is capable of supplying hot metal at a rate sufficient to maintain steady production. The advantage of continuous casting as compared with ingot mould casting will be self-evident:

- higher productivity
- no waste due to cropping ingots
- all steel is killed

On the debit side there are practical problems in operating the machines; the most serious of which is mould break-out. In addition, solute elements and impurities tend to segregate to the central part of the slab, and this has, for example, given rise to local susceptibility to hydrogen-induced cracking. Electromagnetic stirring is used as a means of combating segregation. Even so, carbon, manganese and phosphorus segregate to the centre line of the slab and to meet the most stringent requirements for resistance to hydrogen-induced cracking it is necessary to reduce sulphur and phosphorus levels to below 0.001% (10 ppm) and 0.01% (100 ppm) respectively. This is done by secondary dephosphorisation combined with vacuum treatment and calcium injection. Finally, the steel may be quenched to minimise carbon migration.[9]

Ingot mould casting is still practised for production of mild steel structural sections such as I beams, and the steel is (in the UK) usually semi-killed. Sections are also available in steel with augmented mechanical properties. Steel of extra high quality is produced by treating the product from the basic oxygen converter by vacuum arc degassing. This gives high levels of cleanness and close control of composition. Ingot casting is used for such steel, and also for plate thicker than about 75 mm.

Rolling mill practice

In the period after World War II, and following the catastrophic brittle failures experienced in some of the Liberty ships constructed during the war, it became evident that there was a need to upgrade the quality of mild steel plate and in particular to improve its notch toughness. This could be done by aluminium killing, normalising and by increasing the manganese-carbon ratio: the latter option being favoured in the UK where most steel was semi-killed. At the same time, some European mills obtained better properties by making the last few hot rolling passes at temperatures lower than normal. This was the prototype for controlled rolling, which became the subject for much investigative work in the 1960s. At the same time the beneficial effects of small additions of niobium and titanium – microalloying – were explored.

The object of controlled rolling is to produce a fine grained steel which, in the as-rolled condition, has properties equal to or better than those of normalised plate. Reducing the grain size increases the yield strength. According to the Hall-Petch relation, the yield strength is inversely proportional to the square root of the ferrite grain diameter (Fig. 4.9).[10] It also reduces the ductile brittle transition temperature as measured, for example, by the Charpy impact test.

The grain size of the finish rolled steel is largely controlled by the austenitic grain size immediately prior to the austenite-ferrite transition. In this respect deformation temperatures during hot rolling in the austenite range may be divided into three regions. Deformation above 1000 °C produces coarse recrystallised γ grains which transform into a coarse ferritic-bainitic structure. Within the range 1000–900 °C the austenite refines by repeated recrystallisation, leading to a fine ferrite. Below 900 °C the austenite does not recrystallise but forms elongated grains with deformation bands. On transformation, ferrite grains nucleate internally on the deformation bands, leading to a yet finer structure. Finally, if rolling is continued in the $\alpha + \gamma$ temperature region, the ferrite grains break down into sub-grains. These changes are shown in Fig. 4.10.[10]

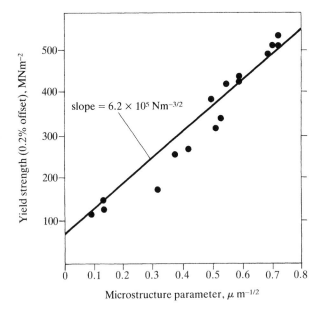

4.9 Yield strength versus microstructural parameter that takes account of both grain size and subgrain or cell size (reproduced by courtesy of The Institute of Materials).[10]

Controlled rolling as first developed did indeed continue in the $\alpha + \gamma$ region. This increased yield strength and reduced transition temperatures down to finishing temperatures of about 840 °C. Lower finishing temperatures, however, caused the transition temperatures to increase again, as shown in Fig. 4.11. This was due to anisotropy associated with the formation of texture combined with elongation of type II sulphide inclusions. In impact testing a crack initiates at the tip of an MnS inclusion and propagates along the banded structure. This results in splitting or delamination in the Charpy test, in low through-thickness ductility, and in a high susceptibility to hydrogen-induced cracking on exposure to wet H_2S. Such low finishing temperatures are now exceptional.

Some of the more recent developments of controlled rolling (known as thermomechanical treatment) are illustrated in Fig. 4.12. The compositions of steels suitable for such treatments, and the properties obtained are listed in Table 4.2 with those of hot-rolled (R) and normalised (N) steel listed for comparison.[11]

Steel A is a case where rolling has been continued below the austenite recrystallisation temperature to obtain fine grain. Additions of small amounts of niobium are advantageous in this type of treatment, because niobium raises the recrystallisation temperature and increases the temperature range in which deformation bands are

4.10 Microstructural changes during the three stages of the controlled rolling process (reproduced by courtesy of The Institute of Materials).[10]

formed. Steel B is a conventionally controlled rolled type, as discussed above.

Type C is a calcium-titanium treated, extra low sulphur steel rolled from a low reheating temperature to give a fine equiaxed structure. The presence of TiN particles inhibits grain growth in the heat affected zone of welds, thus permitting high heat input rates whilst obtaining acceptable impact properties in the heat affected zone.

Type D combines controlled rolling with quenching, so that a high yield strength is obtained with a low carbon content. This in turn permits welding with a low or zero preheat. In type E the final step is

Table 4.2 Composition and properties of Japanese HT 50 steel. plate up to 30 mm as-rolled, normalised and subject to temperature controlled rolling as shown in Fig. 4.12

Rolling programme	Chemical composition, %									CE (IIW)	YP N/mm²	UTS N/mm²	Elongation %	VE −40 °C J	50% VTrs °C	Drop-weight °C	Grain size ASTM
	C	Si	Mn	P	S	Nb	V	Ti	Ni								
A	0.12	0.32	1.36	0.015	0.006	0.019	–	0.013	–	0.36	432	504	29	245	−91	−50	7–9
B	0.13	0.36	1.42	0.02	0.003	–	–	–	–	0.37	368	526	31	139	−92	−95	8–11
C	0.06	0.22	1.32	0.01	0.001	–	0.04	0.01	0.26	0.31	392	471	40	294	−125	−100	11–12
D	0.12	0.27	1.13	0.016	0.004	–	–	–	–	0.31	362	493	30	288	−71	−45	7–9
E	0.05	0.31	1.39	0.019	0.004	0.03	0.05	–	–	0.29	385	469	34	110	−97	−100	11–12
As-rolled	0.13	0.25	1.37	0.013	0.007		REM	0.035		0.36	420	530	22	200	−42	−10	4–6
Normalised											360	500	27	270	−80	−35	6–7

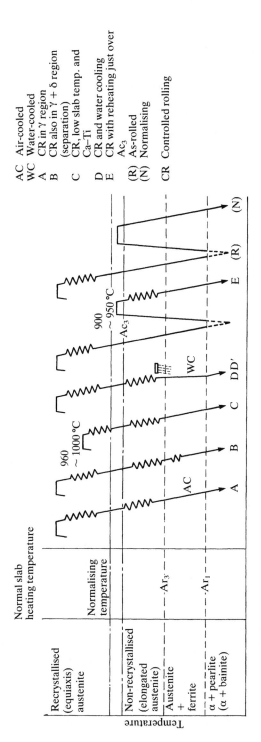

AC Air-cooled
WC Water-cooled
A   CR in γ region
B   CR also in γ + δ region
    (separation)
C   CR, low slab temp. and
    Ca–Ti
D   CR and water cooling
E   CR with reheating just over
    Ac₃
(R) As-rolled
(N) Normalising
CR Controlled rolling

4.12 Comparison of various types of thermomechanically controlled rolling.[11]

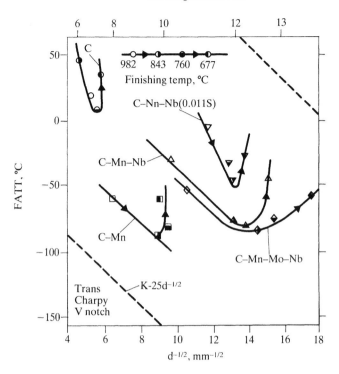

4.11 Effect of finish rolling temperature on transition temperature and grain size steel (reproduced by courtesy of The Institute of Materials).[10]

normalising followed by a finishing pass. Once again good properties are combined with a low carbon content.

There is a current trend towards the use of V-N and V-N-Ti microalloyed steels that are rolled entirely within the crystallisation range from a reheat temperature of about 1100 °C. The presence of TiN and VN particles promotes a fine austenite grain size and gives a fine grained steel with finish rolling temperature as high as 1050 °C. The process is known as recrystallisation controlled rolling and combined with accelerated cooling it is possible to obtain 450 MN/m² yield strength combined with a 40 J transition temperature of −70 °C using a 0.09% C steel.[12] However to achieve such properties the relatively high nitrogen content of 0.018% may be required, which raises questions as to the suitability of such steel for welded applications.[13] Some caution is also required in the use of other thermomechanically treated steels, particularly those that are water quenched in the final stage. The mechanical properties of such steels may deteriorate as a result of heating at 600 °C and above, so that hot forming and post-weld heat treatment must be carefully controlled.[11]

Steel strip

Thermomechanical control is only possible in a reversing mill where it is practicable for example to adjust the interpass time to allow for recrystallisation of austenite, and to control the finishing temperature. In a continuous strip mill the steel passes without interruption through a series of rollers to a run-off table whence it is coiled. The temperature on leaving the last set of rollers depends on the reheat temperature but is generally about 800 °C. At this stage the strip is quenched to about 650 °C and is coiled at this temperature. Thus it is cooled rapidly through the transformation range and then slowly from 650 °C to room temperature. This results in a fine grained ferritic structure, and if microalloying elements are present they are retained in supersaturated solution during the quench and precipitate during the slow cooling period. The precipitation strengthening effect due to titanium and niobium and to vanadium in nitrogen-bearing steels due to this process is more effective than in plate steel.

Strip has a limited application in general structural work but it is used for the manufacture of spiral welded and electric resistance welded pipe as described in Chapter 5.

The development of structural materials

General

Prior to 1939 there was little incentive for any significant improvement in the quality of steel plate. In most countries design stresses were either arbitrary or based on a fraction of the ultimate strength, whilst structures and pressure vessels were, in the main, fabricated by riveting. However in Germany the problem of weldability emerged as the result of the brittle failure of welded bridges referred to earlier. To counter the effects of nitrogen, aluminium was added to the steel, and this had the additional benefit of reducing the ferrite grain size and increasing the yield strength. This was the first step towards the production of weldable structural steels of improved strength and notch-ductility.

After World War II, and particularly after the oil crisis of 1973, the pressure for such improvements has intensified. Suzuki[11] has identified the following market needs:

    (a) heavier structural steel plates for use in larger structures
    (b) higher strength to permit a reduction in weight

    (c) higher toughness to combat the risk of brittle fracture, particularly of thick plate

    (d) improved resistance to hydrogen cracking, making for easier welding procedures and reduced preheats

    (e) improved through-thickness ductility by the removal of harmful non-metallic impurities

    (f) resistance to grain growth in the parent metal adjacent to the fusion boundary, which allows higher heat input rates to be used in welding

    (g) greater resistance to corrosion.

Many of these developments have been made possible by the changes in steelmaking techniques that were discussed earlier. The current trend is towards reducing the content of all non-metallics including carbon, relying on grain refinement and precipitation hardening for increased strength. For example, interstitial-free 'steels' used for roof and door panels in the automobile industry, have carbon contents typically in the range 20–70 ppm, with additions of titanium and/or niobium to combine with residual carbon, nitrogen and sulphur. This is an extreme example and in the steels to be considered here carbon remains an essential alloying element, albeit reduced wherever this may be useful to minimise preheat requirements and improve notch-ductility.

Most of the steel used in structural work is plain carbon and carbon-manganese steel. A high proportion, probably about one-third, is reinforcing bar (rebar) and pre-stressing material for concrete. Most of the remainder, in the form of plate and structural sections, is mild steel with a carbon content up to 0.25%. Such material is to be preferred wherever the design conditions permit on the grounds of cost, availability and weldability. This is the case in welded structures when the governing consideration is fatigue strength (for which, in the welded condition higher tensile strength gives no advantage) or when deflection is the limiting factor and the relevant material property is the elastic modulus. Where tensile and particularly yield strengths are important however high tensile strengths may be required to reduce weight or to limit the section thickness. In 1986 Colbridge[14] estimated that the tonnage of high strength steel used for structural application was about 10% of the total, excluding rebars. For the welding engineer, this 10% provides 90% of the problems, and most of the remainder of this chapter will be primarily concerned with such steel. Aluminium alloys are also used on a significant scale, for example in ship superstructures and military bridges, and these will be considered under the relevant application heading.

Carbon steel

The equivalent carbon content is generally considered to be a measure of the weldability of a steel, and in BS 5135, for example, it is used in combination with section thickness as a means of establishing the required preheat temperature. It will be convenient at this point to consider the means of calculating carbon equivalents. A widely accepted formula is that of the International Institute of Welding:

$$CE = C + \frac{Mn}{6} + \frac{Cu + Ni}{15} + \frac{Cr + Mo + V}{5} \qquad [4.1]$$

where C, etc, represent mass per cent. This is a slightly modified form of the equation developed by Dearden and O'Neill[15] from experimental work. In Japan the quantity C_{eq} was preferred:

$$C_{eq} = C + \frac{Si}{24} + \frac{Mn}{6} + \frac{Ni}{40} + \frac{Cr}{5} + \frac{Mo}{4} + \frac{V}{14} \qquad [4.2]$$

However according to Japanese tests for susceptibility to hydrogen cracking, neither CE or C_{eq} are applicable to steel with a carbon content below 0.17%, for which the Ito and Bessyo[16] formula is appropriate:

$$P_{cm} = C + \frac{Si}{30} + \frac{Mn + Cu + Cr}{20} + \frac{Ni}{60} + \frac{Mo}{15} + \frac{V}{10} + B \qquad [4.3]$$

Other carbon equivalent formulae were developed by Tanaka:

$$P_{NB} = C + \frac{Si + Cu + Mo}{20} + \frac{Mn}{10} + \frac{Cr}{20} \qquad [4.4]$$

and by Mannesmann:

$$CE = C + \frac{Si}{25} + \frac{Mn + Cu}{16} + \frac{Cr}{20} + \frac{Ni}{60} + \frac{Mo}{40} + \frac{V}{15} \qquad [4.5]$$

In practice, and particularly for structural steels, the IIW formula is widely used and there appears to be no evidence of failures resulting from this practice. It must be borne in mind that small scale tests for hydrogen cracking are themselves fallible. For example, the CTS test, used for a long time in the UK, was found to be sensitive to the presence of a small gap at the root of the test weld. Moreover all the formulae are applicable only to a narrow range of alloy contents. For example they would rate a 9Cr1Mo alloy as having a higher carbon equivalent than AISI type 4340 steel, whereas 9Cr1Mo steel is relatively easy to weld and type 4340 is relatively difficult.

Specifications for structural steelwork frequently require a maximum value for the IIW carbon equivalent of, for example, 0.45. In work that is welded in accordance with BS 5135 there is an automatic penalty for a high carbon equivalent in that the required preheat is increased.

As a rule, unalloyed carbon steel with a carbon content of 0.25% or less requires no preheat for thicknesses below about 20 mm. For thicker sections preheat must be considered. The use of basic low hydrogen coated electrodes is unnecessary for mild steel except for thick sections; indeed rutile or cellulosic coated rods are generally preferable because of a better weld profile and reduced risk of porosity. In welding with coated electrodes the cooling rate is high enough to ensure that weld metal of matching composition will overmatch the parent metal in yield and ultimate tensile strength. With high heat input rate processes such as submerged-arc and electrogas welding this is not necessarily the case and alloying additions may be required to maintain properties in the fusion zone. For all processes an addition of 0.5% nickel is required to match the notch-ductility of a normalised mild steel: as used, for example, in subzero temperature duties.

Reinforcing bar may be joined using basic low hydrogen electrodes with a preheat in the range 100 to 250 °C; guidance as to procedures will be found in the AWS standard D12-1.

Alloy steel

Economy dictates that steels used for civil construction should contain the minimum amount of costly alloy additions consistent with achieving the required properties, and in general the alloy steels used are classified as 'microalloyed' or 'high strength low alloy' types. For military purposes the requirements are frequently more severe and the range of alloys used is correspondingly wider; e.g. 18% nickel maraging steel has been employed for assault bridges.[17]

Steels are also classified in terms of heat treatment. For structural work lower tensile grades are usually as-rolled. Fine grain steels of higher yield strength and with specified subzero properties are either normalised or subject to thermomechanically controlled rolling. The highest strength grades may be quenched and tempered.

Steels with guaranteed through-thickness properties are employed to an increasing extent and particularly for the node sections of offshore constructions.

Weathering steel is yet another category. Such steels contain small additions of copper, sometimes combined with chromium, and are intended for atmospheric exposure without painting or other protection. The surface film that forms on this type of material provides adequate resistance to corrosion in favourable climates.

Before discussing these various groups of material it will be useful to consider the effect of individual alloying elements on the processing, properties and weldability of the steel.

The effect of individual alloying elements

Carbon

The carbon content of weld metal has a direct effect on its susceptibility to solidification cracking, as shown in Fig. 4.13. The manganese/sulphur ratio required to avoid cracking increases the carbon content. In practice welding consumables are low in carbon and sulphur so that solidification cracking is a relatively unusual problem. It can occur through dilution from the parent metal however and the related problem of liquation cracking is liable to cause defects in welding higher carbon alloys such as AISI 4340 (0.4C).

Increasing carbon content lowers the transformation temperature of the steel and increases the amount of martensite found in the heat affected zone. This has two effects: it increases the risk of hydrogen cracking and decreases the notch-ductility. Figure 4.14 shows the critical crack opening displacement δ_c for the heat affected zone in a microalloyed steel as a function of carbon content.[18] As already noted

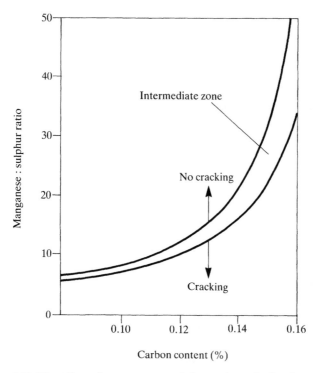

4.13 The effect of manganese: sulphur ratio and of carbon content on the susceptibility of carbon-steel weld metal to hot-cracking.

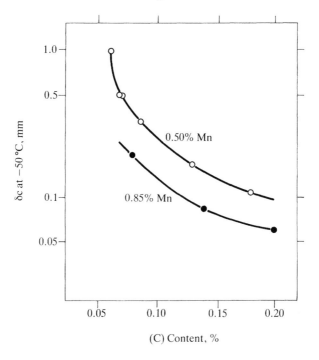

(C) Content, %

4.14 Effect of carbon content on heat affected zone CTOD value at two different manganese levels.[18]

such considerations provide a powerful incentive to reduce the carbon content of weldable structural steels. There is, however, a risk that with low carbon contents the amount of coarse ferrite in the heat affected zone will increase. Therefore any reduction in carbon should be compensated for by an increase in alloy content, with the aim of obtaining an auto-tempered martensitic heat affected zone structure.

Manganese
Manganese combines with sulphur and reduces the risk of solidification cracking; at the same time it forms inclusions that may be detrimental to the through-thickness ductility. It strengthens steel both by solid solution hardening and by grain refinement (through lowering the transformation temperature). The grain refinement leads to increased notch-ductility. However manganese tends to segregate in the ingot and in continuous casting. This can lead to banding and anisotropy in the rolled steel or to localised hard regions in strip. Manganese segregation may also lead to variations in hardness and microstructure across the heat affected zone. Manganese contents are generally held below 1.6% but in extra low carbon steels (C about 0.02%) manganese may go up to 2.0%.[11]

Molybdenum, nickel and chromium

These are classical alloying additions to steel. They act by displacing the austenite-ferrite transformation to lower temperatures, and thereby promote the formation of transformation products that increase in hardness with carbon content and with higher cooling rates. Chromium-molybdenum and nickel-chromium-molybdenum steels used to be called 'air hardening' steels because of their ability to harden after air cooling as opposed to water quenching. The heat affected zone of chromium-molybdenum steels with a carbon content of 0.15% or less is relatively tough in the as-welded condition, probably due to the formation of auto-tempered martensite. This is martensite that forms at a high enough temperature for a degree of tempering to take place during cooling to room temperature. Chromium and molybdenum are added to increase the resistance to high temperature hydrogen corrosion or to improve the creep properties, as discussed in Chapter 5. Nickel additions improve the notch-ductility and have a solid solution hardening effect as well as increasing the quench hardenability.

Titanium, niobium and vanadium

These are the microalloying elements, and they form carbonitrides containing varying amounts of carbon and nitrogen according to the carbon and nitrogen concentrations present in the steel.

Titanium

Titanium nitride is stable at temperatures up to 1350 °C and this, as previously noted, provides a means of controlling austenite grain size in recrystallisation controlled rolling. In continuous casting the cooling rate is fast enough to generate a fine dispersion of TiN particles. The slab is then rolled at a reheating temperature of just over 1100 °C with a finish rolling temperature as high as 1050 °C.

Austenite grain growth is inhibited by the TiN particles and the resulting fine austenitic structure transforms to a fine ferrite. Figure 4.15 shows the effect of V, Ti and N on the grain growth of austenite.[12] Titanium nitride particles are stabilised by the presence of nitrogen, and Fig. 4.15 is illustrative of a steel containing 0.015 to 0.020% nitrogen. As noted earlier it is necessary to consider whether such nitrogen levels are acceptable or not.

Titanium nitride has a similar effect in restraining grain growth in the high temperature region of the heat affected zone of fusion welds. Figure 4.16 compares fusion line toughness for a regular HT 50 steel with those for a titanium treated type and a similar steel containing nickel. The titanium treated steels are better for all heat input rates, including those typical of electroslag welding.[11] Titanium carbonitride precipitates also inhibit grain growth during normalising.

Because titanium carbonitrides are so stable they precipitate mainly

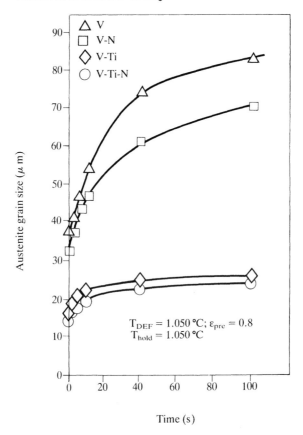

Time (s)

4.15 Effect of vanadium, nitrogen and titanium on grain growth after hot deformation and recrystallisation (reprinted with permission from JOM, formerly Journal of Metals).[12]

at high temperature and therefore do not, as a rule, contribute much to precipitation strengthening at lower temperatures, in the ferrite phase. Such strengthening can however occur under conditions of rapid cooling, as in a strip mill, or when the titanium content is relatively high (say 0.05% or more).

Niobium

Niobium forms precipitates that are stable below about 1000 °C. It raises the recrystallisation temperature of austenite, and increases the degree of deformation of austenite grains. Ferrite is then nucleated from slip bands formed within the grains, giving rise to a fine structure. If the reheating temperature prior to rolling is high enough some niobium will go into solution and there will be an additional precipitation at lower temperatures. Like titanium, niobium prevents grain coarsening during the normalising operation.

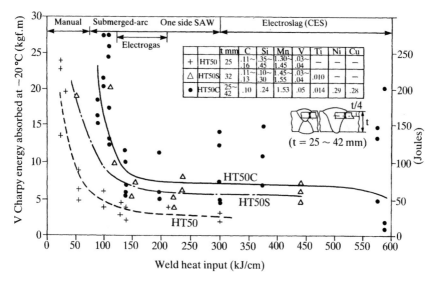

4.16 Charpy V notch impact value in heat affected zone of fusion welds for three types of steel, as a function of heat input rate.[11]

Niobium treated steel is not suitable for welding at high heat input rates. The precipitates are not effective in preventing grain growth and the coarse-grained region transforms to a brittle bainitic structure. With normal welding heat input rates the heat affected zone properties are satisfactory.

Vanadium

Vanadium forms the least stable carbonitrides, and is therefore primarily of value for precipitation strengthening at temperatures below 700 °C. Unlike titanium and niobium, vanadium does not suppress ferrite formation during cooling. Consequently the heat affected zone of welds tends to have a fine intergranular structure which is largely ferritic and has good notch-ductility.[19]

The relative effects of precipitation and grain refinement on strength and notch-ductility of niobium, titanium and vanadium are shown diagrammatically in Fig. 4.17. It is assumed that the steels in question are subject to the same rolling schedule.

In practice high strength structural steels often contain two microalloying elements. By combining niobium or titanium with vanadium it is possible to obtain the benefit of restricting austenite grain growth with precipitation strengthening in the ferrite phase. In particular, titanium/vanadium steels provide good properties in the parent metal with improved toughness in the weld heat affected zone.

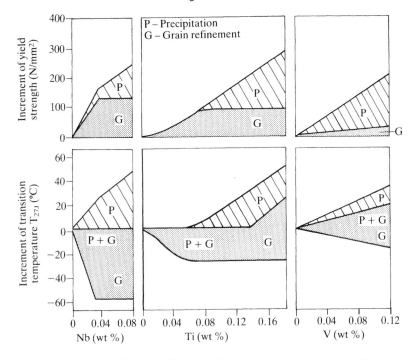

4.17 The relative contribution of precipitation and grain refinement to yield strength and notch-ductility in microalloyed steels containing niobium, titanium and vanadium (reprinted with permission from JOM, formerly Journal of Metals).[13]

Boron

Boron additions to steel are only effective if the nitrogen and oxygen contents are reduced to very low levels, generally by treatment with aluminium and titanium. Free boron then segregates to austenite grain boundaries, where it has the effect of reducing the interfacial energy. This is turn inhibits the nucleation of ferrite at the grain boundaries, and lowers the transition temperature, increasing hardenability. In the coarse-grained region of the heat affected zone boron suppresses the formation of coarse grain boundary ferrite, which is desirable but at the same time increases the hardenability of the grain refined region. Hence for weldability there is an optimum boron content. In the case of low carbon microalloyed steels this lies in the range of 10–15 ppm.[18]

At the present time boron additions are restricted to certain grades of quenched and tempered steels, where there is little risk of embrittlement in the heat affected zone. Titanium plus boron has been added to flux-cored wires to obtain improved notch-ductility at low temperatures.

Copper

Copper has an effect similar to manganese in that it lowers the austenite-ferrite transition temperature, and may produce solid solution hardening. It has been used in alloy boiler plate for precipitation hardening. In structural steel, however such effects are unimportant and it is used primarily for imparting weathering resistance. The amount added is 0.2–0.5%, and it is possible to provide weathering grades in steels of various tensile and impact properties.

Specifications for structural steel

General

In the European Common Market it is proposed to replace existing national standards for structural steels, such as BS 4360 and DIN 17100 with European Standards. These are:

EN 10 025   General steels
EN 10 113   Weldable fine grain steels
EN 10 137   Quenched and tempered steels
EN 10 155   Weathering steels

Individual grades are designated, for example Fe 430 A, where the number is the approximate ultimate strength in N/mm^2 and the letter indicates the impact properties and test temperature. (In BS 4360 the grade number indicates the approximate UTS in kgf/mm^2.) EN 10 025 was published in 1990 and issued by BSI as BS EN 10 025. The remainder are expected to appear in 1991–1993. Prior to publication national standards remain in force. During the transition period there will inevitably be problems because application standards such as BS 5400 for bridges may continue to specify the older standards.

Once a European Standard has been adopted by the European committee for standardisation (CEN), members of the EEC are obliged to withdraw the corresponding national standard and adopt the CEN standard. National standards are a barrier to the free movement of goods within the Community, and the issue of European Standards is seen as a means of removing such restrictions. Progress in this direction has however been meagre, partly because of the lengthy procedures required to obtain agreement on drafts. A number of proposals have been made to speed up the process.

Table 4.3 details the chemical composition and mechanical properties of steels to EN 10 025. In this and other tables referring to standardised materials it is intended only to give a general picture of the steel in question; for details of the current requirements it is necessary to refer to the latest edition of the standard in question.

Table 4.3 Carbon and carbon-manganese steels to EN 10025: plate of thickness between 16 and 40 mm

Designation	Type of deoxidation	Chemical composition, %						YS, N/mm²	UTS, N/mm²	Elongation, %	Impact energy	
		C max	Mn max	Si max	P max	S max	N max				Temperature °C	J
Fe 310 O	Optional	-	-	-	-	-	-	175	290-510	18	-	-
Fe 360 B	Optional	0.25	-	-	0.055	0.055	0.011	225	360-560	26	20	27
Fe 360 B	FU	0.25	-	-	0.055	0.055	0.009				20	27
Fe 360 B	FN	0.19	-	-	0.055	0.055	0.011				20	27
Fe 360 C	FN	0.19	-	-	0.050	0.050	0.011				0	27
Fe 360 D1	FF	0.19	-	-	0.045	0.045	-				-20	27
Fe 360 D2	FF	0.19	-	-	0.045	0.045	-				-20	27
Fe 430 B	FN	0.24	-	-	0.055	0.055	0.011	265	410-560	22	20	27
Fe 430 C	FN	0.21	-	-	0.050	0.050	0.011				0	27
Fe 430 D1	FF	0.21	-	-	0.045	0.045	-				-20	27
Fe 430 D2	FF	0.21	-	-	0.045	0.045	-				-20	27
Fe 510 B	FN	0.27	1.7	0.60	0.055	0.055	0.011	345	490-630	22	20	27
Fe 510 C	FN	0.23	1.7	0.60	0.050	0.050	0.011				0	27
Fe 510 D1	FF	0.23	1.7	0.60	0.045	0.045	-				-20	27
Fe 510 D2	FF	0.23	1.7	0.60	0.045	0.045	-				-20	27
Fe 510 DD1	FF	0.23	1.7	0.60	0.045	0.045	-				-20	40
Fe 510 DD2	FF	0.23	1.7	0.60	0.045	0.045	-				-20	40
Fe 490-2	FN	-	-	-	0.055	0.055	0.011	285	470-610	20	-	-
FE 590-2	FN	-	-	-	0.055	0.055	0.011	325	570-710	16	-	-
Fe 690-2	FN	-	-	-	0.055	0.055	0.011	355	670-830	11	-	-

FU = Rimming steel
FN = Rimming steel not permitted

FF = Fully killed steel with sufficient N-binding elements to combine with N content.

The BS and EN standards cover structural steels for all purposes but in the case of offshore work the Engineering Equipment and Material Users Association (EEMUA) have issued Publication 150, which is an adaptation of BS 4360:1986 for offshore use. The BS grades have been retained but in certain cases the impact energy requirements have been raised and compositions specified in more detail. Where such modifications have been made this is indicated by an additional suffix letter M. The specification includes recommendations for weldability tests, which will be considered later.[20] In the UK, offshore structures are also covered by a non-mandatory guidance document issued by the Department of Energy.[21] The EEMUA document conforms with these guidelines.

In the UK and in many other countries steel for ship construction is specified by the Classification Societies such as Lloyds, and these will be considered later in the chapter.

ASTM publishes separate standards for individual structural grades but also gathers together these standards for particular applications: e.g. ASTM A 709, 'Structural steel for bridges'. This includes grades in yield strengths ranging from 36 to 100 ksi (1 ksi = 1000 lbf/sq in). There are weathering grades with additions of 0.2 to 0.5% copper, designated with a suffix W. Impact test temperatures are indicated by a number, where for example figure 1 indicates that the steel may be used down to 0 °F (−18 °C). Ship steels are also specified by ASTM but these are essentially the same as the ABS (American Bureau of Shipping) grades.

The testing of structural steel plate and sections is normally made to represent a cast, batch or some specified tonnage such as 40 t. Lloyds Rules however require quenched and tempered plates to be tested individually. In the case of boilers and pressure vessels (Chapter 5) the testing of individual plates is the norm.

Steels to EN 10 025

This standard specifies the deoxidation practice for each grade as follows:

FU = Rimming steel, not deoxidised
FN = Rimming steel, not permitted
FF = Fully killed steel with sufficient nitride-forming elements to combine with all free nitrogen present: e.g. 0.02Al.

Type FN would permit semi-killed steel.

The chemical requirements of the standard appear to be designed to make provision for Bessemer steel, where the nitrogen content is a matter for concern. It reflects the practice referred to earlier; namely the use of aluminium treatment to remove free nitrogen and eliminate strain-age embrittlement following welding. The B and C grades are only impact tested when this is specified in the order, but may be useful where the requirements for notch-ductility are not onerous. Fe 510 DD1 and Fe510 DD2 call for a C_v of 40 J at −20 °C, which is adequate for many applications in less cold areas.

For conformity with good practice in the use of steel for welded structures this standard must be used with additional restrictions; in particular, rimming steel is not acceptable, neither is the O grade which has no limitations on chemical composition.

Higher tensile grades to BS 4369

These are listed in Table 4.4, and consist of carbon-manganese and microalloyed types, either normalised or quenched and tempered. In common with EN 10 025 the specification allows thermomechanically controlled rolling as an alternative to normalising, unless otherwise required. For the microalloyed grades it is allowed to use alternatives to niobium and vanadium, or to omit microalloying additions altogether, provided that the mechanical test requirements are met.

Quenched and tempered plate is normally produced using high pressure water jets in a special roller quench unit which gives uniform cooling whilst holding the plate flat.

Microalloying has also been used for some structural sections, mostly intended for North Sea use. Examples are given in Table 4.5 for sections having a ruling dimension up to 50 mm. These niobium and niobium-vanadium treated steels combine high yield strength, good notch-ductility and low carbon equivalent when supplied in the thermomechanically controlled condition; in addition they may be normalised.

Sections may also be given a selective accelerated cooling by means of water jets. In one system, illustrated in Fig. 4.18, the section is subject to thermomechanically controlled rolling and is then passed through the quenching device to bring the flange temperature down to a level near the top of the ferrite range. It is then air-cooled, and during this relatively slow cool auto-tempering occurs. Such treatment gives a very fine grain with, in the case of microalloyed steel, precipitation hardening, and yield strengths up to 700 N/mm² have been achieved.

High strength low alloy steels

The term 'high strength low alloy' has been applied to microalloyed steel, but here it will be used for those alloys containing small amounts of molybdenum, chromium, nickel, vanadium and boron, and normally supplied in the quenched and tempered condition. Table 4.6 lists some typical compositions. These are steels manufactured by Thyssen AG

Table 4.4 Composition and mechanical properties of fine grained normalised and tempered steel plate up to 25 mm to BS 4360

Grade	Normal supply condition	Chemical composition, %							CE (BS 5135)	YP min, N/mm²	UTS, N/mm²	Elongation, %	27 J vTrs °C
		C max	Si	Mn max	P max	S max	Nb	V					
40 EE	Normalised	0.16	0.1/0.5	1.5	0.04	0.03	–	–	0.40	245	340/500	25	–50
43 EE	Normalised	0.16	0.1/0.5	1.5	0.04	0.03	–	–	0.40	265	430/580	22	–50
50 EE	Normalised	0.18	0.1/0.5	1.5	0.04	0.03	0.003/0.10	0.003/0.10	0.48	345	490/640	20	–50
50 F	Quenched and tempered	0.16	0.1/0.5	1.5	0.025	0.025	0.003/0.10	0.003/0.10	0.35	390	490/640	20	–60
55 C	Normalised	0.22	0.60 max	1.6	0.04	0.04	0.003/0.10	0.003/0.20	0.45	430	550/700	19	0
55 EE	Normalised	0.22	0.1/0.5	1.6	0.04	0.03	0.003/0.10	0.003/0.20	0.48	430	550/700	19	–50
55 F	Quenched and tempered	0.16	0.1/0.5	1.5	0.025	0.025	0.003/0.08	0.003/0.10	0.38	430	550/700	19	–60

Table 4.5 Grade E 355 microalloyed structural sections

Steel	Product	Chemical composition, %							CE	Thickness, mm	YS. N/mm²	UTS. N/mm²	Charpy energy J			
													Longitudinal		Transverse	
		C	Si	Mn	P	S	Nb	V					Temp.°C	J	Temp.°C	J
Arctic 355	British Steel	0.10	0.40	1.39	0.019	0.002	0.03	–	0.33 Typical	40	400	512	–		−40	209
															−100	50
Fritenar 355 TZK	Arbed	0.12	0.50	1.60	0.03	0.008	0.04	0.06	0.40 max	≤30	355	470–	−40	70	−40	47
										30–40	345	600				
										40–50	335					
Krytenar 355 TZ	Arbed	0.08	0.50	1.60	0.03	0.008	0.04	0.06	0.36 max	≤30	335	430–	−50	47	−50	32
										30–40	345	550				
										40–50	335					

Composition for Arctic 355 is typical and for Arbed steels maximum.

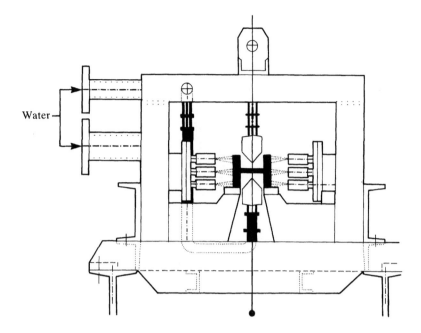

4.18 Quenching arrangement for I beams (reprinted with permission from JOM (formerly Journal of Metals).[13]

Duisburg but most of the qualities are of US origin. Similar steels are specified in ASTM A 709. American high strength structural steels developed during the postwar period, such as T1, were mostly of the high strength low alloy type. T1 is a quenched and tempered NiCrMoVB steel with a yield strength of 100 ksi (690 N/mm²), weldable with a modest preheat. T1 steel has a variety of uses including road tanker vehicles and bridges. It is now possible, as noted earlier, to achieve the same yield strength with a relatively minute amount of alloying element, and current trends favour the microalloy types of steel in the USA as well as in Europe and Japan.[12,13,22]

Weathering steels

The BS 4360 grades are given in Table 4.7. The W grades of ASTM have found a substantial use in bridges and building construction in the USA. In the UK there is less public acceptance of weathering steel and it has mostly been used for road-over bridges, where the supporting structure is unobtrusive. Two other limitations should be noted. Firstly, plate material has impact tests available down to 15 °C, whereas many applications would require testing at lower temperatures. Secondly, the carbon equivalents of grades B and C are over 0.5 and preheating would be required for some welds.

Table 4.6 High strength quenched and tempered steels for structural use

Group	Grade	Thyssen brand	Nominal composition, %								YS N/mm²	Use
			C	Mn	Si	Cr	Mo	Ni	V	Other		
A	ST E 430 V	A537 G2	0.22	1.3	0.4	0.15		0.2		Cu 0.25	430	Storage and transport vessels Penstocks for hydro-electric power Cranes
	ST E 460 V	XABO 47	0.15	1.2	0.4		0.2	0.4			460	
	ST E 500 V	XABO 51	0.15	1.2	0.4		0.2	0.4			500	
	ST E 550	N-A-XTRA 56	0.20	0.8	0.6	0.7	0.2				550	Naval vessels Submarines Heavy road vehicles
B	HY 80	HY 80	0.18	0.3	0.3	1.4	0.4	2.5	0.01		550	Earthmoving equipment Mining equipment
	ST E 620	N-A-XTRA 63	0.20	0.8	0.6	0.8	0.3				620	Cranes
	ST E 690	N-A-XTRA 70	0.20	0.8	0.6	0.8	0.3				690	
C	ST E 690	T1	0.20	0.8	0.3	0.5	0.5	0.9	0.05	B 0.003	690	
	HY 100	HY 100	0.20	0.3	0.3	1.4	0.4	3.0	0.01		690	
	ST E 890	XABO 90	0.18	0.7	0.3	0.6	0.3	1.7	0.07		890	Mobile cranes Mining equipment
	ST E 960	XABO 95	0.18	0.8	0.3	0.7	0.4	1.8	0.07		960	Toothed racks in jack-up drilling platforms

Table 4.7 Weathering steels to BS 4360 (plates up to 25 mm)

Grade	Normal supply condition	Chemical composition. %										CE^x	YP min. N/mm²	UTS min. N/mm²	Elongation. %	27J VTrs °C
		C max	Si	Mu	P	S max	Cr	Ni max	Cu	Al (sol)	V					
WR 50A	As-rolled	0.12	0.25/ 0.75	0.3/ 0.5	0.07/ 0.15	0.05	0.50/ 1.25	0.65	0.25/ 0.55	–	–	0.39	325	480	21	0
WR 50B	As-rolled or normalised	0.19	0.15/ 0.65	0.9/ 1.25	0.04 max	0.05	0.5/ 0.65	–	0.25/ 0.4	0.01/ 0.06	0.02/ 0.10	0.51	345	480	21	0
WR 50C	As-rolled or normalised	0.22	0.15/ 0.65	0.9/ 1.45	0.04 max	0.05	0.5/ 0.65	–	0.25/ 0.4	0.01/ 0.06	0.02/ 0.10	0.55	345	480	21	–15

x Calculated from mid-range of alloy additions.

Steels with improved transverse ductility

Any attachment that is fusion welded to the surface of steel plate causes straining at right angles to the surface and this may give rise to lamellar tearing. Configurations that may be troublesome include web/flange joints, corner joints, nozzle welds and pipe joints such as are found in the node sections of offshore structures. BS 6780 specifies three Z grades of plate according to the reduction of area in the transverse tensile. Three samples are taken to represent either a batch or an individual plate. These are tested and the grade is indicated in accordance with Table 4.8. The standard also gives some guidance as to the grade required for particular types of joint, as shown in Fig. 4.19.

Table 4.8 Grading of plate for through-thickness ductility

<table>
<tr><td></td><td colspan="2">Reduction of area %</td></tr>
<tr><td>Acceptance
Class</td><td>Minimum average
value</td><td>Minimum individual
value</td></tr>
<tr><td>Z15</td><td>15</td><td>10</td></tr>
<tr><td>Z25</td><td>25</td><td>15</td></tr>
<tr><td>Z35</td><td>35</td><td>25</td></tr>
</table>

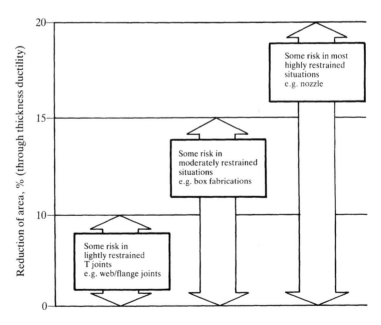

4.19 Guide to assessing the required through thickness ductility of steel plate, according to BS 6780 (source: BSI).

Aluminium alloys for structural work

General

Some of the more important weldable structural aluminium alloys are shown in Table 4.9, together with their nominal composition and typical mechanical properties. The four digit designation is that of the Aluminium Association Inc and it has been adopted by ASTM and BSI, although not by ISO, as will be seen. British standards for wrought aluminium used in general engineering are BS 1470-1475. Castings are specified in BS 1490 but these are little used in structural work.

Aluminium alloys are used in structures primarily to reduce weight, their specific gravity being about one-third that of steel, and to a lesser extent because of their good resistance to corrosion by sea water. Aluminium is easily extruded and is therefore available in a wide variety of sectional forms. This is of particular value in certain types of structural work such as helidecks. An apparent disadvantage of aluminium is that its modulus of elasticity is only one-third that of steel. However the stiffness of a plate is proportional to the cube of its thickness. Thus an aluminium plate is about nine times as stiff as a steel plate of the same weight.

Alloy types

Aluminium alloys may be classified generally into two categories, non-heat treatable and heat treatable. The former category is typified by the aluminium-magnesium alloys, with 3–5% Mg. Their strength is obtained through solid solution hardening and increases with magnesium content. The upper limit is a nominal 5%; above this value a second phase may be formed and the ductility is reduced. Heat treatable alloys gain their strength by precipitation hardening; for example, 6061 is hardened by an incipient precipitate of the intermetallic compound Mg_2Si. Medium strength heat treatable alloys containing zinc and magnesium such as 7005 (Table 4.9) are weldable and although the heat affected zone is softened by welding the strength is recovered by natural ageing over a period of time. This is not the case for the 6061 type alloys and for these it is necessary either to accept a reduced design stress for the structure as a whole or place the welds in a low stress location.

The highest strength aluminium alloys which are commonly used in aircraft construction are unweldable. They suffer liquation cracking, solidification cracking in the weld metal when matching filler rod is used, inadequate strength in the fused zone and irreversible softening in the heat affected zone. These defects can be minimised but not elim-

Table 4.9 Structural aluminium alloys

Designation		Nominal composition, %				Properties		Use
ASTM and BS	ISO	Mg	Mn	Si	Zn	YS N/mm²	UTS N/mm²	
5083	AlMg4.5Mn	4.5				125	275/ 350	Marine structures
5154A	AlMg3.5	3.5				85	215/ 260	Welding wire Non-heat treatable
6061	AlMg1SiCu	1.0		0.6	Cu 0.3	225	295	General, structural
6082	AlSiMgMn					240	310	pipe, plate and
6060	AlMgSi					150	190	extruded sections
6063	AlSiMg0.75	0.7		0.4		180	200	Heat treatable
7005	AlZn4.5Mg1	1.4	0.4	–	4.5	305	345	Structural, military bridges, heat treatable
4043A	AlSi5			5				
4047A	AlSi12			12				Welding filler rod
5356	AlMg5	5						
5554	AlMg3	3						

inated and these alloys (for example 2024: Al4.5Cu1.5Mg0.6Mn0.5Si) are rarely joined by welding. An exception is alloy 2219 (Al6Cu) which was developed in the USA as a weldable alloy for aerospace applications.[23] However a similar alloy was used to construct a welded torpedo boat in 1894, and this failed very quickly due to corrosion.[24] Evidently alloy 2219 would not be suitable for normal structural applications.

Metallurgy

There are three basic problems in welding aluminium alloys. The first is the presence of a refractory oxide film. Prior to the introduction of argon-shielded arc welding processes it was necessary to use corrosive fluxes to remove the oxide. Inert gas-shielded processes with electrode positive remove the oxide film by cathodic action but it is still necessary to avoid procedures that could trap oxide film within the weld. The second is the high ratio of hydrogen solubility in liquid at the melting point and solid at the same temperature. In pure aluminium this is about 14, as compared with about 3 for iron and about 2 for nickel. Consequently aluminium fusion welds tend to be porous if there is a source of hydrogen contamination. The third problem is that many of the alloys have a wide solidification temperature range and may therefore be subject to cracking if welded autogenously (see Chapter 2). The solution is to weld with an alloy that does not have this defect and the filler alloys normally used (aluminium-silicon and aluminium-magnesium) conform to this requirement. The possibility of dilution by the parent metal must however be considered.

The fire risk

Aluminium alloys melt at temperatures within the range 550–660 °C, and suffer a serious loss in strength at temperatures over 300 °C. This means that when exposed to typical fire temperatures of 500–950 °C unprotected aluminium will collapse and melt. Steel, on the other hand, has a softening temperature of 650 °C and a melting point of about 1550 °C so that it is less likely to suffer so catastrophically.

The fire risk problem with alloys has been reviewed by West.[24] Various means are used to reduce the hazard, including a combination of insulation with water sprinkler systems, and the use of steel for strength members. It is evident that special precautions are required in using aluminium for ship superstructures and for accommodation offshore. Likewise in oil refineries and petrochemical plant careful consideration must be given to the use of aluminium piping for carrying inflammable fluids.

Following dramatic incidents during the Falklands War, there was a widespread notion that aluminium ignited during a fire. This, of course is not the case: aluminium will only burn when in a finely divided form.

Welding structural steel

Code requirements: BS 5135

In the UK there is no general code to cover structural steelwork. However BS 5400 'Steel, concrete and composite bridges' contains requirements for fabrication and welding. Steel, whether ordered to BS 4360 or not, is required to meet the minimum quality requirements of that standard. Welding processes and procedures must conform to BS 5135.

BS 5135 defines procedures that will minimise the risk of hydrogen cracking in the weld metal or heat affected zone. The first requirement is to establish the correct level of preheat. This depends on four variables:

the amount of hydrogen injected into the weld, the carbon equivalent of the plate material, the combined thickness of the joint, and the heat input rate.

The hydrogen content of the weld is indicated by the letters A, B, C and D as shown in Table 4.10. As will be seen these scales are consistent with the IIW designations. The diffusible hydrogen content is determined in accordance with BS 6693 and generally as indicated in Chapter 2.

Table 4.10 Hydrogen scales

Diffusible hydrogen content of weld metal ml/100g deposited metal		Hydrogen scale to BS 5135	IIW designation
Over	Up to and including		
15		A	High
10	15	B	Medium
5	10	C	Low
	5	D	Very low

The carbon equivalent value may be calculated using the IIW formula using maxima obtained from the mill analysis certificates. Alternatively, BS 5135 tabulates standard carbon equivalents for BS 4360 grades (but not for EN 10 025 grades). The standard CEs are shown in Table 4.4.

The combined thickness of the joint is, in principle, the sum of the thickness of the plates that meet at the weld. Thus, where plate thicknesses are all equal to t, the combined thickness for a butt weld is 2t and that for a single fillet weld is 3t. Some examples of other configurations are shown in Fig. 4.20.

The heat input rate, or arc energy, is given by

$$\text{arc energy} = \frac{VI}{wx}\ 10^{-3}\ \text{kJ/mm}$$

where V is arc voltage in volts, I is welding current in amps, w is welding speed in mm/s and x is a factor which depends on the welding process as follows:

Submerged-arc welding (single wire)	0.8
GMA welding (solid cored wire or self-shielded)	1.0
Manual metal arc	1.0
GTA welding	1.2

The required preheat is then determined from one of a series of charts which relate the four variables described above. Figure 4.21 is an example. For manual metal arc fillet welds, data obtained from such

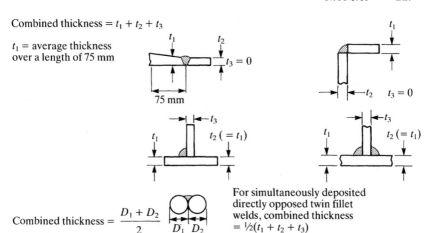

Combined thickness = $t_1 + t_2 + t_3$

t_1 = average thickness over a length of 75 mm

$t_3 = 0$

75 mm

$t_3 = 0$

$t_2 (= t_1)$

$t_2 (= t_1)$

For simultaneously deposited directly opposed twin fillet welds, combined thickness = $\frac{1}{2}(t_1 + t_2 + t_3)$

Combined thickness = $\dfrac{D_1 + D_2}{2}$

4.20 Calculating the combined thickness of a welded joint to BS 5135: typical example (source: BSI).

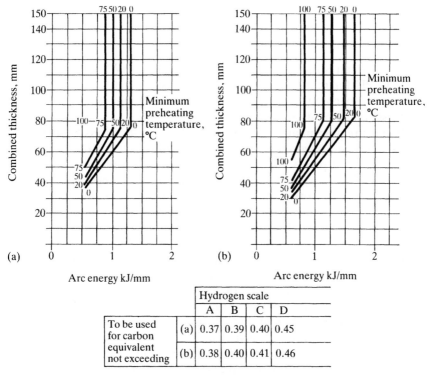

	Hydrogen scale			
	A	B	C	D
To be used for carbon equivalent not exceeding (a)	0.37	0.39	0.40	0.45
(b)	0.38	0.40	0.41	0.46

4.21 Preheat requirements of BS 5135 for steels with carbon equivalents in the range 0.37 to 0.46 (source: BSI).

figures are presented in tabular form. Preheating on large structures is costly and time consuming so there is a strong incentive to keep the hydrogen level down. The required extent and methods of preheating are discussed in Chapter 2.

Post-welding heat treatment, inspection methods and the extent of inspection are as specified in the application code or as agreed by the parties concerned. There is a recommendation (non-mandatory, but generally observed) that because of the risk of delayed cracking a period of at least 48 hours should elapse after welding before inspection. A hydrogen diffusion treatment, as described in Chapter 2, may be used to shorten this interval. There are also recommendations for acceptance standards for welds which may be used if these are not specified in the application standard.

When specified by the purchaser, welding procedure qualification testing is carried out in accordance with BS 4890. Welders are required to be qualified to BS 4872. These two standards are generally similar to the relevant parts of the AWS Code D1.1, which are discussed in the next section.

Code requirements: AWS D1.1

The American Structural Welding Code for steel, AWS D1.1 provides specifically for manual metal arc welding, submerged-arc, gas metal arc and flux-cored, electroslag, electrogas and stud welding. Other processes may be used if qualified.

Preheat requirements cover all the ASTM specifications likely to be used in structural work up to a yield strength of 100 ksi (689 N/mm²). Table 4.11 is a precis giving preheat as a function of steel type and thickness. In the case of mild steel there are two categories, one for coated electrodes other than low hydrogen and the other for low hydrogen rods and other processes. Preheats are lower for the low hydrogen processes. Electrodes other than low hydrogen are not allowed for higher tensile grades. Carbon equivalent values and arc energy are not used in selecting preheat (the governing variables being plate thickness at the weld, steel type and whether or not a low hydrogen process is used). A direct comparison between the preheat requirements of AWS D1.1 and BS 5135 is not therefore possible. Both BS 5135 and AWS D1.1 prohibit welding when the metal temperature is below 0 °C.

Early sections of the code give details of a number of joint types and weld preparations together with acceptable welding processes. These include butt, T, lap and corner joints, both full penetration and partial penetration. Such joints are considered to be pre-qualified, and only require procedure qualification if a non-standard welding process or

Table 4.11 Minimum preheat and interpass temperatures to AWSD1.1

Steel type	Welding process	Thickness of thickest part at point of welding, mm	Minimum temperature, °C
Mild steel with some restrictions	Manual metal arc with electrodes other than low hydrogen type	Up to 19 incl. 19 to 38 incl. 38 to 64 incl. Over 64	None 66 107 150
Mild and medium tensile steel up to UTS of 70 ksi	Manual metal arc with electrodes of low hydrogen type, submerged-arc, gas-metal arc, cored wire	Up to 19 incl. 19 to 38 incl. 38 to 64 incl. Over 64	None 10 66 107
Higher tensile as-rolled or normalised steel	Manual metal arc with electrodes of low hydrogen type, submerged-arc, gas-metal arc, cored wire	Up to 19 incl. 19 to 38 incl. 38 to 64 incl. Over 64	10 66 107 150
Quenched and tempered steel	Manual metal arc with electrodes of low hydrogen type, submerged-arc with neutral flux, gas-metal arc, cored wire	Up to 19 incl. 19 to 38 incl. 38 to 64 incl. Over 64	10 50 80 107

Welding is not permitted when the metal temperature is below 0 °C.
In welding quenched and tempered steel, the maximum preheat and interpass temperature must not exceed 205 °C for thickness up to 38 mm and 230 °C for greater thickness. The heat input must not exceed the steel manufacturers requirements. Preheat is not required for electrogas or electroslag welding.

short circuiting gas metal arc welding is used. Pre-qualified tube joints are included in the section on tubular structures.

For joints that do not conform to these requirements, procedure qualification testing is required. The first step is to specify the procedure, and to do this the code lays down the variables appropriate to specific welding processes. Such variables include material, electrode size, welding current and voltage, welding position, preheat and post-welding heat treatment, and the amount of deviation allowed from the specified quantities is defined in some detail. The welding positions and their designations for pipe are illustrated in Fig. 4.22. BS 4870 (procedure qualification) has requirements that are similar in general but differ in detail. The BS standard positions are also similar but the designations 1G, 2G, etc, are not used.

A sample weld is made in accordance with the specified procedure and is tested by visual inspection, radiography, tensile and bend tests, and impact or other tests when specified. In some instances, for example the node connections for a North Sea oil rig, more elaborate testing up to and including a full scale mock-up of the joint combined with fracture toughness testing may be required.

Records of weld procedure and of the results of testing are made. AWS D1.1 and BS 4870 provide sample forms for this purpose. Manufacturers may also use their own format for records, provided that the essential information is included. As a rule it is desirable to have the tests for both procedure and operator qualification witnessed by an independent inspection agency such as Lloyds, and this will be certified on the test record.

In US practice, checking of procedure qualification records by purchasers is primarily an inspection rather than a technical function. The inspector's task is to ensure that the records conform to the requirements of the code and purchasing specifications, and provided this is the case work can proceed. Only when there is a non-compliance is it necessary to refer the question to the engineer or metallurgist. In the UK, on the other hand, welding procedures are often checked by qualified personnel (BS 5400 for bridges requires that all welding procedures be submitted to the responsible engineer). Even then, BS 5400 does not place too much trust in procedure testing; it requires that tests be made on run-off plates attached to a proportion of welds in tension members.

Both AWS D1.1 and BS 5135 require that welders be qualified. For D1.1 the test position and their designations are the same as for procedure testing (Fig. 4.22). There is a hierarchy of positional welds such that, for example a welder tested on pipe in the 5G position qualifies for the 1G and 2G positions. The same test also qualifies for plate in the flat, overhead and vertical positions.

Completed testpieces are examined visually and either by radiography or ultrasonically, and must meet specified limits for porosity and

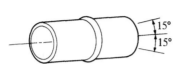

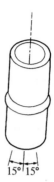

1G Pipe horizontal and rotated.
    Weld flat (± 15°). Deposit
    filler metal at or near the top.

2G Pipe or tube vertical and
    not rotated during welding.
    Weld horizontal (± 15°).

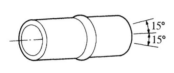

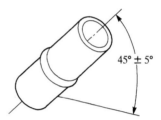

5G Pipe or tube horizontal fixed
    (± 15°) and not rotated during
    welding

6G Pipe inclined 45° and not rotated
    during welding

4.22 AWS designations for pipe welding positions (reproduced by courtesy of the American Welding Society).

inclusions. In addition tensile and bend tests are made. Records are then made on the appropriate forms, samples of which are given in the Code.

BS 4871 (operator qualification) has very similar requirements except that the tensile test is replaced by a macrosection of the weld. A European Standard EN 287 is in preparation and will probably supersede BS 4871 in due course.

Procedure and operator qualification is essential to the control of quality in welding operations, whilst operator qualification is a means of training and advancement for welders. Normally these tests are carried out under workshop conditions but when the actual welds must be made in locations where there is limited access (e.g. a ship's double bottom) the inspecting authority may require the procedures and operators to be qualified under similar conditions.

All codes prohibit downward welding except for repair of undercut or similar defects. In structural work there is always a risk of lack of side fusion and slag traps when using a stringer bead technique and this is most severe in the case of downward runs. Downward welding is however used in pipeline welding and will be discussed in Chapter 5.

AWS D1.1 and application codes such as BS 5400 include provision for the repair of laminations. These are laminar discontinuities that form in plate when gas pores in the ingot mould are flattened during rolling and where the two opposite surfaces fail to weld together, or where there are trapped refractory inclusions or deoxidation products. The AWS code is specific about the action to be taken when a lamination is found (e.g. after edge preparation). Small discontinuities are removed or welded, but those greater than one inch in length and depth are explored by ultrasonic testing. If the area of lamination is not greater than 4% of the plate area it may, subject to detailed provisos, be repaired by welding. This section of the code also covers the case where a lamination is disclosed by ultrasonic testing after welding (see earlier in this chapter). BS 5400 has no detailed provisions and leaves a decision on repair to the engineer. Such decisions are always difficult and AWS are to be commended in providing guidance in this matter. It has been said that because of the widespread use of continuous casting and fully killed steel, laminations are 'a thing of the past'. No doubt their incidence has been reduced and will be further reduced in the future.

Processes

Manual metal arc welding remains the most widely used process for welding structural steel. In the UK rutile-coated rods types R or RR to BS 639 are the usual choice for welding mild steel of moderate thickness that is not subject to severe operating conditions and where the impact requirements for weld metal are not exacting. The designation RR applies to electrodes with heavy rutile coating and as a rule such coatings contain metal powder and give an enhanced deposition rate, as noted in Chapter 1.

Basic low hydrogen coatings (BS types B and BB, US Exx15, Exx16 and Exx18) are used wherever hydrogen cracking is considered to be a hazard and where the weld metal is required to have good notch-ductility. AWS D1.1 requires low hydrogen electrodes for steel grades with a yield strength over 50 ksi ($345 N/mm^2$). BS 5135 has no such limitation but virtually all fabricators would use low hydrogen rods for Grades Fe 510 or BS grade 50, which have similar yield strengths.

The problem of moisture pick-up presents some difficulties. One method of avoiding this, the use of heated quivers, has already been mentioned. A more practical approach is to employ an electrode brand that has been demonstrated by tests to be resistant to moisture pick-up for a period equal to the length of a shift. Any rods left out at the end of the shift are then collected and either redried or destroyed. AWS D1.1 would permit such a procedure for steels having a yield strength below 80 ksi (552 N/mm^2). For 80 ksi and 100 ksi steels there are special procedures for the storage and issue of electrodes.

There is no difficulty in matching the yield and tensile properties of steel with quite modest alloying additions to weld metal. Standard specifications do not spell out these additions but simply define the strength requirements; the means of achieving this being the electrode manufacturer's responsibility. Obtaining adequate notch-ductility is another matter. This question has been discussed in relation to carbon-manganese steels in Chapter 2. Here the addition of, for example, 0.5% Ni can be highly effective, but with HY 80, HY 100 and similar steels the problem is more serious.

Submerged-arc welding is employed in the manufacture of plate girders, box sections, for such applications as attaching stiffeners to panels in shipbuilding and generally wherever the plate thickness and length of weld allow the use of an automatic set-up. In welding of plate girders special purpose machines are available that rectify distortion after welding; otherwise plates are preformed as described in Chapter 2. The two fillet welds joining flange to web are usually made simultaneously with one electrode on each side. Preheat is not required as a general rule but for thicker material when working to BS 5135 it is desirable to limit the carbon equivalent so as to minimise the level of preheat. Radiant heaters have been used to heat the plate ahead of the weld, as described in Chapter 2.[25]

It is possible in submerged-arc welding to produce a fused zone that is deeper than it is wide and centre-line shrinkage cracks may form in such welds. AWS D1.1 requires that neither the depth nor the maximum width of a weld pass should exceed the width at the surface. Figure 4.23 illustrates this condition. Cracks in full penetration web to flange joints may also be caused by bending before the weld metal is fully solidified and in this case is prevented by offsetting the two welding heads.[25]

Gas metal arc welding is tending to displace manual metal arc for shop fabrication, for the most part using flux-cored rather than solid wires. Fume extraction remains a problem with GMA welding however; extractors integral with the gun are the most effective but make handling more difficult. Short circuiting GMA welding is usually restricted to thin sections, say below 10 mm, where problems due to lack of fusion are less likely. Direct current GTA welding is used mainly for shop fabricated tubing. Aluminium is, for the most part, fabricated using the argon-shielded GMA process, with AC tungsten inert gas welding for thinner sections and pipework. Outside work requires effective wind shielding, most particularly for gas-shielded processes but also for manual metal arc welding.

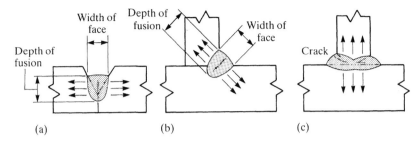

4.23 Centreline shrinkage cracks due to excessive depth/width ratio of weld: a) Groove; b) Fillet; c) T joint. From AWS D1.1 Commentary (reproduced by courtesy of the American Welding Society).

Electrogas welding has been used for welding the vertical seams in large storage tanks and in shipbuilding. Heavy machine tool frames have been fabricated by electroslag welding, but this process has not been used to any great extent in constructional steelwork generally.

Self-shielded welding does not have the disadvantages and complications of gas shield and has had a deal of success in offshore fabrication. The nickel alloyed wire NR 203 Ni produces weld metal with good notch-ductility and resistance to strain-age embrittlement. The disadvantage is the high aluminium content of the deposit, which may be incompatible with weld metal produced by other processes. This, however, is of minor importance.

Shipbuilding

The classification societies

Ship construction is the area where a system of quality control first became necessary. Underwriters had a need to assess the risk in the insurance of a ship and its cargo, and this was done by registering the ships and classifying them according to the condition of their hull and the quality of the equipment. Such a system was first set up at Lloyds Coffee House in London, and the register became Lloyds Register of Shipping, which was first published in 1760. Lloyds Register Office thus became the first Classification Society, and was the model for similar organisations in other countries, such as American Bureau of Shipping, Bureau Veritas and Det Norske Veritas.

The society operated by appointing surveyors who inspected and classified the ships, and by publishing an annual Register Book in which the classifications were recorded. Subsequently rules for ship construction were formulated, and rules for steel ships were published in 1888. Surveyors were then appointed to inspect steelworks and test their products, and Lloyds produced an Approved List of steelworks, which continues to this day. In 1920 an 'electrically welded ship' was registered for the first time, and in the following years Lloyds published a list of approved electrodes.

Originally a single grade of steel was recognised for shipbuilding, but following the loss of Liberty ships during and after the war impact tested grades were introduced. Then in 1959 the International Society of Classification Societies developed unified requirements for ship steels which recognised five grades, lettered A to E. These grades became the basis for the specifications of individual societies. The scantlings of a ship (the section thickness of plate and stiffeners), are based on the minimum mechanical properties listed under the various steel grades.[26,27]

Materials and ship construction

Tables 4.12 and 4.13 give a summarised version of Lloyds requirements for ship steels. Apart from Grade A, which is a general purpose mild steel for use in non-critical locations, all grades are impact tested. Table 4.13 lists high tensile grades which may be quenched and tempered. Where this is so, tensile and impact tests are made on individual plates; otherwise tests are made per cast or per 50 tonne batch. There are also provisions for plate with improved through-thickness properties.

Special steels are required for liquefied petroleum gas or methane carriers. There are basically two systems for the storage and transport of such materials. In the first, the gas is maintained in liquid form at atmospheric pressure by refrigeration. In the second it is kept liquid at atmospheric temperature by holding under pressure. Similar systems may be used for transport of other gases. Normally the liquids are held in containers but provision for spillage may require that parts of the ship also be constructed of similar material. Aluminium-magnesium alloys such as 5054 and 5083 are also used for liquefied gas tanks carried on board ship.[28]

Rules for the construction of such tanks and for the carriers have been formulated by the International Maritime Organisation and by the various classification societies. Table 4.13 lists the steels used in this service, and Table 4.14 shows the Lloyds requirement for impact testing of materials used for low temperature service. Aluminium alloys are not impact tested.

Special materials, including high tensile steel, aluminium alloys and glass reinforced plastic may be needed for small high speed craft such as catamarans, waterplane twin hull ships, hydrofoils and other surface effect vessels. Their primary needs include a reduction in weight combined with corrosion resistance, fire resistance and, for metals, good weldability. Cost is also a factor and, for the present, titanium

Table 4.12 Normal and medium tensile ship steel grades to Lloyds Rules; as-rolled, normalised, normalising rolled or thermomechanically processed

Grade	Type of deoxidation	Chemical composition, %									YS N/mm²	UTS N/mm²	Elongation %	Charpy V Temp. °C	Charpy V J
		C max	Mn min	Si max	P max	S max	Al(sol.) min	Nb	V	Ti max					
A	Any	0.21	2.5x C	0.5	0.04	0.04	-	-	-	-	235	400–490	22	-	-
AH32	K or S-K.	0.18	0.9–1.6	0.5	0.035	0.03	0.015	0.02–0.05	0.03–0.10	0.02	315	440–590	22	0	31
AH36	GR										355	490–620	21	0	34
AH40	GR										390	510–650	20	0	41
B	Not rimmed	0.21	0.80	0.35	0.04	0.04	-	-	-	-	235	400–490	22	0	27
D	K., GR	0.21	0.60	0.1–0.35	0.04	0.04	0.015	-	-	-	235	400–490	22	–20	27
DH32	K., GR	0.18	0.9–1.6	0.5	0.035	0.03	0.015	0.02–0.05	0.03–0.10	0.02	315	440–590	22	–20	31
DH36	GR										355	490–620	21	–20	34
DH40											390	510–650	20	–20	41
E	K., GR	0.18	0.70	0.1–0.35	0.04	0.04	0.015	-	-	-	235	400–490	22	–40	27

Table 4.12 cont'd

Grade															
EH 32	K	0.18	0.9–1.6	0.50	0.035	0.03	0.015	0.02–0.05	0.03–0.10	0.02	315	440–590	22	–40	31
EH 36	K										355	490–620	21	–40	34
EH 40	GR										390	510–650	20	–40	41
FH 32	K	0.16	0.9–1.6	0.50	0.025	0.025	0.015	0.02–0.05	0.03–0.10	0.02	315	440–590	22	–60	31
FH 36	K										355	490–620	21	–60	34
FH 40	GR										390	510–650	20	–60	41

(a) K = killed. S-K semi-killed. GR = grain refined
(b) Grain refined steels may contain the elements indicated singly or in combination. (Nb+V+Ti) = 0.12% max.

Table 4.13 High tensile ship steel grades to Lloyds Rules; quenched and tempered or continuously heat treated

Grade	Chemical composition, %										YS N/mm²	UTS N/mm²	Elongation %	Charpy V Temp °C	Transverse J
	C max	Mn max	Si max	P max	S max	Al(Sol) min	Nb	V	Ti max	Nb+V+Ti max					
DH 42											420	530–680	18	–20	27
EH 42														–40	
DH 46											460	570–720	17	–20	30
EH 46														–40	
DH 50											500	610–770	16	–20	33
EH 50	0.20	1.7	0.55	0.035	0.03	0.015	0.02–0.05	0.03–0.10	0.02	0.12				–40	
DH 55											550	670–830	16	–20	37
EH 55														–40	
DH 62											620	720–890	15	–20	46
EH 62														–40	
DH 69											690	770–940	14	–20	46
EH 69														–40	
FH 42											420	530–680	18	–60	27
FH 46											460	570–720	17	–60	30
FH 50	0.18	1.6	0.55	0.025	0.025	0.015	0.02–0.05	0.03–0.10	0.02	0.12	500	610–770	16	–60	33
FH 55											550	670–830	16	–60	37
FH 62											620	720–890	15	–60	46
FH 69											690	770–940	14	–60	46

Table 4.14 Materials for cargo tanks, secondary barriers and pressure vessels with design temperatures in the range 0 °C to 165 °C: Lloyd Rules

Minimum design temperature °C	Material and heat treatment	Impact test temperature, °C
−55	Carbon-manganese steel with optional low alloy addition, normalised or quenched and tempered	Up to 20 °C below design temperature but not higher than −20 °C
−60	1.5% nickel steel, normalised	−65
−65	2.25% nickel steel, normalised or normalised and tempered	−70
−90	3.5% nickel steel, normalised or normalised and tempered	−95
−105	5% nickel steel, normalised or normalised and tempered	−110
−165	9% nickel steel, double normalised and tempered or quenched and tempered	−196
−165	Austenitic stainless steel 300 series, solution treated	−196
−165	Aluminium alloys such as type 5083 annealed	Not required
−165	Austenitic iron-nickel alloy (36% nickel), heat treatment as agreed	Not required

Impact energy: minimum average 27 J.
Transverse test for plate, longitudinal for sections and forgings.

alloys and carbon fibre reinforced plastic are too expensive to be considered. Panels typically have thin plates and small spacings between stiffeners. Here aluminium has an advantage because such structures can be built up by butt welding extrusions that form both plate and stiffener.

High tensile quenched and tempered steels are also used in naval construction, particularly for aircraft carrier flight decks, side armour, magazine protection and the hulls of submarines. The first of a family of such steels was HY 80 which was developed after World War II in the USA. The composition was based on Krupp armour plate used for German warships in World War I, it is a chromium-molybdenum-nickel steel of high yield strength and good notch-ductility.[22] HY 100, with a yield strength of 100 ksi (690 N/mm²), has a similar composition and is currently used as armour plate for US Naval vessels. The British submarines Valiant and Polaris were fabricated in HY 80; other warships use quenched and tempered armour plate to Ministry of Defence (N) specifications such as Q1N and Q2N.

Present interest lies in the development of hybrids between the Krupp type and microalloyed steel, where the total alloy content is reduced and the properties are maintained by a combination of thermomechanical processing and precipitation hardening. Table 4.15 compares the composition of the HY steels with recent developments, designated HSLA. One object of this development is to improve weldability and reduce preheat requirements. For example the US Navy has approved welding of HSLA 100 in thicknesses up to 25 mm, using solid wire GMA welding with a minimum preheat temperature of 15 °C.[22] Note that the carbon equivalents of most of these steels are out of the BS 5135 range. Welding procedures for such materials are individually qualified and are embodied in (for example) Ministry of Defence (N) specifications.

A steel with 130 ksi (895 N/mm^2) yield strength is under development but there remain problems with welding consumables. It is possible to match the tensile properties but difficult to achieve impact strength equal to that of the parent material.

Aluminium alloys are used for the superstructure of both passenger ships and naval vessels. The liners Bergensfjord, Canberra, Oriana and Queen Elizabeth II each incorporated over 1000 tons of 5083 (Al4.5Mg) alloy plate and extrusions. Royal Navy support ships and frigates were also built with aluminium-magnesium alloy superstructures. The current trend is to combine the use of Al-Mg plate for structural parts with $AlMg_2Si$ (6061, 6063 and 6082) extruded sections for decking. The extrusions incorporate stiffeners with a weld bevel and a profile that gives a backing to the weld at the same time as aligning the joint. Such sections can be longitudinally welded without significantly reducing the strength.

Welding in ship construction

With few exceptions the ship steels listed by Lloyds Register are weldable by the manual metal arc, gas metal arc and submerged-arc processes, as described earlier. Shipbuilding was traditionally an open-air activity but in recent years much of the work has been transferred to indoor workshops where panels are welded using automatic processes, particularly submerged-arc welding. Good improvements in productivity have been obtained by submerged-arc welding the fillets attaching a stiffener simultaneously, as described earlier. The use of robots however has been limited by the fact that many ships are a one-off design, and also by the nature of the structure, which can incorporate a double curvature. Nevertheless some progress has been made; for example in welding of double bottoms. In many cases these consist of numbers of rectangular boxes of generally similar but not identical design. A portal frame (Fig. 4.24) makes possible the transfer of the

Table 4.15 Development of US Navy high tensile steels

| Grade | Nominal composition, % | | | | | | | | | 11W | Yield strength | |
	C	Mn	Si	Cr	Mo	Ni	Nb	V	Cu	CE	ksi	N/mm²
HY 80	0.15	0.25	0.25	1.4	0.4	2.7	–	0.01	–	0.755	80	550
HSLA 80	0.04	0.55	0.30	0.7	0.2	0.9	0.04	–	1.2	0.502	80	550
HY 100	0.15	0.25	0.25	1.4	0.4	2.9	–	0.01	–	0.789	100	690
HSLA 100	0.04	0.90	0.25	0.6	0.6	3.5	0.03	–	1.6	0.812	100	690

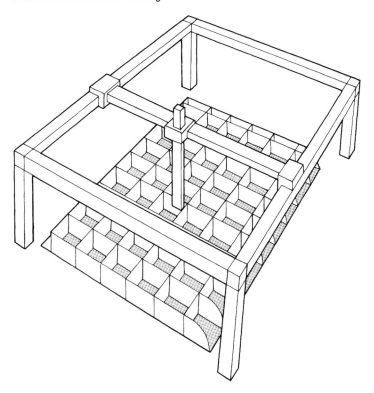

4.24 Portal frame system for robot welding of ship structures.[29]

robot from box to box. Within the box (Fig. 4.25) there are a number of weld patterns which are similar enough to allow macroprogramming of the robot.[29]

In straight runs of submerged-arc welding, increased deposition rates can be obtained by use of multiple electrodes. The simplest system is that of parallel electrodes (so called because they are connected in parallel to a single generator). The electrodes are usually spaced so that they form a single weld pool. However it is also possible to space them so as to make two layers, or to make two parallel runs (in a wide joint preparation). Combinations of single electrodes are likewise possible. AWS D1.1 gives limitations to the maximum current that may be used in multiple electrode welding and this is a useful guide. Such a limitation is necessary to prevent excessive heat input rates and the formation of a coarse-grained heat affected zone.

The most stringent precautions are required for welding of high strength quenched and tempered steels for naval vessels and particularly in submarines, where the plate thicknesses are generally greater. The first step in the control procedure is the qualification of welding

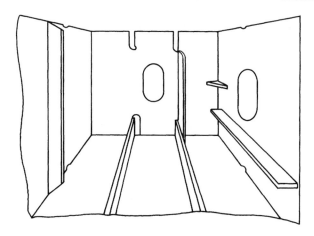

4.25 Double bottom box adaptable to robot welding.[29]

consumables. These must meet the high tensile properties and also be subject to a Pellini bulge test. This test which is a modification of the standard Pellini drop-weight test (see Chapter 3) is illustrated in Fig. 4.26. The testplate which is welded with the consumable to be tested, is expanded into a die of part-spherical form by means of an explosive charge. The test is done at various temperatures and as a rule is required to determine the nil-ductility temperature, as shown in the figure. As in the drop-weight test, a run of brittle weld metal is used as a crack starter.

Having qualified the consumable, the next step in the case of coated electrodes is to apply a rigorous hydrogen control procedure, aimed at ensuring a hydrogen level of less than 5 ml/100g deposited metal. After rebaking to the manufacturer's recommendations, the electrodes are held in heated quivers and are subject to periodic checks for moisture content in the coating. For gas metal arc welding tests of weld metal hydrogen content are used to determine the optimum storage and issue conditions (likewise in submerged-arc welding).

For the welding operation it is necessary to establish control over the heat input rate. This is straightforward for submerged-arc welding and for laying down stringer beads using a manual process. In vertical or overhead welding however, a stringer bead technique increases the risk of lack of fusion defects or slag traps, and weaving is necessary. Control of heat input rate is obtained by either limiting the thickness of the weld deposit, or by limiting the width of the weave in proportion to the core wire diameter.

Finally the seam is inspected by radiography with ultrasonic testing as a back-up. Surface inspection is carried out visually and using magnetic crack detection and dye penetrant testing.[30]

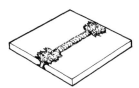

A weld run of brittle alloy
is laid along the centre line
of the test plate and notched
to provide a crack starter.
Ends ground flush to seat
on die.

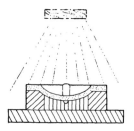

Explosive charge expands test
plate into die. Tests carried
out on plates cooled to
progressively lower temperatures
to determine transition as
indicated below.

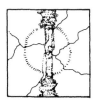

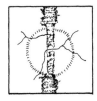

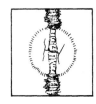

NDT
Nil ductility temperature.
Flat break. Brittle fractures
extend to edge of plate

FTE
Fracture transition
elastic. Fractures are
arrested in elastically
loaded die-supported
region

FTP
Fracture transition
plastic. Fractures are
arrested in plastically
loaded (bulged) region

4.26 The Pellini bulge explosion test[30] (for the Pellini drop-weight
test see Chapter 3).

In submerged-arc welding the welding current is relatively high and,
to maintain the heat input rate level, it is necessary to increase the
welding speed. Sensors or tracking equipment may be required to
guide the welding head.

Aluminium superstructures are almost entirely welded using the
metal inert gas process. When argon is used as the shielding gas the
weld metal contains a fine, uniformly distributed porosity. This is due
to hydrogen derived from water absorbed by or combined with the
oxide coating on the wire. It has only a minor effect on weld properties.
Argon-helium mixtures are also used for shielding. These produce a
'hotter' arc and the burnoff rate is slightly higher in helium than in

argon. It has also been reported that the extent of porosity is reduced in helium and argon-helium shielding.[23]

Traditionally a constant voltage power source is used for manual aluminium GMA welding because the self-adjusting property of the arc is better than with a drooping characteristic source. This means that the arc automatically adjusts to a constant length irrespective of movements of the welding gun. The more sophisticated power sources developed in recent years may be applicable to aluminium welding, particularly for automatic operations, where a constant current source may be advantageous. The GTA process with an AC power source may be used for thin sections.

In vessels with an aluminium superstructure there must be a joint between aluminium and steel. Bolted joints were used at first but proved unsatisfactory because of galvanic corrosion. Aluminium-steel welded transition joints are less vulnerable to this defect and easier to protect. The joint between aluminium and steel is normally made by friction welding but explosion welding has also been used.

Welding procedure and operator qualification testing for aluminium are covered in BS 4870 part 2 and BS 4871 part 2 respectively, and in EN 288 and EN 287.

Welded steel bridges

General

In the postwar period there have been a number of set-backs in steel bridge construction. These included the failure of the Kings Bridge in Melbourne, and the partial collapse of three box girder bridges during construction with accompanying loss of life. All these were avoidable failures, but they undoubtedly had an unfavourable effect on steel usage such that in the early 1980s the market share in terms of number of bridges constructed was steel 10%, reinforced concrete 90%. This trend has now reversed, as shown in Fig. 4.27. In part this recovery is due to the unexpectedly short life of concrete, mainly as a result of attack by de-icing salt. A recent survey[31] found that of 200 concrete bridges studied, 114 were in fair condition and 61 were classified as poor. The worst affected were located in the colder areas, where more salt is used. Apart from salt attack, a number of concrete structures have suffered deterioration due to cracking and exposure of reinforcement, and some concrete buildings have been demolished prematurely.

Steel may also suffer corrosion if the structure is not properly designed and protected. The primary design requirement is to ensure that water from rainfall or condensation drains freely away, particularly in the case of box girders. Figure 4.28 shows the longitudinal seal

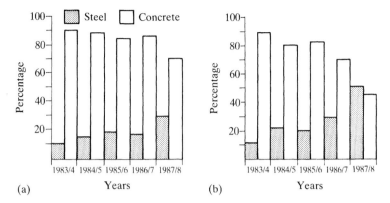

4.27 Relative shares of steel and concrete used to construct highway bridges with spans 15–50 m in England and Wales: a) By number of structures; b) By deck area.[32]

protecting the joint between the box girder and the deck of a recently constructed British Rail bridge.

Coating technology has now reached a point where a life of numbers of years can be expected even when the surface is exposed to salt spray. Such developments have been stimulated by offshore requirements in the North Sea, and many coating systems are available. For example, for the bridge of which a detail is shown in Fig. 4.28 the sequence was:

> Shot blast to SA3;
> Aluminium spray 100 microns;
> Zinc phosphate primer;
> M10 modified phenolic;
> High build Alkyd finish.

Materials for bridges

As a rule, material subject only to compressive stress, or when the governing factor in design is fatigue, will be Fe 430A or Fe 430B (Table 4.3). Tension members will normally be to Fe 510. The impact requirements, which are indicated by suffix letters and numbers such as Fe 510 D1 are determined in the UK from BS 5400.

The first step is to establish the minimum shade temperature for the locality, using isotherms given in the standard and adjusting for height and known local variations. This temperature is further adjusted to take account of the level of stress and the method of construction to obtain the design temperature. The minimum required impact strength C_v at this temperature is then calculated from a formula relating to yield strength and thickness. For example, for a welded member with a design stress over 100 N/mm^2:

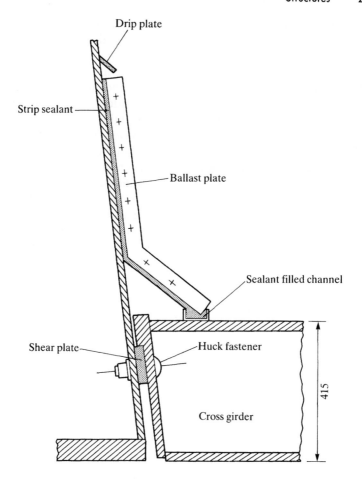

Cross girder connection

4.28 Weatherproof seal for cross girder connection on British Rail bridge (reproduced by courtesy of British Rail Network South East).

$$C_V = \frac{\sigma_y}{355} \frac{t}{2} \qquad [4.6]$$

where C_V is in J, σ_y is the specified minimum yield stress in N/mm² for a thickness of 16 mm and t is the section thickness in mm. Thus, the Charpy requirements for a 40 mm thick steel to Fe 510 would be:

$$C_V = \frac{355}{355} \times \frac{40}{2} = 20 \text{ J} \qquad [4.7]$$

For a design minimum temperature of −20 °C this would be met by Fe 510 D1 (Table 4.3). Lower design temperatures down to −50 °C would require plate to BS 4360: 1990 Grade 50EE. Major steel road bridges in the UK built in the decade prior to 1990 were mainly constructed using BS 4360: 1979 Grades 50 B, C and D.

For military bridges there are special requirements that may, as noted earlier, require the use of materials not normally associated with structural work. These requirements include high strength/weight ratio, good weldability, toughness and reasonable corrosion resistance. Fatigue strength is also important. However, in the UK, military bridges are designed for a life of 10^4 cycles so that the fatigue strength reduction due to welding is less than in permanent bridges.

The strength/weight ratios for some candidate materials are shown in Table 4.16. These have been calculated on the basis of the ultimate stress, which is not an accurate guide to the usable design stress; nevertheless the ranking is about right. The composition and mechanical properties of the high tensile steels are given in Table 4.17. The 18% Ni maraging steels have a good combination of high yield strength,

Table 4.16 Strength/weight ratio of various materials

Material	Tensile strength N/mm^2	Specific gravity	Strength/weight ratio
Carbon whisker	24000	2.2	10909
CFRP	1620	1.55	1403
Steel wire	2500	7.8	321
A1S1 4340	1850	7.8	237
Maraging steel	1435	8.0	179
A14Zn2Mg	380	2.8	136
A14.5Mg	275	2.8	100
BS 4360 55EE	550	7.8	70.5

Table 4.17 Maraging steels

| Grade | Nominal composition, % | | | | | | YS N/mm^2 | UTS N/mm^2 | Elonga- tion % | K_{IC} MN/m$^{3/2}$ |
	C	Ni	Co	Mo	Ti	Al				
18 Ni 1400	0.03	18.0	8.5	3.0	0.2	0.1	1400	1435	12	100
18 Ni 1700	0.03	18.0	8.0	5.0	0.4	0.1	1700	1735	10	90
18 Ni 1900	0.03	18.0	9.0	5.0	0.6	0.1	1900	1935	8	65

toughness and weldability. These steels are alloyed with nickel, cobalt and molybdenum with small additions of titanium and aluminium. The high proof and ultimate strengths are obtained by a martensite transformation followed by precipitation hardening at about 500 °C. The steel is solution-annealed at 820 °C and transforms to martensite on air cooling. The high nickel and low carbon contents both contribute to the good notch-ductility, and the carbon level is such that no preheat is required for welding.

The 18Ni 1400 type steel has been used for a tank mounted assault bridge. This is carried folded in two on the top of the tank and can be unfolded and launched to span a gap of about 24 m. The hinge pins are nitrided 18% Ni maraged steel (Fig. 4.29).

The other alloy that has clear advantages for portability is the heat treatable AlZnMg type, of which a special version, designated DGFVE 232 A has been developed. This contains 4Zn, 2Mg, 0.35Mn and 0.15Zr. It has a proof stress of 380 N/mm^2. On welding the heat affected zone is softened by solution, rather than by over-ageing, and the properties are partly recovered by natural ageing at room temperature. The proof stress obtained in this way is just over 200 N/mm^2. Alternatively the metal may be welded in the solution treated condition and then the whole structure is aged. Stress corrosion cracking may be a problem with welded AlZnMg structures exposed to sea water. The risk of such cracking is reduced by limiting the zinc/magnesium ratio to 2:1 and the total zinc and magnesium content to 6%. The manganese and zirconium additions also help by inhibiting recrystallisation and refining the weld metal.

This alloy is used for bridges of which sections can be manhandled and pinned together to make a span of 30.5 m capable of carrying a 54 ton tank. In this and other recent developments the supporting structure is all below the decking. It was one of the limitations of the Bailey bridge used in World War II that the girders were located on either side of the roadway, and this limited the width of vehicles that could be handled.[17]

Welding of bridges

Welding requirements for bridges are covered by BS 5400, also by AWS D1.5, Bridge Welding Code together with AWS D1.1. The AWS Codes, as noted earlier, contain pre-qualified weld procedures, whereas BS 5400 calls for the submission and approval of all welding procedures before commencing work. This would appear something of a hardship relative to US practice but most responsible fabricators would wish to produce evidence of satisfactory workshop practice, and once a welding procedure is qualified this remains valid indefinitely. BS 5400 requires bend, tensile and impact testing of procedures

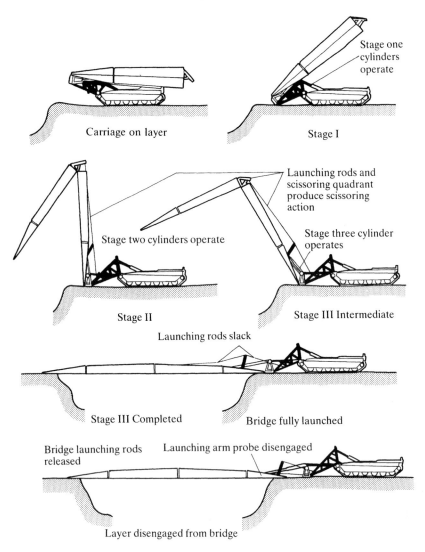

Carriage on layer

Stage one cylinders operate

Stage I

Launching rods and scissoring quadrant produce scissoring action

Stage two cylinders operate

Stage three cylinder operates

Stage II

Stage III Intermediate

Launching rods slack

Stage III Completed

Bridge fully launched

Bridge launching rods released   Launching arm probe disengaged

Layer disengaged from bridge

4.29 Armoured vehicle launched bridge, constructed from 18% maraging steel (reproduced by courtesy of The Institute of Materials).[17]

qualification testpieces, together with a macrosection and hardness traverse. Impact testing is applied both to the weld metal and to the heat affected zone. A proportion of the run-off plates are used as production testplates, with tensile bend and impact testing. Production testplates are not required by AWS Codes.

The welding processes commonly used for BS 4360 steels are submerged-arc and manual metal arc, supplemented to some degree by gas-shielded flux-cored arc welding. Site joints are usually bolted but in some cases they are welded. Welding requires that shop-applied coating be removed from the weld area and the plate surface protected from spatter. An enclosure is built around the joint for weather protection. After testing, including non-destructive checks and testing of a production testplate, the surface is re-coated.

Aluminium alloy and maraging steel for military bridges are welded using GMA welding with solid wires; Al5Mg alloy in the case of the aluminium alloy, and matching filler for the maraging steel. Post-weld ageing at 480 °C is necessary in the case of maraging steel to restore the properties in the HAZ and to develop good strength in the weld metal. This heat treatment is applied to the complete structure.[17]

Repair and strengthening of existing bridges

The first step in developing a weld procedure for repair or renovation is to determine what material was used. If no documentation is available, some guidance can be obtained from the date of construction as follows:

> 1750–1840: cast iron;
> 1840–1890: wrought iron;
> 1880–present: steel.

For the period from 1880 to about 1970 the steel used in the UK and USA would probably be the semi-killed open hearth type, and in Europe it could be rimming steel. After about 1970 basic oxygen steelmaking began to dominate. Continuous casting, using killed steel, has been used to an increasing extent from about 1975. In the case of old bridges, chemical analysis may be necessary, and it is important that this should be done by a laboratory that is engaged routinely in testing of steel. Table 4.1 gives some guidance as to composition.

Cast iron is not weldable and normally cast iron bridges would already have been replaced by steel structures. However, as noted earlier, wrought iron may be butt welded to steel edge to edge. The carbon content, as indicated in Table 4.1, is likely to be low but it may vary and this should be allowed for in setting up the weld procedure. Early steels may be high in carbon and material having more than 0.4% C is of questionable weldability. Rimming steel is high in sulphur and phosphorus, particularly in the mid-section of plates, and is also of questionable weldability. Semi-killed steel is weldable using current welding consumables but BS 5315 does not necessarily apply because of the low silicon content and it is advisable to carry out procedure testing on samples of the original material.

Welded steel frame buildings

At the beginning of the nineteenth century most cotton and woollen mills in Britain had wooden floors and roofs, and there were some disastrous fires. To overcome this problem new designs were developed in which tiled floors were supported by cast iron columns and beams. Later this type of structure was used for office buildings and shops. With the increased availability of iron and then steel, construction of the skeleton frame building became possible. The first skyscraper to be built in this way was the ten-storey Home Insurance Building in Chicago (1885) which had cast iron columns and wrought iron or steel beams. Steel was first used for the complete frame in the Rand McNally building in 1890. Developments in the UK and Europe were more modest in height. Classical examples were the White Star offices in Liverpool and the Ritz Hotel in London, about 12 years later.

Welding had been used for some building work in the 1920s, but was restricted by local regulations. It was first allowed by the London Building Act in 1934. Subsequently most steel framed buildings in England have been welded, one of the first being Simpson's shop in Piccadilly.[33]

Many of the technical problems that attend welding of structural steels for bridges and offshore oil rigs are less important in the case of buildings. Fatigue loading only occurs under special conditions (although vibration may be a significant problem) and exposure to low temperature and corrosion is exceptional. Code requirements for welding are included in BS 5950 and AWS D1.4 and call for compliance with general practice in the qualification of procedures and operators and with BS 5135 and AWS D1.1 respectively. There are in addition local regulations for building construction which must be observed. Materials used are Fe 430 and Fe 510, and welding is manual metal arc for site work and submerged-arc for shop fabrication of beams and columns. However self-shielded cored wire welding was used successfully on one recent project in the UK.

Fire protection is a major concern. Structural elements are required to withstand exposure to specified temperatures; for example a standard time temperature curve for a fire is specified in BS 476. Steel must be insulated such that its temperature does not exceed a stated level. BS 5950 part 8 gives formulae for calculating the performance of elements and for determining the amount of protection required. Welding is not a factor in designing against fire; there is no evidence that welds perform any worse than the parent metal under short term exposure to heat.

Offshore structures

General

Platforms for the drilling and operation of oil and gas wells offshore were installed in shallow water in the Gulf of Mexico nearly 50 years ago. The first rigs (and most of those built subsequently) were essentially flat decks supported by braced tubular legs. This type of design evolved from those traditionally used for piers and jetties. The tubular form is appropriate to a member that is subject to stresses that act in all directions, as opposed to the I form which is most economic in bridges where the load is unidirectional.

The North Sea is a major offshore oil production region and some of the largest rigs are installed there. It also provides a much more hostile operating environment than the Mexican Gulf, and the sea depths are also much greater, up to 250 m. For such depths it is possible to use fixed platforms with the legs secured to steel or concrete structures that are piled to the sea bed. Figure 4.30 shows a typical fixed platform with its support. Where the sea depth is greater than 250 m it becomes necessary to use a floating support which is moored to the sea bed by tubular lines. Figure 4.31 is a sketch of the Hutton tension leg platform. The floating part, known as the hull, consists of six vertical cylinders connected by lateral members. The Hutton platform operates in a depth of only 146 m but its construction has demonstrated the validity of the design. Floating (semi-submersible) platforms that are mobile and held in position by anchors are also used, and one such (the 'Alexander L Kielland') is described in Chapter 6. Concrete structures that are massive enough to retain their position on the sea bed have been employed but these are in the minority.[34,35]

Codes, specifications and materials

Offshore construction in the North Sea is governed by a number of codes and standards. In the UK the Department of Energy issues a guidance document which deals with material selection, design, construction and inspection,[21] and there is a British Standard Code of Practice for offshore rigs.[36] Structures operating in areas within British jurisdiction must comply with the Department of Energy's guidance notes. The classification societies have their own rules, for example Det Norske Veritas publishes 'Rules for the design, construction and inspection of offshore structures'. American codes include the section of AWS D1 on 'new tubular structures' and the API recommended practice 2A.[37] There is no shortage of advice as to how oil rigs should be built.

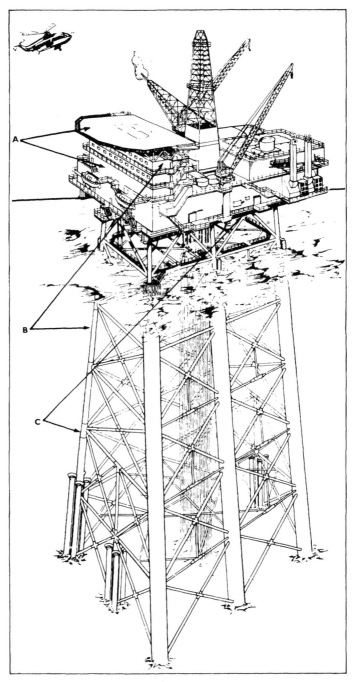

4.30 Typical tubular offshore jacket structure. A) Secondary structural
steel (e.g. Fe 430 D) module walls, decking etc; B) Primary structural
steel (e.g. BS 4360 grade 50 E) jacket legs, module support etc; C) Special
structural steel (e.g. BS 4360 grade 50 E HYZED) node joints (reproduced by courtesy of The Institute of Materials).[34]

4.31 The Conoco Hutton tension leg platform.[35]

The specifications for steel are covered in the UK by EN 10025 and BS 4360. However operating companies often require more rigorous testing than is provided in these standards, and in particular may require additional impact testing. In part these are covered by EEMUA Publication 150, referred to earlier in this chapter. For example in the case of grades corresponding to the former BS 4360 grades 43, 50 and 55, the Charpy V impact properties are enhanced to 40, 50 and 60 J to provide a margin for degradation in the HAZ after welding.

CTOD testing is frequently specified for checking the grain-coarsened region of the heat affected zone. The fatigue crack tip is required to lie within 0.5 mm of the fusion boundary. Alternatively Charpy V tests are made with the notch in this region and typical requirements are 36 J for grade 50 (Fe 510) and 46 J for grade 55 at −40 °C. Another test that may be specified as a control on plate material is a bead-on-plate weld run made under specified conditions. The plate is then sectioned and the maximum hardness determined by a hardness traverse. The results are compared with those obtained from a plate of known acceptability.

More recently there has been concern about a loss of notch-ductility, presumably due to strain-ageing, at a point 5 mm distant from the fusion line, and impact testing of this region may be specified.

Figure 4.30 indicates basic steel grades that are used for major parts of the structure. Steel with improved through-thickness properties (HYZED steel) to BS 6780 grade Z25 or Z35 is invariably required for fabricated nodes. There have been numerous proposals for the use of cast steel for node sections in order to obtain better fatigue resistance at the intersections and to dispose of the through-thickness problem. There is no doubt about the potential advantage of such components but to date they have only been used for special locations such as the intersection between the columns and the square pontoons on the Hutton tension leg platform.[34] In part this has been due to practical considerations such as delivery time, but also because of the greater flexibility of fabricated construction, which can accommodate last minute design changes more easily.

The extension of offshore drilling into arctic regions is likely to pose a further challenge to steelmakers and fabricators. Guaranteed impact properties at −60 °C or lower may be needed. Further development of clean fine grained low carbon steel could make this possible but it might be necessary to consider use of 1–5% nickel steel for critical locations.

The larger offshore platforms support an accommodation block for operators together with a substantial amount of processing equipment and a landing pad for helicopters. A recent development has been the use of aluminium for accommodation module and for the helideck. The alloys used in the structure illustrated in Fig. 4.32 were 5454 (AlMg) and 6082 (AlSiMgMn, heat treatable). The structure was supported on welded AlMg beams in thicknesses of up to 100 mm. Welding

4.32 Aluminium accommodation module for Saga Petroleum's Snorre platform after completion of on-shore fabrication[23] (courtesy of Alcan Offshore).

was carried out using the GMA process with AlMg wire. In the case of 6082 alloy a multirun technique was employed, limiting the heat input rate to minimise the extent of softening.

The oil and gas that is processed on North Sea platforms is generally non-corrosive and carbon steel is the normal material of construction. North Sea crude is sweet but where H_2S may be present welds are specified to have a maximum hardness of 22 Rockwell C. This requirement is not difficult to achieve. Cooling is by sea water which contains dissolved air and is corrosive. Coolers may be plate heat exchangers with commercially pure titanium plates, whilst shell and tube exchangers have 70/30 or 70/10 cupronickel tubes.

Welding

Welding for North Sea offshore work is carried out in some instances to AWS D1.1 but generally in accordance with BS 5135 so far as preheat is concerned. Much of the welding is on fixed position pipe or plate and automatic or robotic welding is limited to prefabricated items that are not too heavy or bulky to manipulate. Almost all welding is done in the shop by the gas-shielded or self-shielded flux-cored process or by MMA with basic low hydrogen coatings. Final assembly welds, for example between jacket and platform, are made *in situ*, again using basic low hydrogen rods. The need to meet high standards both of weld quality and notch-ductility means that welding procedures must be carefully evaluated and controlled. Fabricators like to keep preheat as low as possible and preferably to eliminate it, and where BS 5135 governs, this requires that hydrogen levels and carbon equivalents be kept low.

Rogers and Lockhead of Highland Fabricators reported in 1989 that for an average North Sea jacket of 18 000 tons, 45 000 kg of coated electrodes and 73 000 kg of self-shielded wire would be used.[38] The wire was Lincoln Innershield NR 203 NiC which is suitable for all-position welding and has an impact strength of 27 J at -29 °C (Table 1.3). The self-shielded process is mainly used for node construction. Its success in this critical area is a little surprising particularly as self-shielded welding is little used for shop work elsewhere in the UK. Others have commended the flux-cored gas-shielded process with titanium-boron microalloyed wires, which produces a fine grained acicular ferrite structure in the weld metal with good notch-ductility.[39] This wire has a rutile flux core. Rogers and Lockhead made extensive tests with a basic flux-cored wire on stub-to-barrel and brace-to-barrel welds, but were unable to get satisfactory weld profiles in the transition from overhead to vertical welding. This is a case where a rutile-cored wire would be superior because it gives a fine spray metal transfer whereas the basic-cored wires produce globular transfer.

The use of basic coated low hydrogen electrodes for the connections between platform and hull of the Hutton tension leg platform has been described by Warwick et al.[35] (The welding technique is of considerable interest and therefore it is recorded here in some detail.)

The hull was floated into still water in the Moray Firth and the platform placed in position, supported on jacks. A set of infill plates was welded so as to join the upper and lower parts. The jacks were then removed and a second set of infill plates were welded. The steel was BS 4360 Grade 50E and the weld metal was required to meet the following criteria: 0.25 mm CTOD at -10 °C, 35 J minimum Charpy V at -30 °C, 300 HV 10 maximum and tensile properties at least equal to the plate minimum. A typical weld procedure that met these requirements is shown in Fig. 4.33. Electrodes were tested as-received, after redrying and after several days exposure to air, and in no case did the hydrogen content exceed 5 ml/100g deposited metal.

Preheating at 75 °C was applied by electric heating mats linked to a central control station. Welding of all joints in each phase was started and, as near as possible, completed simultaneously. During this operation welding current, voltage, run-out length and preheat/interpass temperatures were monitored continuously. On completion of each joint the weld was subject to a hot ultrasonic examination at the same time as maintaining a post-heat of 100–150 °C. A proprietary non-hydrocarbon fluid was used. The post-heat was then maintained for 12 hours before cooling to room temperature, after which normal

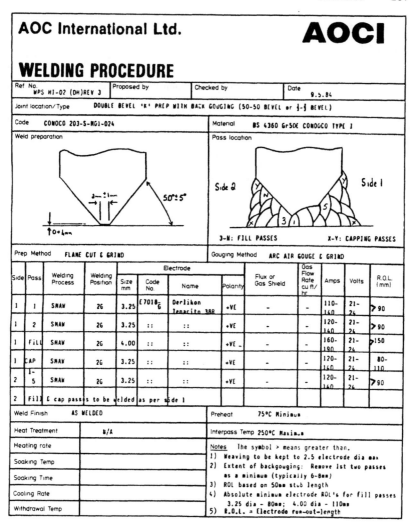

AOC International Ltd.   **AOCI**

WELDING PROCEDURE

Ref No. WPS HI-02 (DH)REV 3	Proposed by	Checked by	Date 9.5.84

Joint location/Type	DOUBLE BEVEL 'K' PREP WITH BACK GOUGING (50-50 BEVEL or ⅓-⅔ BEVEL)

Code CONOCO 203-S-MG1-024	Material BS 4360 Gr50E CONOCO TYPE 1

Weld preparation

50°±5°

3mm ±1mm

↑0+4mm

Pass location

Side 2   Side 1

3-N: FILL PASSES   X-Y: CAPPING PASSES

Prep Method FLAME CUT & GRIND	Gouging Method ARC AIR GOUGE & GRIND

Side	Pass	Welding Process	Welding Position	Electrode Size mm	Electrode Code No.	Electrode Name	Polarity	Flux or Gas Shield	Gas Flow Rate cu ft/hr	Amps	Volts	R.O.L. (mm)
1	1	SMAW	2G	3.25	E7018 G	Derlikon Tenacito 38R	+VE	-	-	110-160	21-24	>90
1	2	SMAW	2G	3.25	::	::	+VE	-	-	120-160	21-24	>90
1	Fill	SMAW	2G	4.00	::	::	+VE	-	-	160-190	21-24	>150
1	CAP	SMAW	2G	3.25	::	::	+VE	-	-	120-160	21-24	80-110
2	1-5	SMAW	2G	3.25	::	::	+VE	-	-	120-160	21-24	>90
2	Fill & cap passes to be welded as per side 1											

Weld Finish AS WELDED	Preheat 75°C Minimum

Heat Treatment	N/A	Interpass Temp 250°C Maximum
Heating rate		Notes The symbol > means greater than.
Soaking Temp		1) Weaving to be kept to 2.5 electrode dia max
Soaking Time		2) Extent of backgouging: Remove 1st two passes as a minimum (typically 6-8mm)
Cooling Rate		3) ROL based on 50mm stub length
Withdrawal Temp		4) Absolute minimum electrode ROL's for fill passes 3.25 dia - 80mm; 4.00 dia - 110mm 5) R.O.L. = Electrode run-out-length

4.33 Extract from a weld procedure for one of the mating joints on the Hutton tension leg platform (source: AOC International).[35]

ultrasonic and magnetic particle testing was carried out. The post-weld heating was intended to diffuse out hydrogen, and justified a final non-destructive test after 12 hours rather than the 48 hours recommended by BS 5135. However retesting was carried out on welds that were potentially subject to strain due to the performance of other welding operations. Some shallow toe defects were found on welds that had previously been cleared. The acceptance criteria are given in Fig. 4.34. On completion of phase 2 welding many plate end terminations were

Type of flaw	Permitted maximum
Internal flaws	
Porosity	Scattered porosity: maximum 1% by projected area Largest pore dimension t/8, maximum 2 mm Cluster porosity: maximum 3% by projected area Largest pore dimension t/16, maximum 1 mm Linear porosity: not acceptable t: thickness of joint

Slag inclusions	Isolated slag: Length t/3
	    Width t/8, maximum 2 mm
	Slag lines: Length (L) 2 t or 50 mm whichever is the lesser
	    Width 1.5 mm
	The width of each line of parallel slag lines (wagon tracks) is
	not to exceed 1 mm
	t: thickness of joint

Lack of fusion	Not acceptable
Incomplete penetration	Not acceptable
Cracks	Not acceptable
Tungsten inclusions	As for scattered and clustered porosity
Hollow bead	Not acceptable
Surface flaws	
Lack of fusion	Not acceptable
Cracks	Not acceptable
Undercut	Intermittent undercut of depth 0.25 mm maximum for an accumulated length of 50 mm in any 300 mm length of weld at each weld toe region
Incomplete penetration	Not acceptable
Craters	All craters to be filled to the full cross-section of the weld

4.34 Defect acceptance criteria for the mating welds on the Hutton tension leg platform.[35]

given a toe burr grinding to a depth of 0.5–1 mm to improve fatigue performance (see Chapter 3).

Specifications for North Sea construction are designed to minimise the risk of brittle fracture by maintaining notch-ductility in the parent metal, the weld and the HAZ, by avoiding hydrogen cracks and by taking measures against fatigue crack initiation. There is no doubt that such precautions, the cost of which is a small proportion of the total cost of developing an oil field, are fully justified, and are beneficial in raising standards of welding practice in general.

Corrosion of welds

Problems due to selective corrosion of weld metal (see Chapter 3) have been experienced in North Sea operations, in particular in sea water and CO_2 injection piping.

Sea water for injection into oil-bearing formations is de-aerated and is nominally non-corrosive. However biocide and other additions may cause the sea water to become aggressive, either directly or by reducing the effectiveness of de-aeration. One way to combat this effect is by small additions (0.5 to 1.0%) of chromium-copper, molybdenum or nickel to the weld metal, either separately or in combination (½Cr ½Cu for example). Alternatively or additionally inhibitors are added. The corrosive effect may be severe such that half or more of the weld metal is dissolved. (See also Chapter 6.)

Non-destructive testing (NDT)

General

In structural work it is possible to be more selective in testing of welds than it is in the case, say, of pressure vessels. Radiography is often not the best means of examination, and more use is made of ultrasonic testing. On the other hand there is a tendency for acceptance standards to become more stringent for critical applications, and in many instances to equal those for pressure equipment.

Normally, welding inspectors and NDT operators are required to be qualified in accordance with recognised standards. The AWS code, for example, specifies that NDT operators should be qualified to AWSQCI level II or should be level I working under the supervision of a level II manager. AWS also recognises other national qualifications that are considered equal. In the UK qualification is to CSWIP standards for the relevant method of examination. In ultrasonic testing much is left to the judgement of the operator, and it is unwise to rely upon a single individual to carry out all the work.

It is most undesirable to make any change in the inspection methods during the construction phase. The AWS code lays down that where such a change is requested by the owner, the contractor is not responsible for the cost of repairs and associated work, except where there is evidence of fraud or gross non-conformance. The methods and extent of weld examination should be clearly laid down in the contract documents.

Methods and extent of examination

Production testplates for steel bridges to BS 5400 are subject to tensile, bend and Charpy impact testing. The impact tests are obligatory for weld metal and may also be required for the heat affected zone. Production testplates are not required for steel buildings or for offshore structures. They are not, of course, practicable in the case of tubular structures. The AWS code does not require production testplates for bridges or any other structural application.

In general codes do not specify the method of examination except that AWS has a positive requirement for visual inspection and lays down the level of acceptability of defects, such as undercut in some detail. In practice visual examination supplemented by magnetic particle or dye penetrant testing is of primary importance. In particular fillet welds must be checked for leg length, throat thickness, excess concavity and excess convexity and undercut. Tools are available for this purpose. Toe grinding is not an easy operation to control, and must be carefully checked.

Magnetic particle testing is almost invariably preferred to dye penetrant for ferrous materials on the grounds that it is capable of detecting subsurface cracks. Ideally the surface is painted white before such testing but in practice this is rarely done. It is of course possible to induce cracking if there is arcing at the prods when this technique is used. Dye penetrant testing is employed in the case of austenitic stainless steel.

It is general practice to examine welds that are in a critical location, such as transverse butt welds in the tension flanges of bridgework, or the node welds in offshore structures, by means of radiographic or ultrasonic testing. The AWS code does not lay down the extent of such examination, this being subject to contractual agreement. BS 5400 however calls for all transverse welds and a proportion of longitudinal welds of tension flanges to be tested, together with butt welds in the web adjacent to the tension flange. The method of testing is left to the engineer. BS 5950, buildings and BS 6235 for offshore structures have no such specific requirements, the extent of testing being laid down in the owner's specifications.

As noted earlier, ultrasonic testing is frequently used for examining structural welds and if any laminations are present in the plate adjacent to the weld these will become evident. The AWS code requires the repair of such discontinuities when they are within one inch of the fusion boundary; otherwise they are ignored. BS 5400 leaves the decision to the engineer.

Acceptance standards

The British Standards for bridges, buildings and offshore structures are not specific in their requirements for acceptability of welds, but all call for welding to BS 5135 whence it can be inferred that the acceptance standard of BS 4870, which covers weld procedure testing, will apply. This standard forbids all planar defects, and gives acceptance levels for cavities, solid inclusions and profile defects such as undercut. For critical areas it may be necessary to tighten the standard. The acceptance criteria for the mating welds for the Hutton field platform, discussed earlier, are shown in Fig. 4.34. Broadly speaking, these are half the maxima permitted by BS 4870; for example, the amount of scattered porosity allowed is 1% of the projected area as opposed to 2% given in the British Standard.

The AWS acceptance levels are detailed and complex and they vary according to the application: i.e. buildings, bridges and tubular structures. The requirements are more severe for bridges and tubular structures than for buildings; for example, a maximum of 0.03 in (0.75 mm) depth of undercut is permitted for stressed welds in buildings, as against a maximum of 0.01 in (0.25 mm) for the other two applications. Ignoring such details, the acceptance criteria for BS 4870 and the AWS structural code are not dissimilar.

Acceptance levels for welds are arbitrary and represent a standard that is achievable with good current practice. They are set to provide a substantial margin over the dimensions at which the relevant defect would represent a threat to the integrity of the structure. By implication, there is a defect size which would pose such a threat, and means of calculating this quantity are discussed in Chapter 6.

REFERENCES

1 Walker F M: 'Iron's contribution to modern shipbuilding' *Metals and Materials* 1990 Vol 6 778–782.
2 Schubert H R: 'History of the British Iron and Steel Industry'. Routledge and Kegan Paul, London, 1957.
3 Wallner F: 'The LD Process' *Metal Construction* 1986 Vol 18 28–32 and 91–94.
4 Morgan P C: 'Steelmaking practice for clean steel production' *Metals and Materials* 1989 Vol 5 518–520.
5 Wightman P and Hengsberger E: 'Cold hearth refining of special metals' *Metals and Materials* 1991 Vol 7 676–681.
6 Pehlke R D and Fuwa T: 'Control of sulphur in liquid iron and steel' *Int Metals Reviews* 1985 Vol 30 (3) 125–140.
7 Nicholson A, Barradell D V and Engledow D: 'Recent developments in steelmaking' *Metals and Materials* 1988 Vol 4 365–371.
8 Islam N: 'Developments in the continuous casting of steel' *Metals and Materials* 1989 Vol 5 392–396.
9 Okaguchi S, Kushida T, Fukada Y and Tanaka S: 'Recent developments of sour gas service line pipe for North Sea use'. The Sumitomo Search, 1990, No 43, 25–34.
10 Tanaka, T: 'Controlled rolling of steel plate and strip' *Int Metals Reviews* 1981 No 4 185–212.

11 Suzuki H: 'Weldability of modern structural steels' International Institute of Welding 1982 Houdremont Lecture.

12 Paules J R: 'Developments in HSLA products' *J Metals* 1991 (1) 41–44.

13 Stuart H: 'The properties and processing of microalloyed HSLA steels' *J Metals* (1) 35–40.

14 Colbridge G B: 'Progress of British Steel Construction' *Metals and Materials* 1986 Vol 2 269–272.

15 Dearden J and O'Neill H: 'Guide to the selection of low alloy structural steels' *Trans Inst Weld* 1940 Vol 3 203–214.

16 Ito Y and Bessyo K: 'Weldability formula of high strength steels related to heat affected zone cracking' *J Jpn Weld Soc* 1968 Vol 37 983–991, *Ibid* 1969 Vol 38 1134–1144.

17 Topp K: 'A survey of military bridges' *The Metallurgist and Materials Technologist*, 1982, 457–465.

18 Grong O and Akselen O M: 'HAZ toughness of microalloyed steel for offshore' *Metal Construction* 1986 Vol 18 557–562.

19 Sage A M: 'Microalloyed steels for structural applications' *Metals and Materials* 1989 584–588.

20 Denham J B: 'EEMUA steel specifications for fixed offshore structures' *Metal Construction* 1987 Vol 19 382–384.

21 'Offshore installations; guidance on design and construction' Department of Energy, London.

22 Holsberg P W, Gudas J P and Caplan I L: 'Welding of high strength low alloy steels' *Weld Met Fab* 1990 Vol 58 492–495.

23 Blewett R F: 'Welding aluminium and its alloys' *Weld Met Fab* 1991 Vol 59 449–452.

24 West E G: 'The fire risk in aluminium alloy ship superstructures *The Metallurgist and Materials Technologist* 1982 395–398.

25 Dewsnap H: 'Submerged arc welding of plate girders' *Metal Construction* 1987 Vol 19 576–579.

26 Annals of Lloyds Register: Centenary Edition 1934.

27 Lane P H R: 'Lloyds Register – surveying for quality' *Metals and Materials* 1985 484–488.

28 Mizuno M and Nagaoka H: 'Application of Al-Mg alloys to large welded structures in Japan' *Int Metals Reviews* 1979 68–81.

29 Petershagen H: 'Trends in design and fabrication of ship structures' in Advanced Joining Technologies (Ed T H North), Chapman and Hall, 1990.

30 Hill P M: 'Cutting and welding in naval construction' *Weld Met Fab* 1991 Vol 60 63–72.

31 Wallbank E J: 'The performance of concrete bridges', 1989, HMSO.

32 Yeo R B G: 'Welded steel bridges' *Welding Review*, May 1990 Vol 9 65–72.

33 Bancroft J and Rogers P: 'Structural steel classics', 1906–1986 British Steel Corporation 1986.

34 Billingham J: 'Materials for offshore structures' *Metals and Materials* 1985 472–478.

35 Warwick P C et al: 'The Hutton TLP mating joint weldout' *Metal Construction* 1985 Vol 7 423–430.

36 Code of practice for fixed offshore structures BS 6235, British Standards Institute.

37 Recommended practice for planning, designing and constructing fixed offshore structures, AP1-RP2A.

38 Rogers K J and Lochhead J C: 'The use of gas-shielded FCAW for offshore welding' *Weld J* 1989 Vol 68 (2) 26–32.

39 Anon: 'Offshore fabrication productivity gains from cored wired welding' *Joining and Materials* 1988 Vol 1 274–276.

40 Musgen B: 'High strength quenched and tempered steels' *Metal Construction* 1985 Vol 17 495–500.

5 Pipelines and process plant

Introduction

The welded equipment that is to be discussed under this heading is used for the transport and processing of hydrocarbons and for raising steam in power stations. The common feature of such plant is the containment of fluids at elevated pressure and at temperatures that may be either higher or lower than normal room temperature. Food and pharmaceutical processing, where the problems are somewhat different, will not be considered.

Steam is used both in hydrocarbon processing and in power generation, and the plant required for raising steam is much the same in both cases. However power plant has its own special problems and will therefore be dealt with separately.

For petroleum refineries and petrochemical plant the feedstock and energy source is either liquid hydrocarbons or natural gas, so it is logical to start with discussing the pipelines that are used for conveying these fluids from the wells to the process area.

Pipelines

General

The techniques for making line pipe and laying pipelines for transporting oil and gas were, for the most part, developed in the USA and the governing codes and standards are US or developments thereof. The code for fabricating pipelines, covering materials, welding, inspection and acceptance levels for welds is API (American Petroleum Institute) 1104. The BS version is BS 4515, which includes provision for underwater welding and for the acceptance of defects either to an arbitrary standard similar to that of API 1104 or by means of an Engineering Critical Analysis (which is discussed in Chapter 6). British Gas has its own requirements embodied in the document PSP2. In the UK there is a Pipeline Inspectorate which is part of the Department of

Energy, and to gain approval from this body the line must be designed and constructed to a recognised code. UK design is covered in BS CP 2010.

In the following sections there is a brief description of methods of making line pipe and the materials used for this purpose, welding techniques and the organisation of welding operators, non-destructive testing, corrosion protection and pipe laying undersea and overland.

Materials for line pipe

The API 5L requirements are shown in Table 5.1. It will be seen that these are not detailed as to chemical composition, it being the responsibility of the steel supplier or pipe manufacturer to meet the tensile minima. For the higher tensile grades the designations are 5L × 42, × 46 etc where 42, 46, etc, are the yield strengths in ksi. Other national and user specifications are more detailed and in particular impact testing is a normal requirement especially for gas lines where it is necessary to design against the possibility of a running ductile fracture.

Table 5.1 API line pipe steel

| Grade | Chemical composition % | | | | | Yield stress | | UTS | Elong- |
	C max	Mn max	V min	Nb min	Ti min	ksi	N/mm^2	N/mm^2	ation %
5L									
× 42	0.28	1.25				42	290	410	25
× 46	0.28	1.25				46	315	430	23
× 52	0.28	1.25				52	360	450	22
× 56	0.26	1.35	0.02	0.005	0.03	56	385	490	22
× 60	0.26	1.35	0.02	0.005	0.03	60	415	520	22
× 65	0.26	1.40	0.02	0.005		65	450	550	20
× 70	0.23	1.60				70	480	560	20
× 80						80	552		

Many different compositions have been proposed for line pipe, but in general they may be divided into three categories. The first is plain carbon-manganese steel, which may be used for strengths up to × 52. Higher tensile steels up to and including × 70 are typically micro-alloyed steels in which the strength is achieved by the precipitation of carbonitrides in a polygonal ferrite structure, often with a reduced pearlite content. Alternately × 65, × 70 and × 80 grades combine a low carbon content, typically in the range 0.03–0.10 with an addition of 0.3% Mo, which in combination with manganese depresses the transformation temperature to form a low carbon bainitic structure, also known as acicular ferrite. This, too, is strengthened by carbonitride precipitation. A similar effect is obtained by the addition of 0.5Ni.

These steels after thermomechanical treatment can achieve a 50% fracture appearance transition temperature of say −70 °C. The highest tensile grade so far contemplated is X 100, for which quenched and tempered low alloy steels of the HY 100 or HSLA 100 types (see Chapter 4 and Table 4.15) would probably be required. However the use of such high tensile grades may be limited by the need for excessively high impact properties.

The requirements for gas pipelines in Arctic regions are necessarily severe. Whereas a crude oil line generates heat due to friction and viscous dissipation, the temperature of an exposed gas pipeline will be close to atmospheric. Good notch-ductility is therefore a prerequisite. Table 5.2 shows the chemical and mechanical requirements for the Canadian standard CSA Z184 Grade 483 (X 70 equivalent) and the actual composition and properties of the two steels supplied to a modification of this standard for the Trans-Alaska Highway project. Both these are microalloyed steels and in both cases very high Charpy V notch energy levels were achieved at −5 °C. Possible exposure to moist H_2S was considered and to avoid stress corrosion or hydrogen-induced cracking the sulphur was reduced to 0.003–0.005%.[1] Low sulphur is often required even for lines carrying sweet (sulphur-free) gas because of the risk of contamination of the source at some later date. Calcium treatment and additions of cerium or misch-metal are also commonly required.

Exceptionally, the fluids handled are corrosive enough to need piping of austenitic stainless steel or Inconel. Neither of these two alloys has an adequate yield strength for pipeline use and they are therefore employed as liners to carbon-manganese steel pipe. As a rule, pipe lengths are made from roll-clad plate which is formed and longitudinally welded.

Plate or strip for the manufacture of pipe is normally supplied in the as-rolled or heat treated conditions; normalising or quenching and tempering of the finished pipe is not practicable because of distortion. Stress relief at 600 °C is possible but rarely used.

The use of quenched and tempered plate is not uncommon even for tensile strengths as low as X 46 (317 N/mm^2 yield strength). This guarantees good notch-ductility and ensures a good result in the drop-weight test. Steel plate may also be normalised or thermomechanically treated. Strip is subject to a quench and temper as part of the rolling process, as described in Chapter 4.

The Batelle drop-weight tear test

This test, which is specified in API 1104, has been much used for pipeline steel since it was empirically found to correlate well with the results of full scale trials. The general set-up for the test is similar for

Table 5.2 Canadian CSA Z184 Grade 483 API 5L×70 type for Alaska Highway gas pipeline[1]

	Chemical composition %												IIW CE	YS N/mm²	UTS N/mm²	Elongation %	Cv −5°C J	BDWTT* −5°C Single heat	Shear J Average
	C	Si	Mn	P	S	Cr	Ni	Cu	Nb	V	Nb+VCe								
Specification	0.18 max	0.5 max	0.6 min	0.03 max	0.035 max	–	–	–	0.08 max	0.08 max	0.12 max	0.018 max	0.50 max	483 min	565 min	22 min	54 min	60	85
Supplier A actual	0.06	0.14	1.73	0.011	0.004	0.07	0.16	0.22	0.065	–	0.065	–	0.43	541	665	33	226	83	99
Supplier B actual	0.06	0.28	1.43	0.010	0.003	0.19	–	–	0.039	0.048	0.087	–	0.37	520	653	33	165	100	100

*BDWTT = Batelle drop-weight tear test (see text)
Pipe sizes 915 × 10.28, 1067 × 12 and 1067 × 16 (mm dia × wall thickness)

the Pellini test (see Chapter 3). The sample is a full thickness bar 3 inches by 12 inches with a central 45 ° notch having a 0.01 inch root radius (Fig. 5.1). This is cooled to the test temperature and fractured by means of the drop weight. The per cent of fibrous or shear fracture is measured, ignoring a length equal to the plate thickness at each end. The test is repeated at different temperatures to determine the temperature at which the specification requirement for per cent shear fracture is met.

Pipe manufacture

General

Line pipe may be seamless or welded, although in practice almost all long distance lines (and even relatively short lines) are welded. Seamless is more costly and more likely to vary in wall thickness around the circumference.

There are three main routes to the production of seamless pipe. The first starts with round billet which is pierced and then hot drawn over a mandrel and through dies to its finished size. Such pipe is used for smaller diameters, say 250 mm diameter and less, and for process piping. It is possible to make seamless pipe up to quite large diameters but apart from other factors this would be too costly for line pipe.

The second method is by extrusion; that is to say by pressing a heated billet through a die and over a mandrel. Extrusion is used to produce aluminium pipe and sections. It has also been employed for nickel base alloys, but only for process duties. The third technique is centrifugal casting. A cylindrical steel mould is spun and liquid metal poured in at one or both ends. Centrifugal forces spread the liquid over the inner surface of the mould and when solidified this forms a tube.

Welded pipe may be made either from plate or strip. When plate is used it is, as a rule, first pressed into a U shape and then formed into a circle. After making the weld the pipe may be expanded. This is done progressively by moving an internal hydraulically operated mandrel into position, expanding it, then reducing the diameter and moving to a new position. Expanding has the virtue that it tests the weld and at the same time rounds up the pipe. In plain carbon-manganese steels this plastic deformation removes the yield point and results in some loss of strength but with microalloyed steels the effect is small.[2] Expanding is an acceptable procedure according to API 5L but is not mandatory.

Pipe made from strip may be wound into a spiral, the weld being made continuously as the spiral is formed. Alternatively it may be formed progressively through a series of shaped rolls so that the seam is longitudinal. Such pipe is normally welded by electric resistance welding. These various routes are described below.

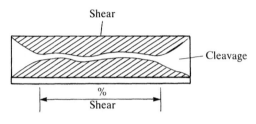

5.1 Batelle drop-weight tear test.

Longitudinal submerged-arc welding
The weld is normally made in a single pass from each side. This places a lower limit of about 250 mm on the pipe diameter, since it must be large enough to accommodate the carriage, wire feed and flux dispenser for the inside weld. To avoid excessive grain growth the heat input rate must be limited. One way in which to increase the deposition rate without increasing heat input is to add an extra electrode. The use of two electrodes is normal but three, four and even six electrode heads have been developed. In using multiple electrodes the arcs may be deflected by their neighbours and it is necessary to ensure that such deflections are favourable to the weld profile. US practice is to use DC for the leading electrode and AC for the remainder. One European manufacturer has proposed AC for all electrodes such that with four electrodes each is 90 ° out of phase with its neighbour (Fig. 5.2). A three electrode system is shown in Fig. 5.3. The wires are spaced so that a single elongated weld pool is formed with a cavity below the leading electrode and weld metal flowing rearward to fill the joint behind the trailing lead. Fused zone profiles obtained by a four electrode arrangement are shown in Fig. 5.4.[3]

The need to obtain adequate notch-ductility in the weld and heat affected zone may impose limitations on such techniques. The best impact properties in the fused zone are obtained by using basic fluxes (see Chapter 2) but the current carrying capacity of the flux decreases as the basicity increases. For the heat affected zone it may be necessary to make a multipass weld to meet impact requirements. Thus the maximum productivity may not be achievable where the pipe is required for, say, arctic conditions. For most purposes however the heat input rates will be sufficiently low.

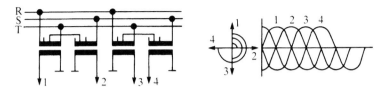

5.2 Connections for a four electrode submerged-arc welding system for pipe.[3]

5.3 Three electrode system for outside longitudinal weld in pipe (courtesy of The Welding Institute).[3]

Electric resistance welded pipe

For line pipe of modest dimensions high frequency electric resistance welding is the lowest cost process. The principle is illustrated in Fig. 5.5. Current in the range 200–5000 A at a frequency of about 500 kHz and 100 V is introduced just upstream of a pair of squeeze rolls. The frequency is high enough to restrict the current to a depth of less than 1 mm so that when the rolls press the edges together they forge without any general distortion and a relatively small amount of metal is

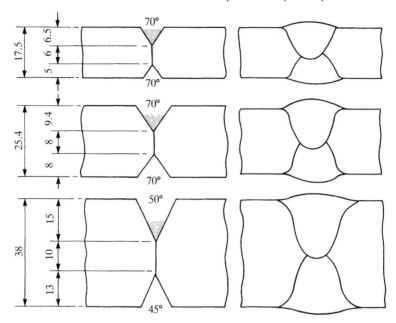

5.4 Fused zone shape for longitudinal weld in various thicknesses of pipe using a four electrode submerged-arc welding system (reproduced by courtesy of Cambridge University Press).[3]

extruded to form a flash. Upstream of the welding station the strip is formed progressively into tubular form by a series of rolls. Downstream there are knives to remove the flash both internally and externally, and a high frequency induction unit reheats the weld to about 850 °C to normalise the structure. Flying shears cut the pipe into standard lengths, typically 12.5 m.

The joint made is a solid-phase weld, any liquid that forms being extruded by the pressure of the rolls. There should be no problems due to slag inclusions, porosity, cracking or the other defects to which fusion welds may be subject. However, in the author's experience, high frequency electric resistance welds that appear under the microscope to be perfectly sound may yet have low impact properties. The welds in question were tested with the V machined normal to the tube surface, so that the fracture surface ran along the centre-line of the weld. Tests on pipe from various suppliers gave results in the range 0–5 J at room temperature. It has been suggested that such problems can be caused by non-metallic films, but no evidence of films or indeed any other defect was found in the samples tested. Many kilometres of ERW pipe have been laid, however, and have performed satisfactorily. It may be that the welds giving low Charpy values were unrepresentative. Alternatively, it may be that a high frequency electric resistance weld behaves well in service even when it has low impact strength.

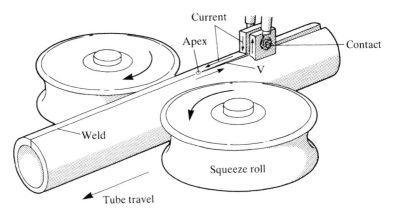

5.5 Principle of high frequency butt welding of pipe.

Spiral weld pipe
The raw material for spiral weld pipe is normally strip although in principle plate welded end-to-end could be used. The strip is uncoiled and then recoiled as a spiral in order to form pipe. A weld is made by the submerged-arc process on the inside as the incoming strip joins the pipe, and then on the outside after it has rotated 180 °. The weld is continuous and it would be possible to make pipe of indefinite length: in practice it is cut into standard lengths. It is normal practice to inspect the weld ultrasonically downstream of the welding station. Any defects found are automatically marked using paint or other means for subsequent further examination by ultrasonics or radiography. Continuous radiography, with the image shown on a screen, has also been used, sometimes as a back-up to ultrasonics.

Strip can be produced to closer tolerances than plate, so spiral weld pipe has some advantage in consistency of wall thickness. However spiral weld pipe has been accepted somewhat reluctantly by US based organisations, and this situation was not improved by the failure of a spiral welded pipe after a short period of service in the Middle East. The failure was due to hydrogen-induced cracking and was associated with the steel quality rather than the way it was welded. Subsequently the developments in steelmaking and sulphide inclusion control have made repetition of such a failure less likely.

Circumferential welding

Stovepipe welding
The technique of pipeline welding, with a string of welders working along the line was first developed in the USA in the 1920s, and the method used for manual welding remains essentially the same to this

day. The electrodes used at first were wrapped with paper bonded with water glass. Later came the cellulosic electrode, EXX10, where the coating consists typically of 40% cellulose (replacing the paper) $25TiO_2$ $20MgSiO_3$ and 15% other including sodium or potassium silicate. This type of rod is suited to pipe welding for a number of reasons:

(a) Drying is not required;
(b) A high volume of shielding gas is evolved by the coating, giving good protection against atmospheric contamination when welding in the open;
(c) The amount of slag produced is small, which enables the weld to be made downwards.

The factor limiting the rate of progress along the line is the linear rate of welding for the root and second (hot) passes. Downward welding is much faster than the upward method for wall thicknesses below 25 mm; moreover with downward welding it is possible to employ two pairs of welders working simultaneously whereas with upward welding only one pair is possible.

In the field the joint is clamped internally to obtain the best possible alignment and the rooters start one (or two) on either side welding from twelve o'clock downwards. When the root is complete the clamps are removed and the hot pass is made. After this come the fillers and finally the capping pass. The root pass is of course the most critical and requires the most skilled operators. For overland pipelines the metal deposition rate is not so important because more welders can be employed for the filler passes if necessary. On lay barges for subsea lines, where space is limited, this may not be the case.

There have been numerous attempts to develop and promote basic coated electrodes for downward welding of line pipe but these have had only limited success. Basic rods do not meet the requirement for a copious gas shield and low slag volume which was noted earlier. Table 5.3 shows that a typical cellulosic rod produces about four times as much gas and nearly one-third the slag as compared with a basic low hydrogen type.

Alloy additions to the weld metal may be necessary to match the tensile properties of API 5LX grades of pipe. For example, in the case of × 70 pipe, electrodes of type E 8010 containing 1.5Ni and 0.25Mo have been used. Proprietary electrodes may also contain microalloying constituents and there is likely to be a significant amount of titanium (~0.015%) reduced from rutile in the coating. The impact properties of weld metal having matching tensile values are usually somewhat lower than that of the parent plate in the higher tensile grades. However in circumferential welding there is little difference between cellulosic and basic rods so far as impact strength is concerned.

Table 5.3 Slag and gas produced from one 5.0 × 350 mm cellulosic or basic electrode

	Electrode type			
	Cellulosic E80109		Basic EY018	
Mass of metal (wire and powder) g	53.8		65.0	
Mass of coating g	9.9		23.8	
Mass of slag formed g	4.6		12.7	
Volume of gas liberated at NTP litres	$CO + CO_2$	2.3	CO_2	1.4
	H_2	1		
	H_2O			
	Total	5.3	Total	1.4

The use of a high hydrogen electrode for welding steel with yield strengths up to 70 ksi is anomalous in welding practice and would be expected, in line with other experience and much experimental work, to result in serious cracking problems.

In fact this is far from the case; the Trans-Alaska pipeline is 1200 km in length and during construction only 28 cracks were found, and these were not thought to be due to welding.[4] The steel used was × 65 and × 70, for which carbon equivalents can range from 0.34 to 0.46, whilst a typical heat input rate for the root and hot passes would be 0.5 kJ/mm. For such conditions BS 5135 would require a 200 °C preheat, whereas the preheat used in practice was 150 °C. Moreover circumferential welds have been made experimentally with 8010 electrodes on × 70 pipe without preheat and with only a small amount of microcracking.[5] One factor that is in favour of the current generation of × 70 pipe is its low hardenability in the heat affected zone. In the tests referred to above the unwelded hardness of the pipe was 214 HV and the peak hardness in the heat affected zone ranged from 214 HV to 245 HV. Such tractable steels have not always been available, however, and the justification for stove pipe welding must rest with experience.

Automatic welding
The imminent demise of manual welding and its replacement by automatic techniques is announced from time to time, but in pipeline work the change is slow to appear. One reason for this is that the work is often done in remote areas in difficult terrain. Another is that fit-up, particularly of larger diameter pipe, is far from perfect and whereas a manual welder can make adjustment for some degree of misfit a machine cannot. Therefore machines need to be equipped with hydraulic gear capable of rounding up and aligning the pipe ends and as a result the types that produce satisfactory welds are bulky, heavy

and expensive. These factors have been a deterrent to the overland use of mechanised welding. For lay barges, on the other hand, the fact that with a machine all welding is done at one station is a real advantage. Subsea lines on the other hand are usually relatively short, and a machine may not be economically viable.

Disadvantages notwithstanding, there has been increased use of machine welding in recent years. Table 5.4 summarises experience from the Alaska Highway project, from which it is evident that the welding rate of the mechanised process was over double that of manual operation, and the productivity nearly three times.[6] For a large project the machine does, after all, have an advantage. The capital cost of existing machines is too high for short lines but development work on lighter, cheaper equipment is in progress.

Mechanised welding operates from the inside of the pipe for the root pass using the CO_2 shielded metal arc process in the short circuiting mode. Short circuiting is required for the all-position operation. Procedures generally follow the same pattern as manual welding; the root pass is followed by an external hot pass, then filler and capping passes. In the CRC Crose machine, which has been used on Canadian pipeline work, there are four internal welding heads mounted on a central rotating ring. One head is lined up with the top of the pipe and one is in the three o'clock position. The ring then rotates clockwise and these two heads complete half the weld. The other two heads are then in position, the ring rotates anticlockwise and the other half of the root pass is completed. Shortly after the internal weld is started an external head starts the hot pass, welding downwards. Up to four external heads are used, with CO_2 shielding and in the short circuiting mode. The capping pass however is made with an argon-rich shield to reduce spatter.

Short circuiting CO_2 welding is subject to porosity and lack of sidewall fusion. The latter defect is not acceptable and mechanised ultrasonic testing has been developed to ensure its detection. The principle is shown in Fig. 5.7; multiple beams are used to strike the expected defect at as closely as possible a right angle.[6] CO_2 welds contain about 400 ppm oxygen and a low nickel alloy wire may be necessary to obtain

Table 5.4 Relative productivity of manual and mechanised pipeline welding in Canada[1]

Project	Process	Pipe diameter mm	Wall thickness mm	Man-power	Welding per day	Man days per weld
Northern borders	Manual	1067	15	217	72	3
Foothills	Machine	1067	12	164	171	0.96
TCPL	Machine	1067	10.3	192	160	1.2

acceptable impact properties. These disadvantages have prompted Canadian investigators to look at the possibility of applying pulsed GMA welding with an argon-helium CO_2 or argon-helium-oxygen shield gas for automatic operations. Pulsed arc welding may also be subject to lack of fusion defects but control over the process variables is better than with short circuiting CO_2.

There is a substantial amount of smaller diameter pipework associated with pipeline construction, for example around compressor stations, and this is normally shop welded using the techniques applicable to process piping and discussed later in this Chapter.

Other welding techniques
Pipeline welds have been made by flash butt welding on a section of the Trans-Siberian pipeline. Welds are completed in 3 minutes at a rated power of 820 kVA, and achieving a production rate of 60 welds per 12 hour shift. This system appears to be slow in making any impact on US or Western European practice.

A method of friction welding in which a ring is rotated between the ends of the pipes to be welded has been proposed. The rotating ring generates sufficient heat for a solid-phase weld to be made when the joint is upset. There do not appear to be any practical applications of this technique in pipelining.

Other one shot methods are electron beam welding and magnetically impelled arc butt welding. These have been proposed specifically for J curve laying of subsea pipe and will be described later in this Chapter.

Overland pipelaying

Pipelaying starts with the establishment of the right-of-way, a strip of land typically 25 m wide along which lengths of pipe are placed prior to starting construction. Where practicable, pipes are doubled ended; that is to say, two lengths are welded together in a temporary workshop to make a single length of about 25 m, which is the upper practicable limit for handling. Figure 5.8 illustrates a typical double ending procedure for pipe up to 120 mm wall thickness.[7] In rough terrain it may not be possible to use double ended pipe.

The field work is carried out by a team of construction workers known as the 'spread'. At the front end of the spread are fit-up engineers who fit the pipes end-to-end using a system of clamps. Then come welders to make the root pass, followed after a short interval by those making the hot pass. Further behind are the filler and capping pass welders, then visual inspection and radiography. Finally the pipe is cleaned, coated with a coal tar enamel or other protective material, wrapped with a protective tape, lowered into the trench and covered with a backfill of earth or other material that will not damage the

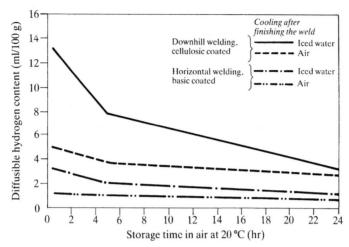

5.6 Diffusible hydrogen content of weld metal in 10 mm thick pipe weld in API 5LX60.

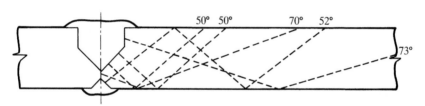

5.7 Location of probes and beam geometry for mechanised ultrasonic testing of girth welds in 8.4 mm thickness line pipe.[1]

Pass	Volts	Amperes	Welding speed, m/min
First	34-36	760-780	1.5-1.8
Second	34-36	550-600	1-1.3
Third	34-36	800-850	0.75-0.9

5.8 Typical procedure for double jointing line pipe using submerged-arc welding process.[7]

coating. When a suitable length of line has been completed it is hydrostatically tested. The tie-in (joining) welds are then made and cathodic protection is installed. The spread may extend for a distance of 10–15 km. In good terrain it is possible to complete 90 welds per day, corresponding to a progress of over 2 km, but the overall average construction rate is likely to be somewhat less than 1 km/day.

Table 5.5 shows the field welding procedures for 30 inch and 32 inch diameter x 60 pipe of ¾ inch (19.05 mm) wall thickness used for the pipeline described in Ref. 7. The root weld was made with a mild steel electrode to minimise the risk of cracking, but remaining passes were with electrodes matching x 60 properties. Preheat was only applied to remove overnight condensation or, as indicated, when the air temperature fell below 5 °C. BS 5135 requirements for preheat of 19 mm × 60 material (with some guesswork as to the carbon equivalent) welded with cellulosic electrodes would be 150–200 °C.

Quality control receives special attention in pipeline work, with several inspectors assigned to the various welding operations, to examining radiographs and to the other back end operations such as coat and wrap. Radiography is carried out with isotope sources, and sentencing of radiographs is usually in accordance with API 1104 (but see the next section for an alternative procedure). The quality of manual welds depends entirely upon the ability of the welders, and this in turn depends on the quality of training and supervision. Mathias[7] reports a repair rate of less than 0.5% on an overland line inspected to API 1104, but Parlane and Still found when examining to BS 4515 a daily reject rate of 10–40% for welding on a lay barge.[8] Much depends on the severity of the radiographic interpretation. Figure 5.9 shows the reject rate for welds made on a lay barge when the radiographs were examined wet (lower line) compared with the reject rate when the radiographs were examined dry at a later date by the owner (upper line). Most of the observed defects were porosity,[9] for which the BS 4515 requirements are more severe than API 1104.

Subsea pipelaying

Methods
Four methods of undersea pipelaying are shown diagrammatically in Fig. 5.10. The first is S-lay, where the line is welded on the barge and launched over a stinger at the stern. In the second technique the line is welded onshore and coiled on a reel, then fed over the stern of the reel barge into position. One such barge can hold up to 80 km of pipe, depending on the diameter, with a maximum diameter of 16 inches (406 mm). In the third system the pipe is again welded onshore but is towed to the required location by a tug. Finally, the J-lay system shown in Fig. 5.10a has been proposed, but not used, for pipelaying in deep water. The pipe is fed vertically through a 'moonpool' in the middle of the ship. This requires a one shot welding method such as electron beam welding. The practicability of such a method has yet to be demonstrated. An alternative technique for J-laying is described later.

Onshore fabrication has self-evident advantages in that all welds

Table 5.5 Stovepipe welding procedure for line pipe in the field[7]

Pass	Electrode		Volts		Amperes		No. of welders	
	In-line	Tie-in	In-line	Tie-in	In-line	Tie-in	In-line	Tie-in
Root	Fleetweld 5 4 mm	Fleetweld 5 3.25 mm	24–28	18–24	140–180	90–140	3	2
Hot	SA85. 4 mm	SA85. 4 mm	26–28	26–28	200–220	160–200	2	2
Filling	SA85. 5 mm	SA85. 5 mm	26–29	26–29	170–220	170–200	2	2
Capping	SA85. 5 mm	SA85. 5 mm	26–28	26–28	160–210	160–210	2	2

Notes: 1 Preheating to 100 °C when ambient temperature is below 5 °C
2 Clamp removed after completion of 75% of root pass
3 Time lapse: 5 min root-to-hot pass. No interruptions to 4th pass
4 All welds completed same day
5 All passes downhill except root pass on tie-in weld
6 Number of filling passes: 1–9, 9.14–19.05 mm wt
7 Joint bevel: 60 ° incl
8 Root gap: 1.5–3 mm
9 Root face: 0.8–1.5 mm
10 Interpass cleaning: brushing and grinding

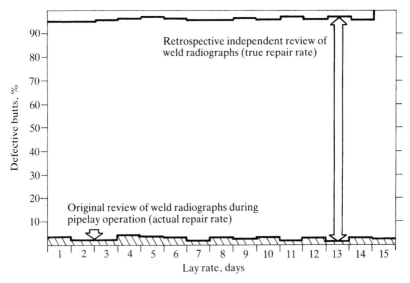

5.9 Results of wet inspection of pipe weld radiograph on lay barge with later dry inspection.[9]

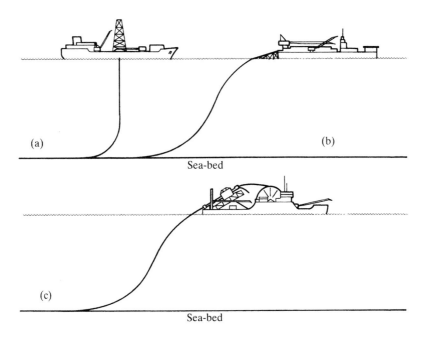

5.10 Subsea pipelaying methods: a) J lay; b) S lay; c) Reel laying (reproduced by courtesy of The Institute of Materials).[8]

can be completed and cleared prior to the laying operation, and is less likely to be affected by unfavourable weather. However the strain caused by reeling is in the region of 1% and may, depending on the type of steel, cause a reduction in the yield strength. The onshore welding operation is similar to that described below for offshore work but space is no longer such a significant constraint.

Most North Sea lines have been fabricated offshore on a lay barge, such as that illustrated in Fig. 5.11. The layout of the various operations is shown diagrammatically in Fig. 5.12. There are facilities for double ending, and in other respects the sequence of activities is the same as in overland welding but much compressed. As in overland welding there is a tendency to increased use of mechanised techniques. Flux-cored self-shielded welding has been introduced successfully for the filler and cap welds. The pipe is supplied wrapped and concrete coated, with bare ends for welding. The concrete coating (Fig. 5.13) provides negative buoyancy to keep the line on the sea-bed and gives protection against corrosion and damage from, for example, fishing tackle. Similar protection is applied to overland lines where they cross rivers under water. It is of course necessary to make good this protection over the welds, as shown in Fig. 5.14. After weld repair (if necessary) and final acceptance, the pipe ends are cleaned, hand wrapped with plastic tape, then the gap is covered with a steel sleeve and filled with mastic.

The number of butt welds completed per day on lay barges using manual techniques is in the range of 150–200; much higher than for

5.11 The Castoro Sei semisubmersible pipelay vessel (reproduced by courtesy of European Marine Contractors Ltd).

Barge stations   Inspection stations

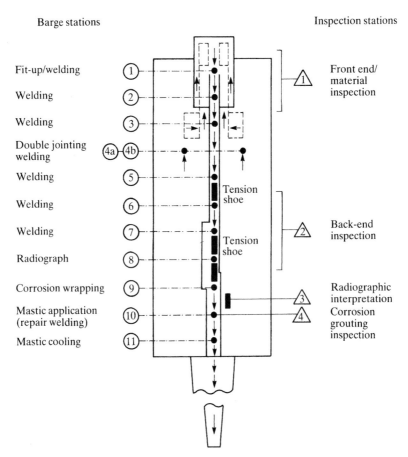

Fit-up/welding   ① ----------   Front end/
                                                material
Welding   ② ----------   inspection

Welding   ③ ----------

Double jointing   ④a ④b ----------
welding

Welding   ⑤ ----------

Welding   ⑥ ----------   Tension
                                           shoe
Welding   ⑦ ----------   Back-end
                                           Tension   inspection
Radiograph   ⑧ ----------   shoe

Corrosion wrapping   ⑨ ----------   Radiographic
                                                  interpretation
Mastic application   ⑩ ----------   Corrosion
(repair welding)   grouting
Mastic cooling   ⑪ ----------   inspection

5.12 The Castora Sei semi-submersible pipelay vessel. (Photo courtesy of European Marine Contractors Ltd) (reproduced by courtesy of The Institute of Materials).[8]

overland work. By using automatic welding, production may be increased still further.

Acceptance standards

Reference has been made to problems encountered in radiographic interpretation in the laying of subsea lines in the North Sea, and illustrated in Fig. 5.9. In order to validate the use of this line it was necessary to recover samples from the sea-bed for CTOD testing, and then carry out an Engineering Critical Assessment (reliability analysis) for all the defects in accordance with BSI PD 6493[10] (see Chapter 6).

For the line in question, the value of CTOD was high enough to allow acceptance of all defects without repair. A similar procedure was adopted for other lines before laying and in one case the acceptance

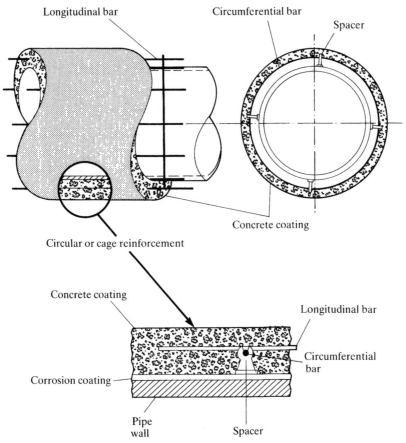

5.13 Typical coating system for subsea pipeline (reproduced by courtesy of The Institute of Materials).[8]

rate was higher than for BS 4515; in others, where the CTOD was lower, there was little difference.[9] In current practice films are processed automatically and are invariably viewed dry.

Alternative welding techniques for J-laying

Various alternatives have been considered for the one shot welding which might be required for J-laying, including magnetically impelled arc butt welding, flash butt welding and friction welding, but the most serious contender at the present time is electron beam welding. Attempts have been made to use this process for normal pipelining, with the pipe axis horizontal, but it is not suitable for positional welding. The electron beam is, however, capable of operating in the horizontal-vertical position required for J-lay.

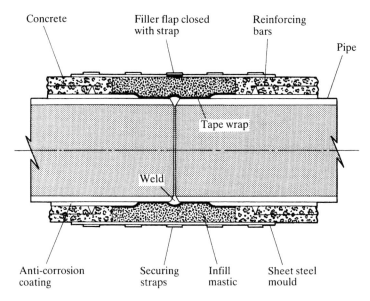

5.14 Coating for field welds in subsea pipe (reproduced by courtesy of The Institute of Materials).[8]

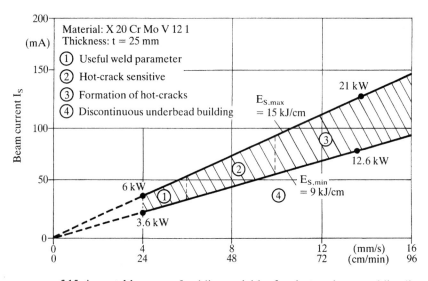

5.15 Acceptable range of welding variables for electron beam welding.[12]

The obstacles to be overcome are many. These include:

(a) Sufficient accuracy of fit-up to accommodate the narrow beam;
(b) Maintaining a vacuum in the joint area;
(c) Achieving a good termination to the weld;

(d) Welding across the longitudinal weld, which has a relatively high oxygen content;

(e) Avoiding planar defects due to shrinkage or cracking.

Probably the best solution to the fit-up problem is that proposed by Sevry and Bonnet[11] namely, to make a wide weld. Good results were obtained in pipe welding with a weld 5 mm wide. The vacuum problem can be solved by using inflatable seals, whilst obtaining a defect-free overlap at the beginning/end of the seam is possible using a programme which modifies the beam variables. The other problems can be overcome by control of gas content and by maintaining the weld variables within prescribed limits (Fig. 5.15).[12]

An alternative to the system of J-laying described earlier is to lay the pipe over the rear of the barge. A 72 metre length is welded up on the deck and then loaded on to a ramp. The ramp is raised to line up near-vertically with the stinger and joined using conventional techniques to the previously-layed length still held in the stinger. Such a method was used to lay an internally-clad 20 inch carbon steel line offshore New Zealand.

Underwater tie-ins and repairs
Underwater joints are needed between transmission lines and risers at the platform, and for repairs to damaged lines. Underwater welding is required for the repair of structural parts, for attaching new sacrificial anodes or lifting lugs and for other maintenance or repair work.

The riser joint may be flanged but welding is also used. Such a joint is required to be of a quality equal to that made on land, and the preferred method is by means of a hyperbaric habitat. The piping is aligned using, for example, the type of rig sketched in Fig. 5.16. The habitat (Fig. 5.17) is then lowered over the pipe ends. Divers enter the habitat from below and fold down gratings to form a work platform. The joint is aligned and seals fitted where the pipe comes through the walls. Pumps then evacuate the sea water and fill the chamber with a helium-oxygen mixture in which welders operate at the pressure corresponding to the depth of water. There is equipment to dry the atmosphere to prevent hydrogen pick-up by the weld. Welding (Fig. 5.18) is performed manually by, for example, making the root and hot passes with the GTA process and welding out with basic coated electrodes in the upward direction.[13] Preheat and/or post-welding heat treatment may be necessary for pipe with higher carbon equivalent and in view of the somewhat unfavourable conditions. Automatic machines that use the TIG process throughout are available. Such a system was used in the Norwegian Trench, which was considered too deep for safe diving operations, and the same system has subsequently been employed at lesser depths. After completion of welding

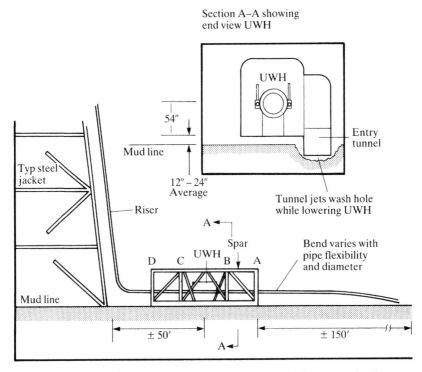

5.16 Typical set-up for hyperbaric tie-in weld between pipeline and riser.[13]

non-destructive testing is carried out in the usual way. Habitats may also be used for repair welds. Where practicable, however, operating companies may prefer to raise the line to the surface and make repairs on board a barge.

Less exacting operations, temporary repairs and attachments are carried out by divers using MMA welding without a habitat – wet welding. Waterproofed coated electrodes are supplied for this purpose. The weld is quenched as it is made and also picks up hydrogen: hardnesses are high and allowance must be made for weld metal having low ductility. Wet welds are therefore almost invariably fillet welds; for example, in the repair of a tubular member a pair of half-sleeves are fillet welded over the original part. One way to circumvent the effects of hydrogen and quenching is to use an austenitic or nickel base electrode. In such cases it must be demonstrated that the corrosion resistance of the joint is adequate; in dissimilar metal welds there is the possibility of selective corrosion close to the fusion boundary.

Other techniques have been developed for underwater welding. The work may be carried out in a chamber which is maintained at one atmosphere. This self-evidently gives better conditions than hyperbaric or

Stern end

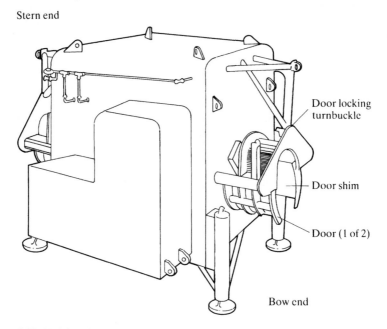

Door locking
turnbuckle

Door shim

Door (1 of 2)

Bow end

5.17 Habitat for underwater welding.[13]

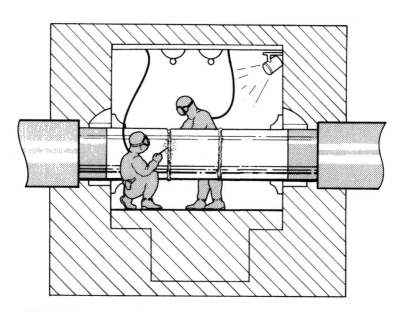

5.18 Underwater welding in a hyperbaric dry habitat.[13]

wet welding, but the engineering complications are such that it has been little used. There is a system in which the diver's head and shoulders are in a gas filled chamber and another in which a gas shield is provided locally around the electrode. Again practical applications are limited.

Requirements for underwater welding are embodied in Det Norske Veritas Rules, in a revision to BS 4515, and in the AWS 'Specification for Underwater Welding D3.6.' All require procedure and welder qualifications, and for Norske Veritas these are carried out under conditions that simulate the extreme of the expected working environment, that is to say at pressures corresponding to the greatest operating depth. Tests are usually in a laboratory facility that simulates hyperbaric conditions. Welders are in most cases qualified for all operations, root, hot pass, filler and cap and for all relevant positions, and approval is based on radiography, visual examination, bend and nick-break tests.

The AWS specification, recognising that different underwater techniques can produce different results, details four levels of weld quality, designated O, A, B and C. Types O and A are equivalent to the quality requirement for welds made under normal atmospheric conditions. Type B is an intermediate level for less onerous service and type C is intended for non-load applications.

Underwater welding has been the subject of many investigations, some of which are collected in the International Institute of Welding book 'Underwater Welding', quoted in Ref. 13. Its practical use has been limited by the high cost of providing a barge, divers and specialist equipment for a relatively small amount of work.

Service behaviour

The avoidance of any risk of a running ductile failure is a prerequisite for any gas pipeline design. This problem is not related to welding; because the longitudinal welds in a pipeline are staggered, they are not considered to have any influence on running fractures. On the other hand such hazards may well limit the tensile properties of pipe for handling gas, and thereby affect welding technology.

The Charpy impact level required to arrest a remaining fracture is obtained by means of the type of plot shown in Fig. 5.19. Here the decompression curve for the gas in question (that is to say, the fracture velocity at which the gas pressure starts to fall) is plotted as a function of pressure. The fracture resistance of the pipe material, which is obtained from the R curve (see Chapter 3) and assumed to vary with the Charpy properties is plotted on the same chart. The crack will arrest in those cases where the R curve lies below the decompression curve. In the plot shown (Fig. 5.20) $\times$ 70 steel would need an impact value of 160 J to arrest the crack, a figure which was achieved in the

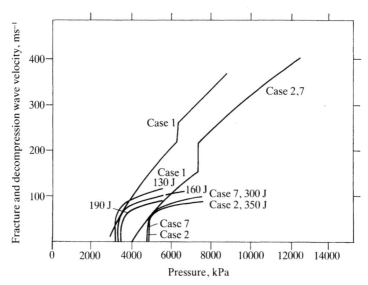

5.19 Fracture arrest predictions for ×70 and ×100 pipe in relation to impact properties for a design temperature of −4 °C and conditions as below (reproduced by courtesy of The Institute of Materials).[1]

Case	OD mm	Wall thickness mm	Operating pressure kN/mm²	Grade	Y.S N/mm²
1	1219	13.72	8687	×70	483
2	1219	13.72	12411	×100	689
7	914	10.28	12411	×100	689

case of steel for the Alaska Highway project (Table 5.2). On the other hand, × 100 steel (case 2) would require 350 J, which is not obtainable in commercial practice. To use such high tensile steel for gas transmission it would be necessary to insert crack arresters at intervals along the line.[1]

Cracking may occur in lines handling gas or oil containing H_2S if moisture is present. For such conditions the accepted upper limit for yield strength of the steel is 550 MN/m², but in practice the steel used for sour service has considerably lower strength. Weld hardness is held to a maximum of 200 Brinell. The trend to clean steel favours resistance to H_2S cracking, and such precautions together with those outlined in Chapter 3 are capable of controlling the problem.

Stress corrosion cracking may occur also from the outside of the pipe. At points where the coat-and-wrap has failed ('holidays' in the coat) the action of the cathodic current may, in certain soils and in the presence for example of a varnish coat on the pipe generate alkaline conditions. This has caused carbonate-bicarbonate stress corrosion cracking in some lines. There is no positive evidence that limiting pipe

5.20 An aerial view of an ethylene plant (reproduced by courtesy of M W Kellogg Ltd).

or weld hardness is of value in this case, and maintaining the integrity of the coating, in particular by effective shot blasting before applying the first coat, appears to be the best defence.

Process plant

Introduction

The plant to be considered under this heading is primarily that which uses a hydrocarbon feedstock. Inorganic chemical, metallurgical and nuclear power plant are not covered and fossil-fired steam plant is dealt with later in this chapter.

The primary purpose of structural steelwork is to support platforms that may be used for transport, manufacture, merchandising or other useful activities. Process plant, on the other hand, is for the containment and treatment of fluid to produce useful substances. Such processing often requires elevated temperatures and pressure, and exposure to corrosive substances. The materials used in this field therefore include almost all engineering alloys, ferrous and non-ferrous, and a number of corrosion-resistant or heat-resistant non-metals.[14] There is some common ground with structural work; structures are required to support pipework and structural steel is employed in the construction of storage tanks for hydrocarbons that are liquid at room temperature.

Petroleum or rock oil is generally considered to have been derived from the remains of living organisms deposited over millions of years in sedimentary layers. It migrates to porous formations and becomes trapped below impervious strata, often floating on a layer of water and held under pressure, below methane (natural gas). In some parts of the world it leaks out at the surface, and has been used in various ways for thousands of years. The modern oil industry was born in 1859 when Colonel Drake drilled the first oil well at Oil Creek, Pennsylvania. Drake, whose military connections were somewhat obscure, adapted the methods that had been developed during the nineteenth century for drilling through rock to obtain salt. At that time oil was used primarily as kerosene for heating and lighting. The kerosene was obtained by distilling the crude oil in a pipe still, and the lighter fraction, gasoline or petrol, provided the fuel that made possible the development of the internal combustion engine. ·

Oil technology also laid the foundations for the development of petro-chemicals, which started in the 1920s in the USA with the separation of isopropanol from petroleum fractions and the production of ethylene by thermal cracking of hydrocarbons. This made possible the manu-facture of polymers such as polyethylene in its various forms. Later a petrochemical route was found for the production of some inorganic chemicals such as ammonia and nitric acid. Originally the hydrogen needed for these products was obtained from coke oven gas but steam-methane reforming proved a more economical and cleaner method of hydrogen production, and liquid or gaseous hydrocarbons are now the feedstocks for the bulk of the chemical industry. An exception to this rule is found in South Africa, where Sasol produces gasoline and numbers of petrochemicals from local supplies of cheap coal.

Plant layout: contractors and codes

Traditionally, process plant is laid out in two areas, on-site and off-site. Processing equipment is located on-site, whilst the off-site area is used for storage of feedstock, products and essential services such as cooling water supply, auxiliary steam generation, water treatment and so forth. The off-site area is generally considered to be exposed to a lower degree of hazard than the on-site one, but this does not necessarily apply to the storage of liquefied gases such as ammonia or propane. Figure 5.20 is an aerial view of a process plant and Fig. 5.21 is a process flow diagram for an atmospheric crude oil distillation unit in a refinery. The general arrangement is typical of many hydrocarbon processing operations. The oil comes from storage and is first heated in a furnace, then transferred to a distillation tower at a temperature of, say 350 °C. Here it flows over a succession of trays, each of which operates at a particular temperature, with the maximum at the bottom of the tower and the minimum at the top. Sidestreams are taken off at the appropriate levels, cooled and sent off to storage. The overhead vapours are

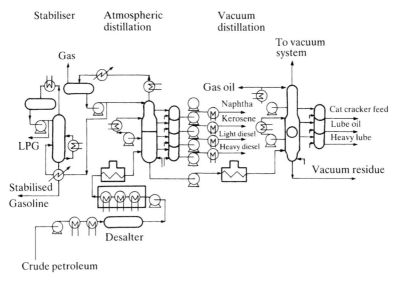

Stabiliser   Atmospheric   Vacuum
                  distillation   distillation

5.21 Process flow diagram for crude oil distillation.

condensed and the liquid fraction is recycled; the gas is stored or used as a fuel. The residue from the bottom of the tower goes to a vacuum distillation unit from which heavier fractions such as lubricating oils are obtained.

The design, purchasing and construction of process plant is normally carried out by contractors, who work to specifications, more or less detailed, laid down by the operating company concerned. Most contractors are capable of making the process and mechanical design, purchasing equipment and piping, and constructing the plant on-site. Some, however specialise; for example in process design or site construction. Operating companies occasionally design and construct their own plant but this applies only to a small proportion of plant construction.

Most of the major contractors for hydrocarbon processing plant are US based and the basic codes and standards used for specifying equipment and piping are American (Table 5.6). However UK based operating companies may specify British Standards, and there are also countries such as Germany where there is a legal requirement that pressure vessels conform to the national code.

Materials: general

The materials used for process plant are dictated in part by the processing conditions, and in part by the nature of the equipment such as heat exchangers, pressure vessels and the like. Details will be discussed below firstly for particular types of service, secondly for individual

Table 5.6 Codes relevant to the construction of equipment for process plant

Item	Code	
	US	UK
Pressure vessels	ASME code section VIII	BS 5500
Heat exchangers	TEMA*	
Structures	AWSD1.1	BS 5135
Piping	ANSI B31.3	
Tankage:		
Ambient temperature	API 650	BS 2645
Low temperature to −50 °C	API 620	BS 4751
Low temperature to −196 °C	API 620	BS 5387

*Standard formulated by the Tubular Exchanger Manufacturers Association

items of equipment and thirdly for processes where special materials and welding techniques are used.

The changes in steelmaking methods described in the previous chapter have also been of benefit for pressure equipment. Various impurity elements such as arsenic, antimony and tin promote temper embrittlement, which, in turn, can result in the long term deterioration of certain steels used at elevated temperature. The same elements, plus sulphur, contribute to strain relaxation cracking (see Chapter 3). The development of clean steel helps in avoiding these and other problems.

Austenitic chromium-nickel steels find an extended use in petrochemical plant and these, too, have been profoundly affected by processing changes. The first change was the introduction of the argon-oxygen decarburisation process. By injecting oxygen into the melt, carbon may be reduced to low levels, but with severe loss of chromium. However, the lower the level of partial pressure of carbon monoxide in equilibrium with the melt, the higher the retention of chromium (Fig. 5.22). Diluting the oxygen with argon makes it possible to achieve an appropriate CO partial pressure. The steel is first melted in an arc furnace and then transferred to a refractory lined vessel, where the argon-oxygen mixture is injected. Carbon levels below 0.01% can easily be obtained in this way and at the same time lead and other impurities are vaporised, hydrogen contents reduced and, by employing a suitable slag, sulphur is removed. The second step was the introduction of continuous casting, which results in a yield at least 10% higher than ingot casting. These developments, together with the use of more efficient rolling mills, have resulted in the reduction of the cost of stainless steel relative to that of other materials (Fig. 5.23) whilst at the same time making possible an improvement in quality. In particular, extra low carbon steels that are required for welds exposed to some aggressive media have become more readily available at lower cost.[15] Table 5.7 lists commonly used austenitic chromium-nickel steels.

Table 5.7 Austenitic chromium-nickel steels

| AISI no. | Composition (%) | | | | | | Electrode type* |
	C, max.	Cr	Ni	Mo	Ti	Nb	
304	0.08	18–20	8–11	–	–	–	E308
304L	0.035	18–20	8–13	–	–	–	E308L
304H	0.04–0.10	18–20	8–11	–	–	–	E308
316	0.08	16–18	11–14	2–3	–	–	E316 or E16-8-2 †
316L	0.035	16–18	10–15	2–3	–	–	E316L
316H	0.04–0.10	16–38	11–14	2–3	–	–	E316
321	0.08	17–20	9–13	–	$5 \times C, 0.07$	–	E347
347	0.08	17–20	9–13	–	–	$10 \times C, 0.10$	E347
309	0.15	22–24	12–15	–	–	–	E309
310	0.15	24–26	19–22	–	–	–	E310

* To AWS A5.4–69.
† For high temperature creep resisting applications.

In process plant metallic materials are frequently required to operate in tensile loading at elevated temperature. In US practice permissible stresses for various materials are listed in the relevant section of the ASME code. The properties of new materials, suitably certified, may be submitted to ASME for inclusion in the Code. If approved, a 'Code Case' is issued giving permissible stresses and any restrictions on use. BSI operate a similar system in conjunction with the BS code for unfired pressure vessels, BS 5500. However BS and European codes do not publish allowable stresses; instead the codes give a formula which is applied to properties listed in the relevant materials standards: e.g. the permissible stress may be 2/3 yield stress at design temperature.

As noted earlier, plate for boilers and pressure vessels is subject to one or more tests per plate. The testing for tensile properties is carried out at room temperature and it is implicitly assumed that the elevated temperature properties, both in the welded and unwelded condition, will correspond with those given in the standard or code. For critical vessels and steam drums however it is customary to specify testing of welded samples at the design temperature as part of the procedure test.

The specified room temperature properties of different product forms (plate, forgings, pipe, etc) of any specific grade may be slightly different. Such differences, where they exist, tend to diminish with increasing temperature.

Material for specific services

Subzero or elevated temperatures
A number of petrochemical processes operate in part at subzero temperatures. In ethylene plants, for example, the product is so separated

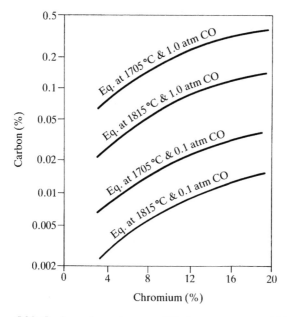

5.22 Carbon-chromium equilibrium curves at 1705 °C for partial pressures of 1.0 atm CO and 0.1 atm CO (reproduced by courtesy of The Institute of Materials).[15]

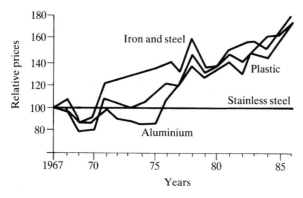

5.23 Price of various materials relative to that of austenitic stainless steel 1967–1985 (reproduced by courtesy of The Institute of Materials).[15]

by distillation at low temperature and equipment may be exposed to temperatures as low as −120 °C. Ammonia is separated from synthesis gas at −35 °C. The distillation of liquid air for oxygen and nitrogen production requires temperatures down to −196 °C. The storage of liquefied ammonia, propane, butane, ethane, ethylene, methane, oxygen and nitrogen is done at low temperature. Table 5.8 shows temperatures and materials for storage of these gases at atmospheric pressure. The use of impact tested, thermomechanically treated or normalised carbon

Table 5.8 Storage temperature and materials for liquefied refrigerated gases at atmospheric pressure

Gas	Storage temperature °C	Material
Butane	0 to −10	Impact tested
Ammonia	−33	carbon steel
Propane	−45 to −50	
Ethane	−90	9% Ni steel,
Ethylene	−105	aluminium,
Methane	−162	or austenitic
Oxygen	−183	Chromium-nickel
Nitrogen	−196	steel

steel for temperatures down to −50 °C is well established and is generally successful, but the failure of a tank at Qatar in 1977 has caused some users to take a more cautious approach. Japanese steelmakers are offering 2½% and 3½% nickel steel with crack arrest properties and such proprietary alloys may find greater use in the future. In the lowest temperature range 9% nickel steel has the best combination of strength and notch-ductility, and is the normal choice. Aluminium and stainless steel may be viable for small tanks.

In process plant of moderate thickness impact tested carbon steel is used down to −50 °C, 3½% nickel steel from −51 to 150 °C, 9% nickel, aluminium or stainless steel down to −196 °C and aluminium or stainless steel below this temperature. Aluminium is employed for a special design of regenerative heat exchanger and for some piping in ethylene and liquid oxygen plant, otherwise 9% Ni or stainless steel is the normal choice. Copper has been a traditional material of construction in air separation plant but finds little use in recent large scale units.[14, 16]

Alloys for elevated temperature use, and specifically those with augmented creep strength are not so easily categorised. The most widely used low alloy creep-resistant steel is the 2½Cr1Mo type, and this is employed for pressure vessels and piping for temperatures up to about 550 °C. Recently the Materials Properties Council in the USA has sponsored research into a 2–3 Cr1Mo¼V with titanium and boron additions and at 550 °C this has a 10^5 hour rupture strength, about twice that of the conventional alloy (Fig. 5.24). Matching weld properties have been obtained and the steel has been tested for susceptibility to reheat cracking. ASME Code Case 1960 has been issued to cover this alloy.[7] Numbers of proprietary alloys have been developed for heavy-wall pressure vessels and steam-drums and these are reviewed later in this chapter.

For higher temperatures up to 600 °C 9Cr1Mo and 12Cr1Mo with their vanadium treated variants may be used. Above this temperature it is necessary to specify austenitic chromium-nickel or nickel-base

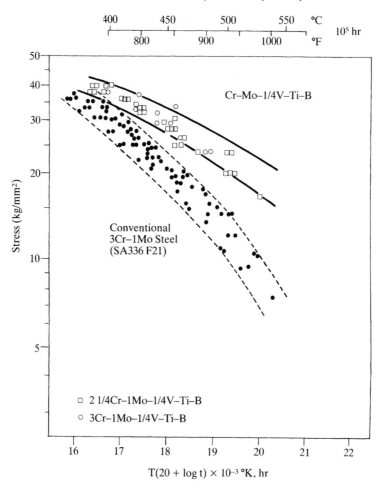

5.24 Creep rupture strength of boron-treated Cr-Mo-V steels.[17]

alloys. Some of these are applicable primarily in steam power plant and will be discussed later in this chapter. Special high temperature alloys are used for ethylene pyrolysis and steam-methane reformer furnaces and these are also considered later in this chapter.

Scaling in oxidising atmospheres
Steel which is exposed to an oxidising furnace atmosphere, to air, CO_2 or steam will oxidise and there is a temperature above which the oxide scale ceases to be protective. Table 5.9 gives these upper limiting temperatures for various alloys. Such limits need to be considered for furnace parts such as tube supports or fins. The temperature of furnace tubes,

Table 5.9 Limiting temperature for exposure to oxidising atmosphere

Material	Scaling limit*, °C	Typical use
Carbon steel	540†	General
C–0.5Mo	540†	Elevated temperature and hydrogen service
1Cr–0.5Mo	565	
1.25Cr–0.5Mo	565	High pressure steam service and
2Cr–1Mo	580	hydrogen service
5Cr–0.5Mo	620	Used for resistance to sulphur
7Cr–0.5Mo	635	corrosion in liquid hydrocarbons or
9Cr–0.5Mo	650	for hydrogen service
18Cr–8Ni austenitic steels	900	General corrosive service; high-temperature S and H_2
Incoloy 800 18Cr–35Ni	1100	High temperature pyrolysis and reformer furnaces
High carbon, centrifugally cast 25Cr-20Ni	1100	High temperature pyrolysis and reformer furnaces
Proprietary centrifugally cast alloys	1100	High temperature pyrolysis and reformer furnaces

* Temperature above which scaling in ordinary flue gas, air, CO_2, or steam will be excessive; temperatures may also be limited by corrosion.
† Graphitisation may occur at temperatures above 400 °C (carbon steel) or 450 °C (C − 0.50Mo steel) and this possibility must be considered in design.

piping and other items may well be limited by corrosion from the process fluid. In addition, carbon and carbon ½Mo steel may suffer from graphitisation at temperatures over 400 °C and 450 °C respectively. Cementite (Fe_3C) is a metastable compound and at elevated temperatures may decompose, forming graphite nodules. The tensile properties are thereby reduced. In the case of welds graphitisation may occur preferentially in the heat affected zone, giving rise to the possibility of rupture. The graphitisation temperature limits given above are therefore observed for welded pressure equipment.

Hydrogen attack
Cementite is also broken down by hydrogen dissolved in steel at temperatures over about 250 °C. There are many process units in which it is necessary to contain hydrogen at elevated temperature and pressure, and for these hydrogen-resistant materials must be used.

Hydrogen at elevated temperature may damage steel in two ways. The first is surface decarburisation, which, like decarburisation by oxygen, leaves a weak surface layer. The second is by decomposing cementite to form methane, which precipitates at the grain boundaries and causes the steel to disintegrate. The heat affected zone of welds may be preferentially attacked. Figure 5.25 shows a cover plate on a 5Cr½Mo vessel which had been exposed to hydrogen at too high a

5.25 Cover plate of 5Cr½Mo steel welded with 25Cr20Ni electrodes after exposure to hydrogen at elevated temperature.

temperature. The metal as a whole was attacked, but it has fractured at the weld boundary.

Stabilising the carbides by chromium and molybdenum additions raises the temperature at which hydrogen attack occurs, whilst austenitic chromium-nickel steels are resistant under all practicable conditions. Materials are selected according to the Nelson Chart,[18] which shows the pressure temperature limits for a number of alloys. This chart takes into account the susceptibility of welds (Fig. 5.26).

There is an incubation period before fissuring and permanent damage occurs and API publish charts giving information on this subject in the document which contains the Nelson Chart. Such information may be useful in cases where there has been an accidental excursion above the safe operating limit.

Mercaptan attack

Some crude oils contain sulphur in the form of organic compounds known as mercaptans. When the sulphur content exceeds 0.5% the crude is 'sour', below this level it is 'sweet'. When mercaptans are heated in contact with steel they break down releasing hydrogen sulphide. This happens in petroleum refining, and when the temperature exceeds 300 °C carbon steel is attached by the H_2S. For higher temperatures 5Cr½Mo steel is satisfactory for piping, but distillation towers are lined with 12Cr steel and distillation trays are likewise 12Cr. 18Cr8Ni stainless steel is completely resistant to attack by organic sulphur compounds but crude oil normally contains a proportion of salty water, and unless this is effectively removed the austenitic stainless steel may suffer stress corrosion cracking.

It is generally thought that the corrosive effect of mercaptans in contact with carbon steel is due to their catalytic breakdown to form H_2S. The presence of 5% or more chromium in the steel inhibits the catalytic action and the corrosion rate is sharply reduced.

The ease with which mercaptans decompose depends on their molecular weight. The lighter compounds, which occur in the lower boiling fractions, decompose more easily and vice versa. Thus, the bottom outlet of a crude atmospheric tower, which carries the heavy residue, may be made of carbon steel even though it operates at about 400 °C. At the other extreme, where it is necessary to reheat a light sulphur-bearing fraction such as naptha, it may be necessary to use austenitic stainless steel. In practice most operating companies use 5Cr½Mo, 7Cr1Mo or even 9Cr1Mo for all operating temperatures over 300 °C.

Welds do not appear to be preferentially attacked in sour crude service. There is some indication that sulphur and other impurities in the steel accelerate the attack, and in this respect weld metal, which normally has a low sulphur content, would be expected to behave well.

Hydrogen/H_2S mixtures

Sulphur is removed from hydrocarbons by hydrogenation, which converts mercaptans to H_2S. The resulting mixture of hydrocarbon, hydrogen and H_2S is corrosive to ferritic alloy steels. In this instance there is no advantage in using 5Cr½Mo steel, which is consistent with the idea that chromium prevents the catalytic breakdown of mercaptans in

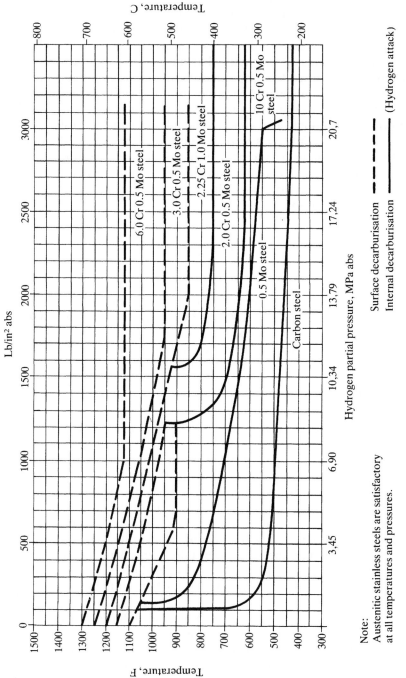

5.26 Operating limits for steels in hydrogen service according to API publication 941 June 77 edition.

Note:
Austenitic stainless steels are satisfactory at all temperatures and pressures.

Surface decarburisation - - -   (Hydrogen attack)
Internal decarburisation ——

crude. In H_2/H_2S mixtures no catalytic breakdown is required to produce the corrosive medium. Charts are available to estimate corrosion rates as a function of temperature and H_2S concentration but as a rule temperatures above 280 °C require austenitic chromium-nickel steels.

Welds are not preferentially attacked by H_2/H_2S mixtures. However if austenitic stainless steel welds that have been sulphided by such mixtures are exposed to damp atmospheres acidic conditions are set up. Steels that are not extra low carbon or fully stabilised may then be subject to intercrystalline corrosion in the vicinity of the welds. This is known as 'polythionic acid attack' although the actual corrosive agent appears to be a sulphurous acid. A neutralising wash after shutdown and prior to opening up is sometimes recommended. It is worth noting that sulphided surfaces produced in this way and by sour crude attack are pyrophoric when exposed to air, and may cause fire in contact with inflammable substances.

Shop fabricated equipment

Classification
It is customary to classify types of equipment by a letter and individual items by a number; thus a heat exchanger may be designated 101–C where 10X represents the process unit, the last figure is the item number, and C represents exchanger. Most equipment is purchased from specialist vendors; some however is site fabricated. Site fabrication is discussed later in this chapter.

Heat exchangers
Heat exchangers are of two main types: plate heat exchangers and shell-and-tube exchangers. In the first type heat exchange is between streams of different temperature flowing on either side of a plate; in the other it is between fluids inside and outside a tube.

Plate heat exchangers are used on North Sea platforms for sea water cooling. This type of cooler is a development of the devices used for pasteurising milk. They consist, in principle, of a set of parallel plates separated by gaskets and with a hole in each corner. When the plates are clamped together the holes form tubular headers, the flow being directed by the gaskets. Plates for sea water duty are pure titanium. Welding is used for fabricating the frames but not to any significant degree for the plates.

Another plate type is the core exchanger used for low temperature duties in ethylene and air distillation plants. In these a set of parallel aluminium plates are separated from each other by fins which project at right angles to the surface. The pure aluminium plate and fin material is coated with an Al–Si brazing alloy, and the whole assembly is dipped into a salt bath, thus joining the tips of the fins to the opposite plate. A large number of joints are thus made simultaneously.

For the great majority of process plant applications, however, the shell-and-tube exchanger is used. The three main types, floating head, fixed tubesheet and U tube are illustrated in Fig. 5.27. These three types represent different means of accommodating the differential expansion between tubes and shell. In the floating head type the tube bundle has its own head, which is free to float outwards when the tubes expand or contract. A division plate at the inlet end directs the flow outward through one-half of the bundle and back through the other half. In the fixed tubesheet type there is an expansion joint in the shell, whilst in the U tube type the bundle, within limits, is free to expand within the shell.

The materials for heat exchangers are those required for the relevant duty at the operating temperature. However coolers and condensers with salt or brackish water on the tube side employ copper base alloys. For land based service tubes are aluminium brass or admiralty metal, and tubesheets naval rolled brass. Under conditions where risks must be minimised, e.g. North Sea platforms and some seaborne applications 90/10 or 90/30 cupronickel is used for both tubes and tubesheets. In both cases tubes are expanded and not seal welded.

Titanium tubes are also used for sea water coolers. Titanium is resistant to most of the corrosion problems associated with sea water. The relatively high cost is minimised by using thin-wall welded tube (0.7 mm wall as compared with 1.2 mm for copper base alloys). Tubesheets are titanium clad carbon steel or aluminium bronze. The channel, for this and other sea water coolers, may be carbon steel cathodically protected by sacrificial anodes, or carbon steel coated with epoxy resin or rubber.

Where there is a requirement for austenitic chromium-nickel or other high alloy steel this is normally specified for the tube side. In such a case the tubes are (for example) stainless steel, the channel, bonnets and tubesheet stainless clad, and the shell is carbon or low alloy steel. This type of arrangement is necessary to minimise costs but it also avoids welds in thick austenitic material. The welding of shells is conventional and typically employs coated electrodes for root passes with weld-out using the submerged-arc process. The tubesheet is normally forged steel and to radiograph the tubesheet to shell weld it is necessary to machine a stub at the other circumference of the tubesheet (Fig. 5.28). To avoid lamellar weakness this stub may be formed by upset forging.

The traditional method of making a leak-proof joint between tube and tubesheet is by expanding. An expanding tool consists of a set of rollers that are expanded inside the tube until the torque required for the expansion reaches a predetermined level. The tube is deformed

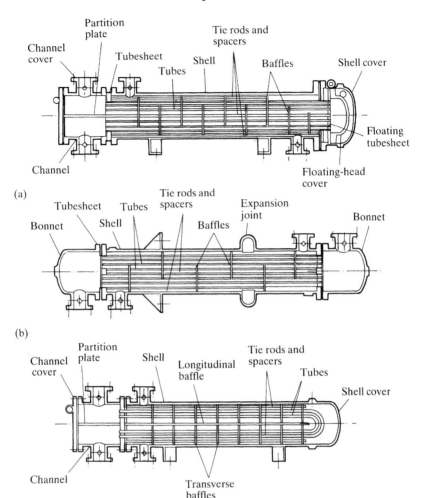

a Floating-head type: *b* Fixed tubesheet type: *c* U tube bundle type

5.27 Shell and tube heat exchangers.[19]

plastically whilst the surrounding tubesheet expands elastically; elastic recovery of the tubesheet then grips the tube tightly. This type of joint has served industry well but to an increasing extent more onerous service conditions, higher pressure for example, require a more positive type of seal.

One way to do this is to weld the end of the tube to the face of the tubesheet, as noted in Chapter 1. Typical joint configurations are shown in Fig. 1.11. Such joints may be welded manually using coated electrodes or the GTA process, or by a combination of the two. Single

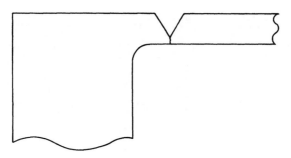

5.28 Inspectable tubesheet shell weld.

run welds are the norm but for severe duties a double run may be preferable. This may consist, for example, of a root run by the tungsten arc process without filler wire followed by MMA welding.

Manual welding can be done with the tubesheet vertical, and the use of automatic techniques was at first limited by the need to place the tubesheet horizontal, which in turn requires special working facilities. However, the use of pulsed GTA or GTA with current modulation makes it possible to control the weld pool when operating automatically in the vertical position. Thus, automatic machines of the type illustrated in Fig. 1.12 are used to an increasing extent, with a substantial improvement in productivity. An electronically programmed power source may be used, or a conventional source with a controller on the power input side. A typical set of welds is shown in Fig. 5.29.

As a rule, the tube is expanded after welding. Expansion starts 25 mm below the weld and is continued to 5 mm short of the inside face of the tubesheet.

Tubes may also be explosively welded to the tubesheet. The technique is shown in Fig. 5.30. The outer end of the tube hole is countersunk to a depth of say 12 mm with an angle of 10–20 °. A detonator is exploded centrally inside a polythene insert. A jet of liquid metal forms at the line of contact between tube surface and bore; this has the effect of cleaning the metal such that a good solid-phase weld is made. Explosive welding has the disadvantage that there must be a relatively large ligament between the tubes,[20] and that productivity is low. It has been used for emergency repairs to leaking joints.

Where conditions are severe or the penalty for a tube failure is very great it may be necessary to weld the tube to the inner face of the tubesheet by internal bore welding. This is a costly procedure and it has been used primarily for high pressure feed water heaters in power stations. It will however be convenient to describe the technique here.

Figure 5.31 shows the various possible configurations for bore welding. In each case the tubesheet is machined so as to provide a spigot and socket joint into which the end of the tube fits. Figure 5.32 shows

5.29 Automatic face welding of heat exchanger tubes to tubesheet (reproduced by courtesy of Magnatech).

details for the fit-up used by the Steinmüller company. The weld is made by the GTA process from the inside. Gas is supplied through the tube and also on the outside through a channel that clamps around the joint (not shown in sketch). Welding is done with the tubesheet vertical, using pulsed current to make the positional weld possible. Figure 5.33

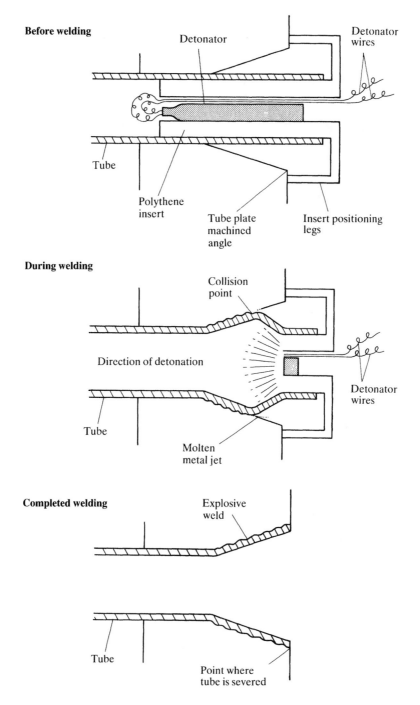

5.30 Explosive welding of tube to heat exchanger tubesheet.[20]

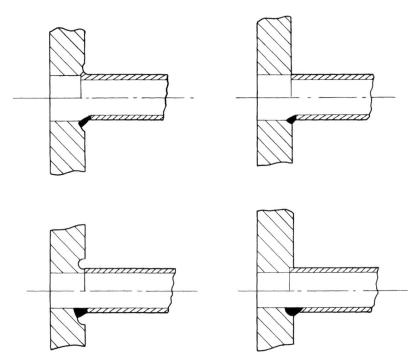

5.31 Various configurations for the internal bore welding of tubes to tubesheet.[21]

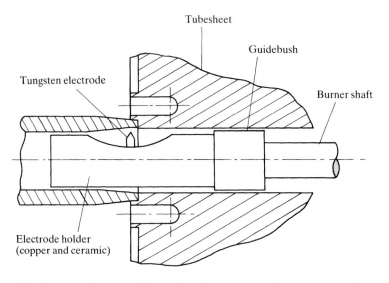

5.32 Internal bore welding of tubes to tubesheet: Steinmuller design.[21]

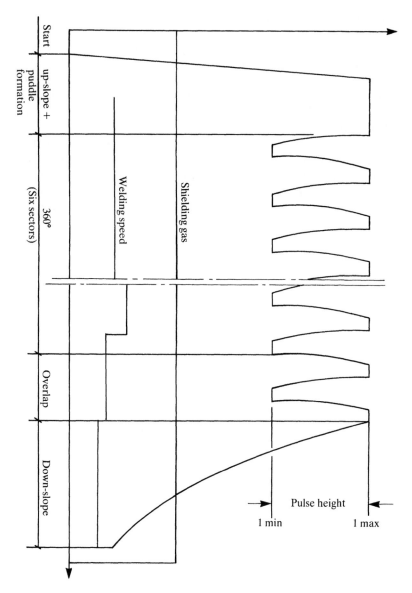

5.33 Typical current time sequence for internal bore weld to tube to tubesheet.[21]

shows a typical current time curve for one weld. This is essentially similar to that used for face welds; there is a controlled rate of current rise at the start (slope-up), a hold period to establish the weld pool, then pulsing during the rotation of the electrode. The rotational speed varies according to position. Finally there is a controlled decrease of current and a further delay before argon flow ceases.[21]

The advantages of such joints are that the crevice between tube and tubesheet (which may cause corrosion problems) is eliminated and that it is 100% inspectable by radiography. Also, in the case of waste heat boilers taking an inlet of hot gas (as in ethylene quench boilers) the weld is removed from the hot inlet and placed in a cooled region.

Extreme cleanliness is necessary for successful internal bore welding, and the work is done in a clean room, as for the fabrication of nuclear plant.

For all types of tube to tubesheet weld a procedure test is made by welding a number of tubes, usually six, into a plate of, say, 50 mm thickness. The welds are examined non-destructively, as specified for the work, and destructively. Destructive tests include measurement of push-out strength, a peel test on a strip of weld, and macro-examination of weld sections.

Non-destructive testing may include radiography. In the case of internal bore welds this is a straightforward panoramic shot with an isotope in the bore. Explosive welds may be ultrasonically tested for lack of fusion. Face welds are, in addition to other tests, examined for cracking and surface porosity either by the magnetic particle or by the penetrant dye method. Defective face welds are cut out and remade. The common defect in internal bore welds is porosity, and this can sometimes be removed by re-running the weld. Otherwise it is cut out.

Pressure vessels: codes
In the previous chapter the evolution of construction codes in the shipbuilding industry was described. As the use of steam power for ships became general, so these codes provided rules for the manufacture of steam boilers. At the same time, the number of accidents and fatalities associated with the use of land based boilers became a matter of public concern. In the USA this led ASME to formulate what later became part 1 of the Boiler and Pressure Vessel Code. The development of the oil industry gave rise to a need for constructing unfired pressure vessels, and the American Petroleum Institute formulated rules for this purpose. Eventually these two codes were merged to become, firstly, the API-ASME code and later the ASME Code Section VIII as it exists today. In the UK rules were provided by Lloyds Register; also by the Associated Offices Technical Committee, which is the technical organisation serving a group of insurance companies. Later the British Standards Institution took responsibility for code writing and produced BS 1500 and the associate material specifications, BS 1501-6.

Both the ASME Code and BS 1500 listed possible stresses for design based, at room temperature, on one-quarter of the ultimate tensile strength. In Germany, Holland and other Continental countries however design was based on 2/3 the yield strength, which gave higher design

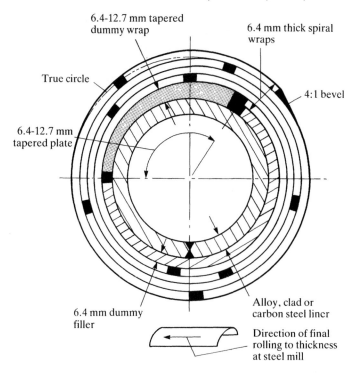

6.4-12.7 mm tapered dummy wrap

6.4 mm thick spiral wraps

True circle

4:1 bevel

6.4-12.7 mm tapered plate

6.4 mm dummy filler

Alloy, clad or carbon steel liner

Direction of final rolling to thickness at steel mill

5.34 Spiral wound multi-wall pressure vessel as fabricated by the Nooter Corporation.[22]

stresses. The relevant authorities also required suppliers from abroad to demonstrate that their materials met the local requirement. There was pressure to revise the basis for design in the UK, and the British Standards Institution produced BS 1515, which aligned with Continental practice. Eventually BS 1500 and BS 1515 were replaced by BS 5500, where only the BS 1515 design basis is used. Similar pressures induced ASME to write a 'high stress' version of the code for unfired pressure vessels based on one-third the ultimate tensile strength. This became ASME Section VIII Division 2, the original design basis being retained in ASME VIII Division 1. Division 2 contains extra quality requirements so that it is used primarily for critical high pressure vessels.

In British practice boilers and pressure vessels are given a quality classification. This follows the Lloyds Register system for ships, and it appears in Lloyds and AOTC rules and in BSI documents for various types of boiler and pressure vessel. Class 1 vessels are subject to the highest level of examination, such as 100% radiography or ultrasonic testing of main welded seams, whilst classes 2 and 3 have more relaxed requirements. Class 1 is normally specified for all high pressure duties, and for vessels handling toxic or inflammable substances. ASME codes for boilers, vessels and piping do not include such classes, it being argued that the standard of safety must be uniform. However ASME VIII Division 2, which calls for severe inspection is often specified for higher duty vessels and represents a superior construction class.

ASME and BS codes include detailed provision for welding and provide valuable guides as to the proper use of welding in the manufacture of pressure equipment.

Pressure vessels: categories

Pressure vessels for hydrocarbon processing fall into three categories: drums, towers and reactors. Drums are used for separation or mixing of fluids, and in the main are light-wall vessels in carbon steel, sometimes clad for corrosion resistance. In processes employing high pressures, drums may be thick-walled vessels, and steam drums, which operate at elevated temperatures, are made of thick-wall low alloy steel.

Towers are used for distillation, absorption or stripping and normally operate at modest temperature and pressures. They may be clad with ferritic or austenitic stainless steel in corrosive duty; for example a crude oil atmospheric distillation tower treating sour crude is clad with 13Cr steel from the bottom up to a level where the temperature has fallen to 315 °C. In fabricating the majority of towers and drums, the welding processes are MMA and submerged-arc, sometimes flux-cored arc welding, and no special procedures or processes are required.

With the exception of high pressure and elevated temperature conditions, alloy is generally required for resistance to hydrogen corrosion. Table 5.10 shows the steels listed by Lloyds for unfired pressure vessels, together with some of the elevated temperature properties used for design. This list does not include the two carbon-molybdenum grades, namely C½Mo and the German DIN steel 15Mo3. In practice these steels when used for pressure vessels have little cost advantage over 1Cr½Mo. Also, there have been unexplained cracking failures of C½Mo vessels in thicker sections, and 1Cr½Mo or 1¼Cr½Mo steel is often the better choice in hydrogen service. Alloy steels in Table 5.10 show properties in the normalised and tempered condition; quenched and tempered 2¼Cr1Mo is used for high pressure vessels such as hydrocrackers.

Reactors are the vessels in which the essential chemical changes take place and often elevated temperature and pressure require the use of thick-walled vessels. Such vessels may be solid wall or the thickness may be built up using a number of layers, as described below.

Table 5.10 Lloyds carbon and low alloy steels for boilers and pressure vessels
(a) Composition

Grade	Deoxidation	Chemical composition, max or range %							
		C	Si	Mn	P	S	Al	Cr	Mo
410	Killed or Semi-killed		0.35	0.50–1.3					
460			0.40	0.80–1.4	0.05	–	–	–	–
490	Killed	0.20	0.10–0.50	0.90–1.6					
410 FG	Killed, Fine-grained		0.35	0.50–1.3					
460 FG			0.40	0.80–1.5	0.04		0.018	–	–
490 FG			0.10–0.50	0.90–1.6					
1Cr½Mo	Killed	0.10–0.18	0.15–0.35	0.4–	0.04		0.02	0.7–1.3	0.4–0.6
2¼Cr1Mo		0.08–0.18	0.15–0.50	0.8				2.0–2.5	0.9–1.1

(b) Mechanical properties for design: thicknesses between 63 and 100 mm

Grade	Tensile strength N/mm²	Yield or proof stress N/mm²							
		Temperature °C							
		20	100	200	300	400	450	500	550
410	410–530	205	178	170	150	138	136		
460	460–580	235	206	197	176	162	158		
490	490–610	255	222	212	192	177	172		
410 FG	410–530	220	200	175	152	134	130		
460 FG	460–580	255	233	202	177	158	153		
490 FG	490–610	280	253	218	192	173	168		
1Cr½Mo	410–560	255	210	192	150	130	127	124	120
2¼Cr1Mo	490–640	275	261	245	230	218	205	189	167

(a) Solid wall

Vessels may be fabricated from plate, from forged rings, or they may be made in one or more pieces from hollow forgings. Plate fabrication is the most common but Japan Steel Works have the capability of forging heavy-wall rings of large diameter thereby eliminating longitudinal welds. Forgings are used for low density polyethylene production: pressures are very high and the material is a high carbon quenched and tempered alloy steel. The reactors are of a bolted construction with special flange connections for piping. Hollow forgings were at one time employed for ammonia converters but such very large forged vessels are unlikely to be made in the future.

Heavy-wall vessels have the disadvantage that it is difficult to obtain good tensile and impact properties in thick sections. Moreover brittle behaviour is inherently more likely in thick plate. Nevertheless, most heavy pressure vessels are of solid wall construction, using either plate or forged rings.

(b) Multilayer

A number of pressure vessel manufacturers offer multi-wall construction for high pressure duties. There are three main types; multilayer, where there is a core tube over which formed sheets are layed and welded longitudinally. Weld shrinkage tightens the layers, which may be concentric or, in the case of Plywall vessels, made by the Nooter Corporation, in spiral form (Fig. 5.34).[22] The second type is coilayer, where strip is wound continuously on the core tube up to the final thickness. Thirdly there is the Struthers-Wells method, where rings about 25 mm thick are longitudinally welded and then machined. These rings are heated to normalising temperature and then slipped over the core tube or the previously applied ring to obtain a shrink fit.

With few exceptions heads are solid wall. For high pressures the head is of hemispherical form and therefore the stress is half that in the cylindrical portion, and the thickness is correspondingly reduced.

Multilayer vessels have the advantage that thin plate has inherently better notch-ductility, whilst the laminated structure is by nature less prone to brittle failure. In hydrogen service there is a risk of hydrogen evolution between the layers, causing blisters to form. This risk is minimised by boring weep holes that penetrate to the outer surface of the core tube. Also, welding a nozzle into a multilayer shell is not usually acceptable, so that all nozzles must be placed in the heads.

Nozzles, to which pipe connections are welded, may be cylindrical forgings but are preferably flanged such that the connections to the shell or head is a fully inspectable butt weld.

Multilayer vessels are not stress-relieved after welding. They nevertheless have a much better record than solid wall vessels so far as brittle fracture is concerned. Multilayer vessels in urea plants are discussed later in this chapter.

Pressure vessel materials

Steels suitable for high pressure petrochemical vessels and steam drums are listed in Table 5.11. The last two, HT 60 and HT 80, have been used in Japan for storage spheres. The remainder are mostly

Table 5.11 High tensile steels for solid wall vessels

Type of steel	Designation	Country	Heat treatment	Chemical composition % max or range								Yield strength N/mm²		UTS N/mm²
				C	Mn	Cr	Mo	Ni	Cu	Nb	V	20 °C	350 °C	
MnMo	ASTM A302B	US	As-rolled, N or N and T	0.25	1.15–1.5	–	0.45–0.6	–	–	–	–	345	290	552
MoNiV	Ammo 65	France	N and T	0.15	1.55	0.25	0.45–0.55	0.6–1.0	0.25	–	0.10	462	353	621
MoNi	BHW 38	Germany	N and T	0.18	1.0–1.65	–	0.2–0.6	0.5–1.2	–	–	–	421	363	568
CrMoV	BS 1501-271 Ducol W30	UK	N and T	0.17	1.5	0.7	0.28	0.3	0.20	–	0.10	414	354	553
CrNiMoV	Asera 60N	Italy	N and T	0.20	1.5	0.6	0.25	0.6	–	–	0.20	462	345	586
MoNiCu	WB 36	Germany	N and T	0.17	0.8–1.2	–	0.25–0.4	1.0–1.3	0.5–0.8	–	0.10	431	353	607
MnMoNi	ASTM A533 Cl.1.1	US	Q and T	0.25	1.15–1.5	–	0.45–0.6	<1.0	–	–	–	345	290	552
2¼Cr1Mo	ASTM A542 Cl.1	US	Q and T	0.15	0.3–0.6	2.0–2.5	0.9–1.1	–	–	–	–	586	–	724
2½CrMo	24CrMo10 2.5FDO	Germany France	Q and T	0.25	0.5–0.8	2.3–2.6	0.1–0.3	0.8	–	–	–	441	352	634
Microalloy	WES 135 HT60 QT	Japan	Q and T	–	–	–	–	–	–	–	–	618	–	687
Microalloy	WES 135 HT80 QT	Japan	Q and T	–	–	–	–	–	–	–	–	687	–	785

steels that were originally developed for steam drums, where a traditional requirement in Europe is for a yield strength of 36 kgf/mm² (353 N/mm²) at 350 °C.

Strain relaxation cracking has been experienced in the heat affected zone of nozzle welds in thick-wall (over 75 mm) pressure vessels and steam drums, mostly in steels with 0.02% max V.[23] The compositions of most steel drum shells have been adjusted to avoid this problem and, bearing in mind the increasing standard of cleanliness in steel, it might be hoped that the strain relaxation problem is a thing of the past. However, the recently developed 2¼Cr1Mo¼VB steel (ASME Code Case 1960) has a composition which previous experience would indicate to be extremely susceptible to this defect. Also the ½Cr½Mo¼V steel used in UK power stations is likely to continue to give trouble in the future. Adding vanadium is a cheap way to improve the creep-rupture properties of low alloy steel so that the risk of strain relaxation cracking is likely to persist.

Pressure vessel welding: main seams
The processes most appropriate to pressure vessel manufacture are manual metal arc and submerged-arc welding. These are supplemented in particular cases by gas metal arc and electroslag welding.

A common procedure for butt welds is to use a double J preparation, make root passes from one side using coated electrodes, back-chip and weld up with coated electrodes from the reverse side, then complete the joint with numerous runs of submerged-arc welding. The head for welding from the outside of circumferential seams is located at the twelve o'clock position, that on the inside at the six o'clock position, with welders manually controlling the weld bead location. Other operators ensure that all fused slag is detached and removed. Slag inclusions and porosity are occasional defects in such welds. A potential problem in welding thick plate is transverse cracking. The manual root passes are made under conditions of severe restraint combined with 3-D cooling, and under unfavourable conditions transverse cracks may form. As subsequent weld runs are made the tensile stress in the central parts of the weld is converted into a compressive stress, which squeezes the crack faces together and may, as noted later, make them difficult to detect.

The time spent in laying down metal for the longitudinal and circumferential seams of a heavy-wall pressure vessel is considerable, and much effort has been put into means of reducing this. Single pass electron beam welding is a tempting prospect but in spite of a great deal of development work, both in making large evacuated chambers and in welding within a locally evacuated volume, the practical use of this process for large pressure vessels would still appear to be a remote possibility.

Much work has also been done on narrow gap welding. This process, described in Chapter 1, employs a J weld preparation with, for example, a 4 mm root radius and a 1° taper. The weld is therefore made in a narrow slot, and special provision must be made to avoid arc strikes between the welding head and the side walls, to obtain adequate sidewall fusion, to ensure alignment of the head and to compensate for longitudinal drift of the rotating vessel. Gas metal arc welding has been used but spatter is deposited on the nozzle and in the groove, such that it is difficult to avoid interruptions in welding. Gas tungsten arc welding is spatter-free, but the deposition rate is low. The most practical solution therefore may be to use submerged-arc welding. Ellis[24] has described an application of narrow gap submerged-arc welding to the fabrication of high pressure feed water heaters at the GEC plant in Larne, Northern Ireland. The equipment, which is sketched in Fig. 5.35, was designed primarily to make circumferential seams in thicknesses up to 350 mm with a gap width of 18 mm or greater. The weld metal is deposited in two runs per layer and the nozzle is directed slightly towards the flank along which the run is being made. On completion of the run a hydraulic device directs the nozzle towards the opposite flank. This system ensures good sidewall fusion.

The nozzle is guided by a probe which tracks one sidewall and a roller which runs over the top of the deposited weld metal and controls height. At the same time it feeds a signal to the central control system that adjusts rotational speed so as to keep the linear speed constant. All welding variables are automatically monitored and maintained within specified limits. An essential feature of the process is that slag detaches easily, even in such a deep, narrow groove.

Ellis suggests that although the time to complete the joint is reduced, the main benefit of this process is the guarantee of a metallurgically sound joint. This is made possible by the close control over the location of the weld beads and the welding variables. Avoiding defects is of special importance in welding thick-wall vessels because the cost of a repair, and the associated risk, is so high. The use of narrow gap welding for hydrocracker reactors is considered later in this chapter.

Electroslag welding is another means of increasing productivity in the welding of thick plate, and has been applied for example to the longitudinal welds of ammonia converters. The material used was 24CrMo10 (Table 5.11). Individual tiers or strakes (cylindrical rings) were made up with either one or two welds; these were set up in a furnace with the axis vertical and heated to about 840 °C. The heated rings were then transferred to a set of rolls where they were rounded up and at the same time quenched by means of a water spray. This system failed in some cases due to an inadequate quenching rate but the welding was satisfactory. A small number of seams were rejected for shrinkage cavities; the majority however were virtually defect-free.

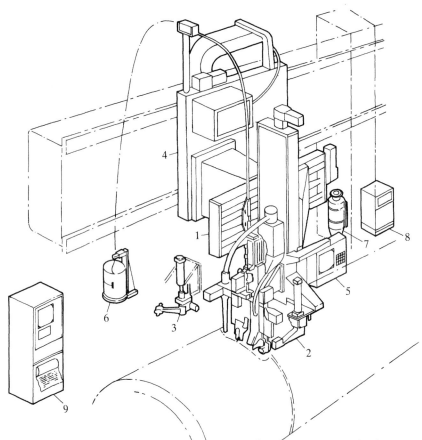

1–Welding head with cross slide; 2–Scanning device; 3–Wire cutting device; 4–Carriage;
5–Control box; 6–Wire feed system; 7–Flux feeding system; 8–Welding set;
9–Central data computer.

5.35 ESAB submerged-arc narrow gap welding machine.[24]

The disadvantage of electroslag welding is that it cannot be applied to circumferential seams. It is difficult to obtain a defect-free overlap of start and finish of the weld. Moreover most specifications require that electroslag welds be normalised or quenched and tempered, and it is not possible to maintain dimensional tolerances when the heat treatment is carried out with the axis horizontal.

Flux-cored welding is also effective for shop and field fabricated pressure vessels. For this purpose rutile-cored low hydrogen wire is especially useful. The rutile-core ensures spray transfer and makes vertical (3G) and horizontal-vertical (2G) welds practicable. At the same time the hydrogen content of the deposit can be maintained below 5 ml/100g. Higher impact properties can be achieved using a wire giving a 1% Ni steel deposit.

A good example is the pressurised fluid bed combustion vessel fabricated by the Swedish company, Uddcomb.[25] This, together with the main seam joint preparations, is illustrated in Fig. 5.36. The material selected was BS 1501-271 (Table 5.11) which, in spite of an early setback, has a good record for weldability and freedom from strain relaxation cracking.

The vessel was shop fabricated with nozzles attached in four parts. These parts were shipped to site and welding was completed. After hydrotest the top head was cut off, the internals were loaded and the head rewelded.

Welding was entirely manual and carried out partly with coated electrodes and partly with the Filarc rutile-cored wire PZ6138. Typical weld procedures for horizontal-vertical and vertical-up welds respectively are given in Table 5.12. The welds were of satisfactory radiographic quality and in particular the flux-cored welds were free from lack of fusion defects. With a basic flux-cored wire it would have been necessary to use a dip transfer technique for the positional welds, with a corresponding risk of lack of fusion. This is one of the cases where a basic flux may produce a joint which is inferior to that associated with a rutile flux.

In principle it is possible to mechanise vertical-up welding using rutile flux-cored wire. Attempts at such mechanical welding have not always been successful however.

Pressure vessels: preheat and post-welding heat treatment
Preheat and post-welding heat treatment have already been discussed in Chapter 2. It is important, in the case of heavy-walled vessels, to minimise the time of heat treatment so far as possible. On the one hand, it may for example be desirable to carry out an intermediate heat treatment on a seam when it is half completed, so as to diffuse out hydrogen and to minimise the stress build-up in the outer parts. On the other hand, increased exposure to the stress relief temperature may be damaging, for example by lowering creep strength, promoting creep embrittlement and in the case of hydrocracker reactors, promoting disbondment. All such possibilities need to be studied, and the risks assessed, before determining the final heat treatment programme.

Table 5.13 lists ASME and BS requirements for preheat and post-welding heat treatment temperatures.[26] These are given in general terms; there may be detailed provisions for which the Code must be consulted.

Pressure vessel welding: clad steel
In many pressure vessels the shell or cylindrical portion, and the heads are clad with corrosion-resistant steel. Crude and vacuum stills handling sour crude are normally clad with 13Cr steel. Such ferritic alloy stainless

Table 5.12 Typical welding procedures for a fluidised bed combuster vessel[25]

Plate quality	BS 1501-271						
Gas	80%Ar+20%CO$_2$, 23 litre/min						
Joint preparation:	35 ° symmetrical, 3 mm gap, 4 mm land						
Position:	Horizontal-vertical						

						Heat treatment (recorded)	
						Heat	62 °C/hr
						Hold	615–635 °C, 115min
						Cool	61 °C/hr

Pass No.	Consumable	Dia. mm	Welding current, A	Arc voltage, V	Travel speed,* mm/min	Interpass temperature, °C	Arc energy, kJ/mm
1	PZ6138	1.2	190	26	~170	150	~1.7
2	PZ6138	1.2	190	26	~280		~1.1
3–5	PZ6138	1.2	190	26	~300	180	~1.1
			Backgouge				
6	E7018-G	3.25	140	22	~130	165	~1.9
7–8	E7018-G	3.25	145	22	~280	170	~1.0
9	E7028-G	4	200	29	~260	180	~1.5
10–20	E7028-G	4	200	29	~290	185	~1.6
21–29	PZ6138	1.2	200	25	~300	180	~1.0

Table 5.12 cont'd

Joint preparation: 50 ° symmetrical
3 mm gap, 4 mm land

Position: Vertical-up

52 mm

PZ6138

Pass No.	Consumable	Dia, mm	Welding current, A	Arc voltage, V	Travel speed,* mm/min	Interpass temperature, °C	Arc energy, kJ/mm
1	PZ6138	1.2	150	24	~ 90	150	~2.4
2	PZ6138	1.2	170	26	~135	170	~2.0
3–7	PZ6138	1.2	185	25	~190	190	~1.5
			Backgouge				
8–16	E7018-G	3.25	120	22	~140	160	~1.6
17–19	E7018-G	3.25	115	22	~130	165	~1.6
20–22	PZ6138	1.2	185	25	~190	170	~1.5
23–25	PZ6138	1.2	175	25	~190	195	~1.4

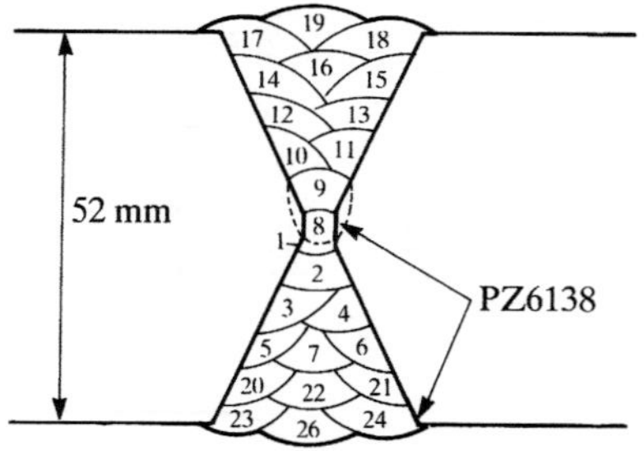

* Runout, mm. for MMA electrodes.

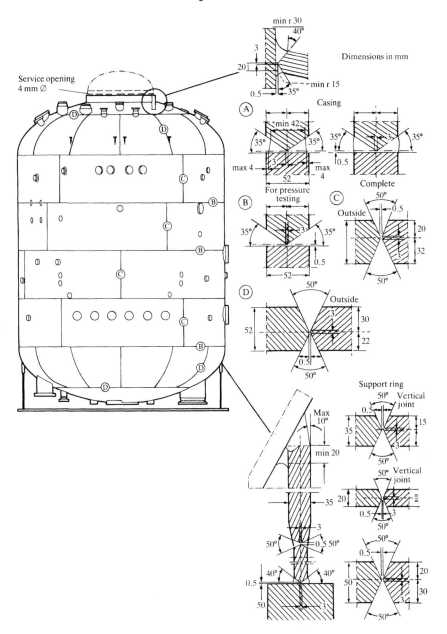

5.36 Schematic drawing of a 19 m high 42 m diameter 500 t pressurised fluid bed combustor.[25]

Table 5.13 Code requirements for preheat and post-welding heat treatment of pressure vessels[26]

Alloy	Preheat			Post-weld heat treatment			
	ASME VIII Dw1 Temperature, °C	BS 5500 Thickness, mm	Temperature, °C	ASME VIII Dw1 Thickness, mm	Temperature, °C	BS 5500 Thickness, mm	Temperature, °C
C½Mo	79	<12	20	<15	–	<20	–
		>12	100	>15	593	>20	630–670
1¼Cr½Mo	121	<12	100	<15	–	Max strength	630–670
		>12	150	>15	593	Max softening	650–700
2¼Cr1Mo	204	<12	150	All	677	Max strength	630–670
		>12	200			Max creep	680–720
5Cr½Mo	204	All	200	All	677	All	710–750

Table 5.13 lists ASME and BS requirements for preheat and post-welding heat treatment temperatures.[26] These are given in general terms: there may be detailed provisions for which the Code must be consulted.

steels are not normally considered to be weldable because of a combination of grain growth and embrittlement in the heat affected zone. In the form of cladding, however, the embrittlement is not significant and welds do not deteriorate in service. The procedure frequently specified for welding 13Cr clad steel is illustrated in Fig. 5.37. The cladding is stripped back to avoid embrittlement of the deposit by chromium pick-up when making the carbon steel weld. The cladding is made good using 25Cr20Ni electrodes to compensate for dilution and obtain a ductile layer (procedure qualification may require bend tests).

In exceptional circumstances it may be required to weld clad steel (usually 18Cr8Ni clad) from the backing side. One possible technique is illustrated in Fig. 5.38. A V preparation is made and the cladding is welded with a GTA root pass. This is followed by one or two passes with a nickel base electrode such as Inconel 182, which is compatible with both stainless steel and carbon steel. The weld is then completed with electrodes that match the backing material.

Another exceptional case is sketched in Fig. 5.39; welding titanium clad steel. Pick-up of iron in the titanium must be avoided since it can severely damage the corrosion-resistant properties. In this case the cladding is stripped back, the backing metal welded up and ground flush on the clad side. Then a strip of titanium or copper is inserted to fill the gap in the cladding, and a titanium cover strip is fillet welded over the joint.

Pressure vessels: procedure and operator qualification
In the US, procedure and operator qualification tests are prescribed in the ASME Code Section IX. The UK standards are BS 4870 and 4871 and a European Standard is in preparation. The ASME Section IX requirements are similar to those of AWS D1.1 except that there are no pre-qualified procedures and a test must be made to cover each change in an essential variable. With this exception operator and procedure qualification tests for pressure vessels are similar to those described for structural work in Chapter 4. In procedure testing for heavy-wall pressure vessels, however, Code requirements are usually trivial compared with those specified by purchasers. Testing may well be carried out near the surface, at quarter thickness and at mid-thickness. Tests may include hardness traverses, yield, tensile and impact properties, and CTOD measurement. Suppliers of heavy-wall pressure vessels will often carry out very extensive testing of welded samples prior to submitting a bid. All such testpieces must, as already noted in Chapter 2, be subject to a heat treatment simulating that given during fabrication. It may also be desirable to include provision for additional heat treatments which may be required after weld repair or after a period of service.

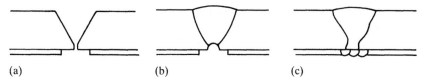

(a)   (b)   (c)

5.37 Welding of 13Cr clad plate: a) Strip back cladding; b) Weld backing plate and back-gouge; c) Complete weld of backing plate, grind finish and restore cladding using 25Cr20Ni electrodes.

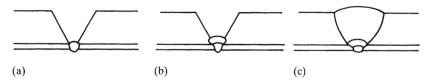

(a)   (b)   (c)

5.38 Welding clad steel form the backing side a) GTA root pass; b) Cap with nickel-base electrode.

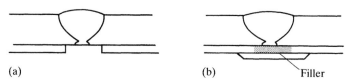

(a)   (b)   Filler

5.39 Welding titanium clad steel: a) Strip back cladding and weld up backing plate; b) Insert titanium or copper filler and fillet weld titanium cover strip using GTA process.

Pressure vessels: NDT

Codes require radiography, or radiography and ultrasonic testing as primary methods for the examination of main seams. Neither method is fully satisfactory. Radiography is insensitive to the most serious type of defect: namely a crack oriented at right angles to the plate surface. Ultrasonics, on the other hand, will in most cases detect such cracks, but cannot unequivocally determine the size and nature of other types of discontinuity such as pores and slag inclusions. These limitations are accentuated in the case of thick-walled vessels, where for radiography it is necessary to use high voltage equipment, or a gamma ray source such as Cobalt 60, and where the contrast obtained in the radiographs is relatively low. Ultrasonic testing is also adversely affected by attenuation of the beam but not quite to the same extent. Moreover, the central regions of a thick weld are in a state of compression prior to stress relief, and transverse cracks present in this region are invisible both to radiography and ultrasonics. Stress relief normally, but not invariably, relaxes transverse cracks so that they reflect an ultrasonic beam. In spite of such limitations, fabricators that are skilled in the art are not likely to overlook significant defects, and main seams in heavy pressure vessels are normally subject to both 100% radiography and 100% ultrasonic testing. It is necessary to carry out non-destructive testing after completion of post-weld heat treatment. However, the disclosure of defects at this stage may create problems since a further heat treatment will be required after repairs, with possibly damaging effect on properties. Therefore it is common practice for critical vessels to carry out testing both before and after heat treatment and for some vessels after hydrostatic testing.

An additional security is provided by acoustic emission testing. Acoustic probes are attached to 'wave guides' that are welded to the vessel at selected points. The output from these probes is fed to a computer which filters out extraneous noise and analyses the acoustic emissions that occur when the metal is strained during hydrotest. Pressure is applied in steps, and any region where there is evidence of abnormal acoustic activity is investigated by normal ultrasonics and possibly by radiography. Records are kept of the emissions that result from the progressive application of pressure and these may be compared with the results of similar tests made after a period of service. In this way it may be possible to detect the initiation or extension of cracks.

Magnetic crack detection may be applied to the main seams after completion and also to nozzle welds. Penetrant dye and powder is used to test austenitic chromium-nickel steel overlays. Chloride-free materials are commonly specified. Hardness testing of welds, using a portable tester, is sometimes required.

Site fabrication

General
A high proportion of fabrication work, most particularly welding, is carried out on-site in the case of process plant. This work includes structural steelwork, furnaces, piping, tankage and pressure vessels that are too large or heavy for transportation. Site conditions are less favourable than those in a normal workshop; nevertheless the quality of workmanship is not allowed to suffer.

Furnaces
Furnaces are used for two main purposes; firstly, to preheat a fluid prior to distillation, and secondly to provide the heat input for an endothermic chemical reaction, in particular, ammonia synthesis and ethane pyrolysis. The second type of furnace is considered later in this chapter.

A sketch of the first type of furnace is shown in Fig. 5.40. Fluid normally enters through the convection section and then passes through the main body of the furnace, where heating is primarily by radiation. Tubes are usually supported horizontally on brackets made from heat resistant (typically cast 23Cr12Ni) steel. Tubes may be connected in series by U bends, as in the diagram, or in parallel by headers. Tube material ranges from carbon steel through CrMo grades to austenitic chromium-nickel, according to duty.

Return bends and headers may be forgings but are usually cast. They may be shop welded to a pair of tubes to form a hairpin, the welds at the opposite end being made on-site. Access to site welds is necessarily restricted and welding is by MMA, using the same procedures as for the same grade of piping. However, in the case of alloy material a GTA root run may be made, giving a much improved internal finish particularly at return bends. In order to supply backing gas a paper dam is placed internally on either side of the weld and gas is introduced by tube through a hole in the edge of the preparation. The dams are washed or blown out of the hairpin or completed coil.

Piping: general

Piping conveys fluid between the various items of plant such as furnaces, reactors, heat exchangers and distillation towers, and is of necessity convoluted in form. It also connects one process unit with another, and here there may be long straight runs. The design of piping was at one time a considerable art. After completing the general layout the drawings were broken down to 'isometrics'; these show sub-assemblies small enough to handle with a light crane. These sub-assemblies, sometimes called spools, are then assembled and welded together in the field. Currently, the design is done by computer, which also prints out the isometrics.

Individual sub-assemblies are fabricated in a workshop; sometimes by off-site contractors but more often in a site workshop under the direct control of the construction manager. Figure 5.41 shows a typical flow chart for a large site fabrication shop. In the stock yard each length of pipe is marked along its complete length either with colour bands or a code number, or both, which should maintain the material identity until final inspection. Pipe is cut, bevelled and passed to the tackers who tack up sub-assemblies that can be rotated and welded into the 2G position. For carbon steel this is normally done by dip transfer CO_2 welding, which is particularly well adapted to pipe welding because it is suitable for both root and filler passes. Lack of side-wall fusion is not a problem with thin-wall pipe but can result in severely defective welds when the wall thickness exceeds about 18 mm. The rotationally welded sub-assemblies are then tacked together, supported by frames or 'horses' and welded with stick electrodes into spool

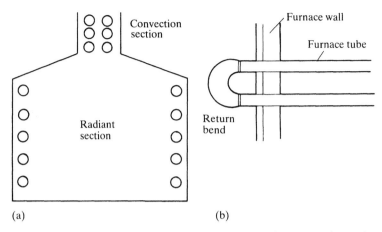

5.40 Furnace for distillation plant: a) Diagrammatic section; b) U bend connection.

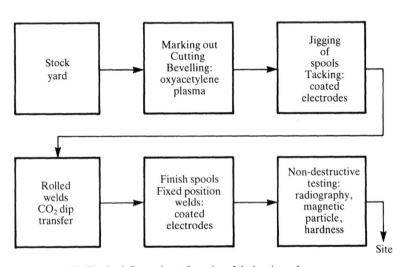

5.41 Typical flow chart for pipe fabrication shop.

assemblies. At this stage the spools are subject to visual inspection and any non-destructive testing that may be specified. Finally they are transported to site and the final tie-in welds are made.

Piping: welding and NDT
As indicated above, semi-automatic GMA processes may be employed for rotatable welds. Submerged-arc welding is employed for double ending in the case of long straight runs. (Generally this is done in a small temporary workshop close to the pipe rack.) Otherwise the

welding is manual with stick electrodes. Stovepipe welding is used only exceptionally: the time required for fit-up is such that a fast welding process would provide little advantage. All welding is therefore upwards, both for root and filler passes. Gas tungsten arc welding is employed for the root pass where slag or internal spatter is objectionable; otherwise for carbon steel the root pass is made either with a cellulosic (US practice) or a rutile (UK and European practice) electrode. As noted earlier, a root run made with a basic rod is likely to be porous. Filler passes are made with basic coated or nickel alloy electrodes where impact testing is specified. Otherwise filler passes are made with either cellulosic or rutile-coated rods.

Alloy welds are made with a GTA root and weldout with matching basic coated rods, or with basic coated rods throughout. One method of obtaining a consistent root fusion when making a GTA root pass is by the use of a fusible insert. Such inserts take various forms; in the simplest a ring of metal is placed between the root faces of the preparation, and fused without addition of filler rod. A more elaborate arrangement is illustrated in Fig. 5.42. The J preparation is appropriate for wall thicknesses in excess of say 5 mm and once again the root pass may be made without filler rod additions, but with argon gas backing. Machining the bore is necessary to obtain an accurate fit-up but it must not reduce the wall thickness below the allowable minimum. Fusible inserts are of particular value in making high quality welds in austenitic chromium-nickel steel pipe. Table 5.14 shows code requirements for preheat and post-welding heat treatment temperatures, whilst the method of applying such treatment is usually by electric resistance heaters such as the finger elements shown in Fig. 2.33.

NDT is specified in accordance with the line class, which in turn is determined by the required service. For example, welds in carbon steel exposed to moist gas containing H_2S or cyanides will be required to meet a maximum hardness value of 200 Brinell. Various types of hardness tester are available but the one most frequently used is the Tellerbrineller. The instrument holds a hardened steel ball which is placed in contact with the weld to be tested. It is then struck with a hammer, and the ball simultaneously indents the weld (or rather a flat area ground thereon) and a standard bar held inside. Comparison of the diameter of the two indentations gives the required hardness. The impression left by a Tellerbrineller is relatively large but it is nevertheless used also for testing the heat affected zone, where this is required. In practice there is rarely any difficulty in meeting the 200 Brinell limit, and failures due to H_2S stress corrosion cracking or hydrogen-induced cracking are rare in petroleum and petrochemical plant.

Magnetic particle testing is sometimes specified for ferritic steels. Austenitic welds are frequently examined by penetrant dye testing. In addition there is likely to be a limitation of ferrite content within, say, 3

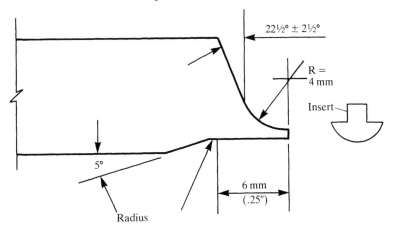

5.42 Typical edge preparation for GTA root pass in pipe over 5 mm wall using a fusible insert.

and 10%. This is best established by procedure testing, but a field check using a magnetic gauge may be desirable. Ultrasonics are rarely used for piping in hydrocarbon processing plant; radiography is the primary NDT method. In severe service 100% of welds are so tested, but for the majority of duties 5% or 10% radiography will be specified. ANSI B31.3 lays down the procedure to be followed when a weld subject to spot radiography is found to be defective.

In the field radiography is made using radioactive isotopes, preferably Iridium 192 but also in some cases Cobalt 60. A double wall single image technique is normally employed; that is to say, the isotope is located so as to radiograph part of the circumference of the weld by shooting through the opposite wall of the pipe. Radiographs are sentenced in accordance with the governing code, and the rejection rate provides an important measure of the effectiveness of welder training and supervision.

The final steps in acceptance of a piping system is 'punch-out' and hydrostatic testing. Punching-out requires inspectors to walk the lines and note any deviation from drawings or specifications; these discrepancies are then rectified. One of the persistent problems in field fabrication of pipework is material mix-ups, and these are not necessarily disclosed by visual inspection. It may therefore be required to make a positive material identification on critical lines (e.g. for elevated temperature service). Such testing is done using a portable spectroscope, by chemical spot test and sometimes by a system that polishes and etches an area of pipe or weld so that it can be examined microscopically *in situ*. A portable spectroscope, in particular an instrument known as the 'Metascop', is most commonly employed.

Table 5.14 Code requirements for preheat and post-welding heat treatment of piping in hydrocarbon service[26]

Alloy	Preheat				Post-weld heat treatment			
	AWSI B31.3		BS 3351*		AWSI B31.3		BS 3351	
	Thickness, mm	Temperature, °C	Thickness, mm	Temperature, °C	Thickness, mm	Temperature, °C	Thickness, mm	Temperature, °C
C½Mo	<12.7 / >12.7	10 / 80	<12.5 / >12.5	20 / 100	<19 / >19	– / 595–720	<12.5 / >12.5	– / 630–670
1¼Cr½Mo	All	150	<12.5 / >12.5	100 / 150	<12.7 / >12.7	– / 705–745	<12.5 / >12.5	– / 630–670
2¼Cr1Mo	All	175	<12.5 / >12.5	150 / 200	<12.7 / >12.7	– / 705–760	Max strength / Max softening	680–720 / 700–750
5 to 9 CrMo	All	175	All	200	≤12.7 / >12.7	– / 705–760	All	710–760

*Assumes use of low hydrogen electrodes.

Piping: hot tapping
The term 'hot tapping' is the technique whereby it is possible to modify a piping system without interrupting the flow of gas or liquid. It is used in hydrocarbon processing plant when a change in the process requires realignment of the piping, or on pipelines for by-passing a damaged length or for re-routing the line.

The first essential step is to weld a fitting with a set-on branch and flange around the existing pipe. The fitting, known as a 'T' is in two halves and fits snugly around the pipe. The welds between the two halves are made first, then the fillet welds connecting the fitting to the pipe. After non-destructive and hydrostatic testing a valve is fitted, then a device which bores the pipe through the open valve. When this operation is completed the valve in the branch is closed, the boring equipment is removed and the modified line may be connected.

The two main areas of concern are the possibility of burnthrough, and excessive hardening due to the cooling effect of the fluid flow in the line. In practice, provided that the pipe wall thickness is greater than 4 mm, the risk of burnthrough is minimal, whilst hardening can be reduced by preheat and laying temper beads.

Cassie[27] has described a technique used by British Gas on high pressure gas lines. The fitting is cylindrical in form with the inside diameter equal to the outside diameter of the pipe. The preparation for the longitudinal welds is shown in Fig. 5.43. A backing bar is placed in the groove at the root of the preparation, and this extends beyond the fitting to allow run-on and run-off; when the weld is completed these projections are removed. Preheat for this stage is 250 °C. Then buttering runs are laid on the pipe surface, as shown in Fig. 5.44. The pipe is preheated with gas torches to 250 °C and the buttering is applied in blocks, a block being completed before a preheat falls below 100 °C. Welding is with low hydrogen electrodes running downwards to minimise penetration. Finally a temper bead is laid over the last run. The same technique is used to build up the filler welds as shown in Fig. 5.45. Non-destructive testing is by the magnetic particle technique, both for the buttering runs, fillet and butt welds; the combined thickness of the joint is too great for effective radiography, and ultrasonic testing is impractical.

Fittings of spherical form are also employed and less exacting techniques have been used for hot tapping in refineries. In spite of apparent hazards this operation has a good safety record and has been in use for a number of years.

Cylindrical storage tanks operating at ambient temperature
This category covers the great majority of tanks used to store crude oil, intermediate and final products. The design of such tanks evolved during the development of the oil industry in the USA, and was

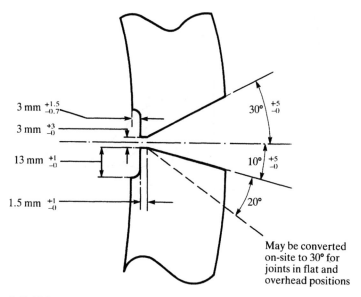

5.43 Edge preparation for longitudinal weld in cylindrical fitting for hot tapping of pipe.[27]

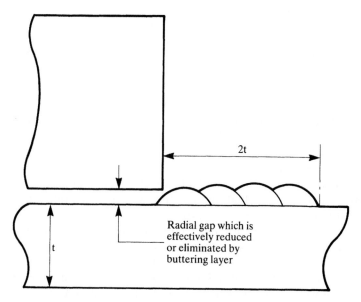

5.44 Buttering layer on pipe prior to fillet welding.[27]

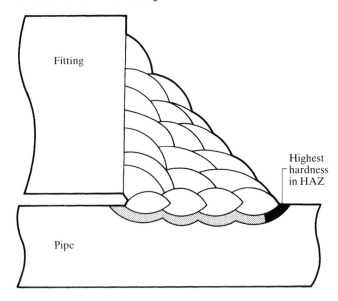

Fitting

Highest
hardness
in HAZ

Pipe

5.45 Completed fillet weld. The hard zone may be tempered by making
an arc heating run.[27]

standardised by the American Petroleum Institute, in API 650. Subsequently other countries have produced tankage standards, notably in
the UK BS 2654, in Germany DIN 4119 and in Japan JIS B8501. These
codes generally follow API; however there is a difference in allowable
stress that parallels the difference in unfired pressure vessel codes. BS
2654 allows up to 2/3 yield stress but API 650 has a limit of 3/8 times
the ultimate stress. Higher design stresses are permitted by API 650 but
with restrictions as to material and thickness.

All codes require stress relief above a limiting plate thickness: e.g. 30
mm for BS 2654. Since it is virtually impossible to stress-relieve a completed tank, this limits the thickness of the lowest course. Some nozzle
connections and the manhole in the bottom course may exceed the
limiting thickness because of the need for reinforcement: here the nozzle
and plate are supplied as a pre-fabricated stress-relieved assembly.

There are two main types of ambient temperature atmospheric storage
tanks, fixed roof and floating roof, and these are illustrated in Fig. 5.46.
The floating roof design eliminates the potentially hazardous roof
space but is mechanically more complex. The wall thickness of each
course or tier is sized such that the stress is about constant except for
the top one or two courses where stability dictates a minimum thickness, typically ⅜ in (9 mm). Large tanks may require high tensile steel
for the lower courses to keep within thickness limitations.

Plate for the cylindrical shell is structural grade, for example ASTM

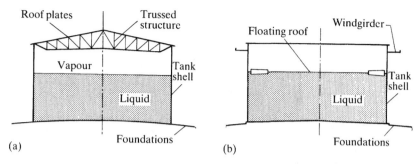

5.46 Cylindrical storage tank design: a) Fixed roof; b) Floating roof.[29]

A283 or EN 10 025 Fe 430. The higher tensile grade to BS 2654 is BS 4360 50D, but API 650 (and possibly BS 2654 in the future) allows the use of quenched and tempered steel. The British Standard calls for impact testing of both plate and weld metal, whilst API 650 requires impact testing for higher stress design and high tensile steel.

Welding of vertical seams is as a rule carried out with coated electrodes, but where impact properties allow, and for thickness over say 20 mm, it may be possible to use electrogas welding for the bottom course. The circumferential welds are made with coated electrodes or by GMA welding. Submerged-arc welding may also be employed. After a course has been tacked in position and the vertical welds completed, a special purpose carriage is mounted on top of the plates. An outside and an inside head, with platforms for the operators, are suspended from the carriage, which then travels round the circumference of the tank. Using a simple dam to support the flux, the submerged-arc process operates well in the horizontal-vertical position. Figure 5.47 shows some typical joint details for vertical and circumferential joints.

Non-destructive testing consists of radiography of a proportion of T joints: that is to say, joints between vertical and circumferential welds. In addition the vertical welds, or a proportion of them, may be examined by radiography.

The tank bottom is made of ¼ in (6 mm) thick carbon steel plates that are overlapped and fillet welded. Where three plates overlap the top plate is joggled, as shown in Fig. 5.48. The bottom course of shell plate is fillet welded to a ring of butt welded annular plates, whilst the annular plate is connected to the bottom by means of a 'sketch' plate, as shown in Fig. 5.49.[28]

Welds in the tank bottom are inspected visually and by means of a vacuum box. This is a rectangular box with a glass top. The joint is painted with soap solution and the device is sealed over the area to be tested. The box is partially evacuated and any leaks show up as bubbles in the soap film.

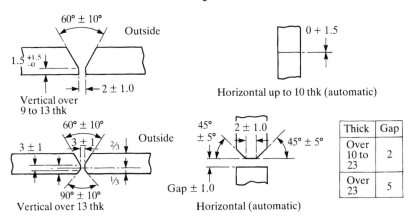

5.47 Joint details for tank shells.[28]

The most severe loading conditions occur during hydrostatic testing of tankage, since water is more dense than hydrocarbons. The maximum stress is at the fillet weld between the shell and the annular plate, where yielding occurs. The degree of overstress depends upon the code but is at least 125% of the normal operating stress. As in other cases such as the expanding of line pipe and pressure testing of boilers and vessels, this process preconditions the structure such that, barring corrosion, mechanical damage, repair welding and the like, it will always be subject to lower stresses in service. This conditioning process, also known as 'shakedown' is an important means of promoting safe operation. Indeed, cylindrical tanks working at ambient temperature have a good service record.

Cylindrical tanks operating at subzero temperatures
There are two ways in which gases like ammonia, propane and methane can be stored economically. These are indicated in Fig. 5.50, which shows the pressure at which ammonia becomes liquid as a function of temperature.[30] The practical alternatives are either to refrigerate to −33.5 °C and store at atmospheric pressure, or to pressurise up to nearly ten atmospheres and store at room temperature. Intermediate conditions are possible and are sometimes used.

Present concern is with refrigerated cylindrical tanks operating at atmospheric pressure. These may be similar in construction to the fixed roof design of crude oil tanks, with a lap welded bottom, butt welded shell and domed roof, but in addition they are insulated and connected to a refrigerating system (Fig. 5.51).

Cold storage is covered by API 620, which has various appendices relating to different temperature levels, by BS 4741 for temperatures down to −50 °C and by BS 5387 for temperatures down to −196 °C.

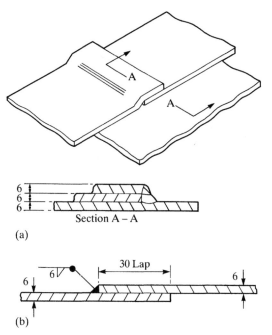

5.48 Lap welding of tank bottom showing section where three thicknesses overlap.[28]

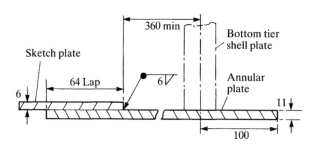

5.49 Tank bottom: joint between sketch plate and annular plate.[28]

Traditionally, the material used for service down to −50 °C is impact tested carbon steel. However, as noted earlier, the failure of liquid propane storage tanks at Qatar in 1977 led to a re-appraisal of requirements and an increase in the level of notch-ductility. Therefore alternatives to BS 4360 Grades EE or F have been sought. Japanese steelmakers are offering a 2½% nickel grade in the thermomechanically rolled or quenched and tempered condition for this duty. The intention is to provide a steel that is capable of arresting a running crack. It is difficult to determine whether this is so because there is no generally agreed test

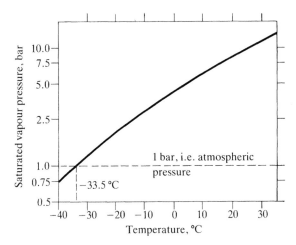

5.50 Relationship between temperature and pressure for equilibrium between liquid ammonia and its vapour.[30]

for crack arrest.[29] Furthermore reliance on material quality alone does not provide a sufficient assurance of safe operation. Account must be taken of the possibility of accidental impact or of terrorist attack; also of the potentially lethal consequences of a large scale spill of liquid ammonia or liquefied hydrocarbon gas. Therefore in recent years refrigerated tanks have in many cases been surrounded by a reinforced or prestressed concrete wall, capable of containing the complete contents of the tank should it fail. Such arrangements have been taken a step further in the recent EEMUA recommendations sketched in Fig. 5.52.[31] In this the concrete wall is protected by an earth embankment and the insulated roof extends over the wall. This design, combined with material selection in accordance with BS 4741, API 620 Appendices R and Q is to be commended.

Welding of materials for −50 °C duties is normally by MMA with 1% or 2½% nickel alloy electrodes, as required to meet impact specifications. Material for −196 °C service is either 9% nickel steel, aluminium alloy or austenitic stainless steel. In practice, economics favour 9% nickel steel. This is welded using nickel base electrodes, selected to match the tensile properties of the parent metal.

Spheres and bullets
Pressure storage at atmospheric temperature is used for all the liquefiable gases. Special problems arise with ammonia because at room temperature (but not at −33 °C) it may cause stress corrosion cracking of ferritic steel. These problems are considered later in this chapter, in the section on ammonia plants.

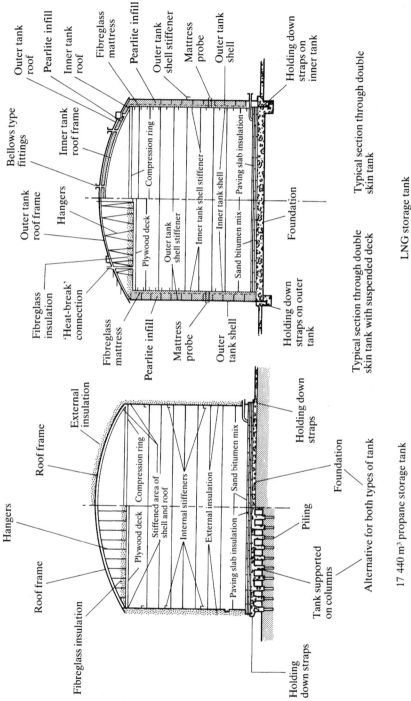

5.51 Refrigerated storage tank construction.[29]

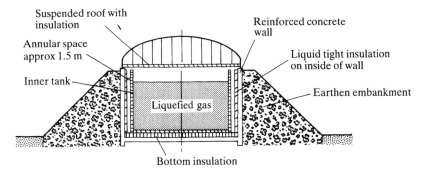

Suspended roof with insulation

Annular space approx 1.5 m

Inner tank

Reinforced concrete wall

Liquid tight insulation on inside of wall

Earthen embankment

Liquefied gas

Bottom insulation

5.52 Engineering Equipment Manufacturers and Users Association recommendation for safe containment of liquefied gases.[31]

Storage spheres are employed for city gas in Japan, where they are usually fabricated from quenched and tempered steel. A typical alloy is HT 60 CF (crack-free) of which the typical percentage composition is:

C:0.06,Si:0.22,Mn:1.3,P:0.012,S:0.005
Cr:0.16,Mo:0.16,V:0.03

and mechanical properties are:

Yield stress 562 N/mm²; UTS 626 N/mm²;
Elongation 27%; Charpy V energy at −40 °C = 167 J.

This is welded with low hydrogen electrodes of matching strength. It is claimed that HT 60 CF does not suffer hydrogen cracking in the heat affected zone when welded without preheat. However, fine cracking can occur in the weld metal so a preheat of 50 °C is recommended.[32]

Special welding problems associated with individual processes

Ammonia

Problem areas
Figure 5.53 is a process flow diagram for a high output single train ammonia plant.[33] The hydrocarbon feed (usually methane) is desulphurised and passes through reformer tubes where, mixed with steam, it is converted to a mixture of hydrocarbon, CO, CO₂ and hydrogen. Air is injected in the secondary reformer, to add nitrogen and convert CO to CO₂. The CO₂ is then absorbed and a hydrogen plus nitrogen synthesis gas is formed. This passes over a catalyst in the converter to form ammonia which goes to storage.

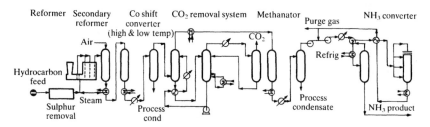

5.53 Process flow diagram for a single train Kellogg low pressure ammonia synthesis unit .[33]

For welding, the chief areas of concern are the primary reformer, the ammonia converter and the storage containers, particularly those operating at or near to room temperature. Problems with the converter do not differ significantly from those of heavy-wall pressure vessels in general, as discussed earlier.

Reformer furnace

A top-fired furnace with catalyst tubes welded directly to the outlet manifold is shown in Fig. 5.54. Furnaces may also be bottom or side-fired, and the catalyst tube is often attached to the outlet manifold via a pigtail. This is a short length of wrought Incoloy 800 tube which can flex to accommodate movements due to expansion. Typical operating conditions are: pressure 400–500 psi (2.75–3.4 N/mm²) and temperature 1275 °F (670 °C) at the inlet and 1600 °F (870 °C) at the centre but design temperature can go up to 1750 °F (950 °C). Tubes are 62.5 mm to 125 mm inside diameter and up to 25 mm thick.

Materials suitable for such duties are listed in Table 5.15. The catalyst tubes are directly fired and operate at the highest temperatures; these are centrifugally cast. Headers can be insulated and here the lower strength wrought alloys such as Incoloy 800 H are suitable.

The catalyst tubes consist of several centrifugal castings joined end-to-end by circumferential welds. As cast, the inner 1–2 mm of metal is porous, so this is either machined out or counter-bored adjacent to the weld preparation, as in Fig. 5.42. The root pass is a GTA weld (with filler wire or fusible insert) and weld out is with coated electrodes or automatic GTA. The tube material originally used was HK 40, and there were a number of failures of welds made with coated electrodes. The rods were made with a 25Cr20Ni core wire of commercial purity, and the creep strength of the weld metal was about half that of the cast material, due to the presence of sulphur, phosphorus or other impurities. No such problems occurred with the automatic GTA welds, which are made with matching filler wire.

More recently the modified HP (Manaurite 36X) type alloy has been

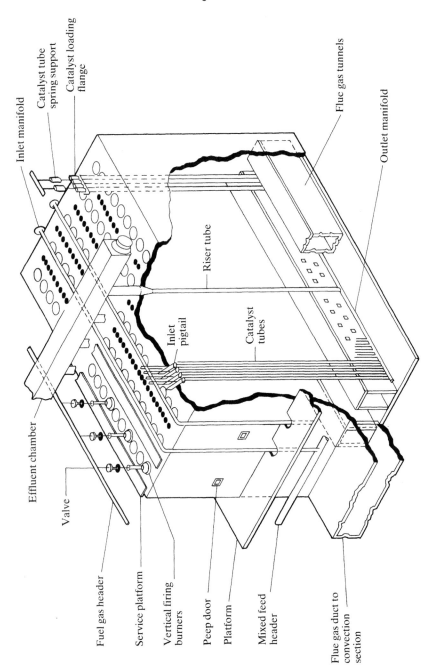

5.54 Top-fired steam-methane reformer furnace.

Table 5.15 Cast, forged and wrought alloys for service at 800–1200 °C in petrochemical plant[33]

Alloy designations*		Chemical composition (approx values, %)								Stress to produce rupture in 1000,000 h Stress, psi			
		C	Si	Mn	Cr	Ni	Cb	W	Other	1400 °F	1600 °F	1800 °F	2000 °F
ACI†, others	Cr/Ni, others												
		Cast alloys											
HK-40	25/20	0.40	1.50	1.50	25	20	–	–	–	4 500	2 250	875	–
IN-519	24/24Cb	0.35	1.50	1.50	24	24	1.5	–	–	6 550	3 050	1 160	–
HP Mod. 36X	25/35Cb	0.40	1.50	1.50	25	35	1.5	–	–	6 550	3 650	1 450	360
HP Mod. 36XS	25/35WCb	0.45	1.50	0.70	25	35	present	present	–	–	3 500	1 310	350
G4868	30/30Si	0.55	2.00	1.30	30	30	present	–	–	5 250	2 500	975	320
NA 22 H	28/48W	0.45	1.50	1.50	28	48	–	5.0	–	–	3 200	1 450	440
Supertherm	28/35WCo	0.50	1.20	1.20	28	35	–	5.5	Co = 15	–	3 650	1 600	610
H110	30/33W	0.55	1.20	1.20	30	33	–	4.5	MT‡	–	2 700	1 300	570
Transfer lines and manifolds													
		Cast and forged alloys											
800 H cast	20/32Cb	0.10	1.50	1.50	20	32	1.0	–	–	5 100	2 200	680	–
800 H forged	20/32	0.10	1.00	1.00	20	32	–	–	Al+Ti<0.7	4 200	1 600	490	–
AISI¶, others	Cr/Ni, others												
		Wrought alloys											
310 SS	25/20	0.15	0.75	2.00	25	20	–	–	–	4 000	1 600	–	–
800	20/32	0.05	0.35	0.75	20	32	–	–	Al+Ti<0.7	5 200	2 100	620	–
800 H	20/32	0.10	0.35	0.75	20	32	–	–	Al+Ti>0.85	8 300	3 200	765	–
802	20/32	0.35	0.75	0.75	20	32	–	–	Al=0.5, Ti=0.75	–	–	950	–
800 DS	18/37	0.15	2.20	0.75	20	37	–	–	Al=0.3, Ti=0.3	–	–	–	–

* The following tradenames are included:
Manaurite 36X and 36XS (FAM, Pompey)
Maerker G4868 and H110 (Schmidt und Clemens GmbH)
Na22H and Supertherm (ABEX)
Incoloy Alloy 800, 800H. 802, 800Ds (INCO)
IN 519 (INCO)

† Steel Founders Society of America (Des Plaines. Ill), formerly Alloy Casting Institute
‡ Metallurgical treatment performed
¶ American Iron and Steel Institute (Washington. DC)

used for catalyst tubes in the UK and in Continental Europe. The higher creep properties of this alloy allows the use of a thinner tube. There have been no problems due to preferential cracking of welds in this or in the other more highly alloy materials listed in Table 5.15; indeed there have been cases where failures were arrested by a weld.

The high carbon centrifugally cast alloys have low ductility at room temperature in the as-cast condition and are further embrittled in service. Repair welding of tubes may therefore present difficulties due to cracking. In one case it proved necessary to heat treat the used tubes at 1225–1275 °C for one hour followed by rapid cool prior to welding.[34] In other locations however extensive repairs have been made without heat treatment. The wrought alloys used for headers and transfer lines suffer only minor embrittlement and provided that cracking is not extensive may be repair welded without difficulty.

Weld inspection is primarily by radiography, supplemented by dye penetrant testing. As with austenitic chromium-nickel steels of the 300 series, ultrasonic examination of welds is not possible due to the coarse grain of the weld metal, which reflects the ultrasonic beam and obscures the reflections from defects.

Storage

Atmospheric pressure, atmospheric temperature and intermediate regime storage is illustrated diagrammatically in Fig. 5.55.[35] Refrigerated cylindrical tanks are used for large scale storage over about 5000 t whilst spheres and bullets are employed for intermediate and atmospheric temperature storage. Stress corrosion cracking may occur in bullets and spheres for two reasons; firstly because of the higher operating temperature and secondly because such vessels are periodically emptied and refilled and so are more subject to oxygen contamination. The cracking is often very fine and difficult to detect with certainty. It occurs preferentially near welds and in cold worked material; it propagates slowly outwards and in one case was reported to have contributed to the failure of a road tanker.[30] More often the discovery of cracking requires decisions about repair, stress-relieving and so on.

It has been demonstrated by laboratory tests that ammonia stress cracking is inhibited by the addition of 0.2% of water, and it is normal practice for bulk suppliers to make such additions. However cracking has been found in spheres even when the ammonia was inhibited, possibly because the water is depleted in the vapour phases, and possibly because the water content was not to specification. Therefore water additions cannot be relied upon to prevent cracking.

Reducing the tensile strength of the steel reduces residual stress and is therefore beneficial. Following accidents with road tankers made of ASTM A 517 Grade F (Tl) steel, the use of quenched and tempered steel has been banned for road transport in some countries. In general,

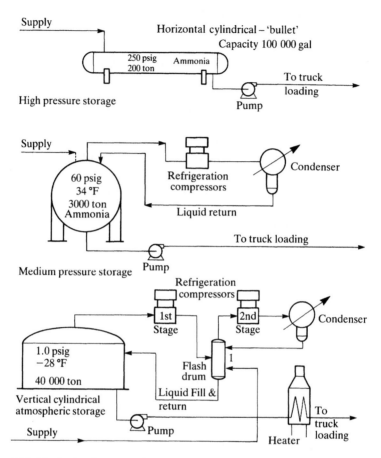

5.55 Methods of storage for ammonia (reproduced by permission of the American Institute of Chemical Engineers. c 1974 AIChE).[25]

codes of practice call for stress relief where this is practicable, as for road tankers and bullets. In the case of spheres it is recommended in the UK that steels conform to the requirements of BS 5500, in particular for low temperature operation, that the yield strength should not exceed 350 N/mm², and that weld metal strength should be minimised. Steels should have sufficient notch-ductility to withstand the presence of at least shallow cracks, and PD 6943 (see Chapter 6) may be used to assess the failure risk.[30]

Inspection for cracks is carried out using the fluorescent magnetic particle method with an AC yoke, and periodic inspection of spheres and bullets is recommended. However, experience shows that not all cracks are disclosed by this technique so that the main effort needs to be directed towards prevention.

Hydrocracking and coal liquefaction

General

These two processes are lumped together because they are basically similar in character and because the critical item in both cases is a 2¼Cr1Mo heavy-wall vessel clad with austenitic chromium-nickel steel. However coal liquefaction processes of the type requiring a high pressure reactor, and which like hydrocracking include hydrogenation as an essential step, will only become economically viable if there is a serious oil shortage. Gasifiers at the Sasol oil-from-coal plants in South Africa are relatively low pressure vessels where coal is burnt in a mixture of air and steam, and do not fall in the same category.

Hydrocracking is a means of converting heavy refinery products, such as vacuum residues or creosote, into lighter fractions which are generally more useful. Heavy oils are used for firing boilers in ships and power stations but in some locations this outlet may not be economic.

Operating conditions in hydrotreaters vary according to feedstock and process route but pressures are frequently up to 25 N/mm² and temperatures up to 460 °C. The process fluid is a mixture of hydrocarbon, hydrogen and H_2S, and the material of construction must resist corrosion by the hydrogen/H_2S component. Coal liquefaction plants, if and when they are constructed, will operate under similar conditions but probably at higher temperatures, up to 500 °C.

Reactor design and fabrication

Figure 5.56 is a simplified process flow diagram for a hydrocracker upstream of the fractionation section,[26] whilst Fig. 5.57 shows a typical design for a solid wall hydrocracker reactor. The base material is 2¼Cr1Mo steel, which for the shell is in the form either of plate or forged rings. Thinner sections are normalised and tempered, thicker sections quenched and tempered to give a room temperature yield strength of 350–450 N/mm². The inner surface is weld-deposit clad with type 347 stainless steels for most applications but where organic acids are present the cladding is type 316L. Internals match the cladding.

The circumferential welds may be made using a conventional submerged-arc welding procedure with a preheat of 150 °C and interpass temperature of 200 °C. The weld is completed from the inside by back-chipping and manual welding with coated electrodes. However there is increasing use of narrow gap welding for such applications, particularly by Japanese fabricators. Figure 5.58 shows the weld preparation, which is parallel-sided. The first few passes on to the backing ring are made with a single electrode and two electrodes in tandem are used for the filler passes. On completion of the joint the

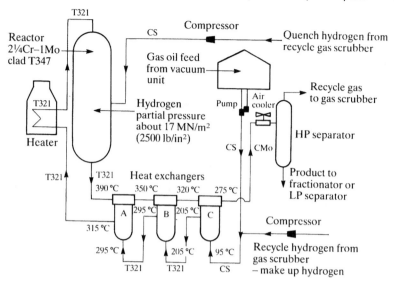

5.56 Simplified process flow diagram for a hydrocracker.[26]

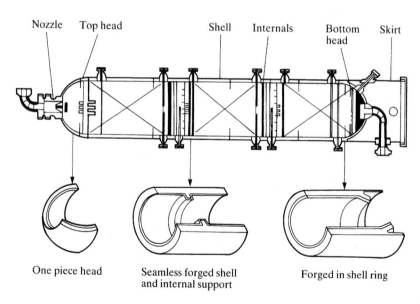

5.57 Typical hydrocracker reactor vessel.[36]

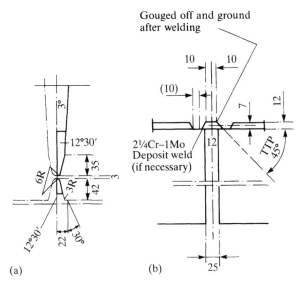

5.58 Hydrocracker reactor weld preparations: a) Longitudinal seam;
b) Circumferential seam (parallel-sided narrow gap weld made from
one side with backing plate).[36]

backing bar is removed by air-arc gouging, the surface ground flush
and clad with stainless steel. Such welds are made without an intermediate heat treatment, but are subject to a dehydrogenating treatment
at 300–400 °C for 2 hours after welding.

Cladding is by submerged-arc welding using either multiple rod or
strip electrodes, and by MMA for areas not accessible to the
submerged-arc process. GMA and GTA welding may also be used, for
example for flange facings.

The final post-weld heat treatment is in accordance with code
requirements and job specifications, and is typically in the range 630–
680 °C. Procedure testpieces are subject to all intermediate plus final
heat treatments, and in some instances with additional treatment to
allow for repair welding after a period of service. Such heat treatment
is beneficial to welds in the low alloy steel. The cladding may however
be embrittled in two ways. Firstly, any ferrite present in the stainless
steel deposit may be partly converted to sigma phase. Both types 347
and 316L deposits normally contain 3–10% ferrite. Type 309
(23Cr13Ni) however is less susceptible to sigma phase formation and is
often used for the first layer of cladding. Secondly carbon diffuses
from the ferritic alloy steel into austenitic deposit, resulting in carbide
precipitation and embrittlement adjacent to the ferrite-austenite
interface.

Service problems that have occurred with hydrocrackers, particularly with earlier generations, include embrittlement and cracking of

the 2¼Cr backing, and cracking and disbonding of the overlay cladding.

The first hydrocracker reactors were fabricated in the US from 2¼Cr1Mo steel quenched and tempered to give a yield strength of about 700 N/mm². These vessels suffered cracking at nozzle and main seam welds after a relatively short period of service. The cracking was thought to be hydrogen-assisted, and to be associated with the high yield strength. Subsequently the yield strength for hydrocracker reactor steel has been maintained in the range 350–450 N/mm², and this type of cracking has not recurred. However there is a current trend towards the development of 2¼Cr and 3Cr steels with properties enhanced by vanadium, titanium, niobium and boron additions, as noted earlier, and careful testing will be required to validate their use in high pressure hydrogen service.

Regardless of the levels of yield strength, 2¼Cr1Mo steel may suffer temper embrittlement when held for a period of time within the temperature range 375–575 °C. Temper embrittlement increases the impact transition temperature, and the susceptibility of a steel to this deficiency is measured by the increase in either the 50 J or the 50% fracture appearance transition temperature after exposure to step cooling. A typical step cooling programme is shown in Table 5.16.

Table 5.16 Typical step cooling programme to simulate temper embrittlement in service[36]

Holding temperature		Holding time, hours	Cooling rate	
°F	°C		°F/hr	°C/hr
1100	593	1	10.6	5.6
1000	538	15	10.6	5.6
975	524	24	10.6	5.6
925	496	60	6.0	2.8
875	468	100	50.0	27.8

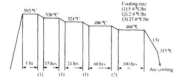

Temper embrittlement is made worse by increasing contents of silicon and manganese, and by the presence of tramp elements, in particular P, As, Sn and Sb. Silicon contents are minimised by vacuum deoxidation (which would normally be practiced for heavy plate steel), whilst typical limits for tramp elements are:

P   0.012% max   Sn 0.018% max
As 0.025% max   Sb 0.004% max

This type of embrittlement is due to the segregation of tramp elements to the prior austenite grain boundaries. It may be removed by heating for a period of time at 600 °C. Such de-embrittling treatments are not used in practice for hydrocracker reactors. The degree of temper embrittlement must be estimated when carrying out a reliability analysis of a 2¼Cr1Mo vessel.

Cracking in the overlay welding is due to a combination of embrittlement due to dissolved hydrogen and sigma phase formation and tensile stress. The tensile stress, which is close to yield point levels, is present after placing the weld deposit and is not removed by post-weld heat treatment. In addition there is a service loading on catalyst support brackets. In early reactors these consisted of rings welded to the inside of the cladding, and were particularly disposed to crack; more recently 2¼Cr1Mo rings are welded to the ferritic alloy shell and then the whole is clad. Such an arrangement is much less vulnerable.

Disbonding is likewise caused by a combination of hydrogen and metallurgical embrittlement. Two types of cracking have been identified, depending on the microstructure at the interface. Type I occurs when there has been relatively high dilution of the cladding near the interface, such that a martensite layer is formed in the carbon migration region. Type II cracks are associated with low dilution, and form intergranularly in the austenite adjacent to the interface. Type II structures are much more prone to disbonding than Type I, and therefore cladding procedures that favour dilution at the interface are preferred.

Disbonding occurs in round or oval shaped patches which grow with time. The rate of spread is greater with reactors that are subject to frequent shutdowns. Although on the face of it this would appear to be a severe condition, there have been no records of any catastrophic results such as collapse of internal supports.

Fabricators of hydrocracker reactors are required to carry out disbonding tests over a range of temperatures and pressures, representing startup and shutdown conditions. Such tests must be sufficiently long in duration to ensure that the testpiece is fully saturated with hydrogen.

The use of PD 6493 to assess the integrity of a hydrocracker reactor is considered in Chapter 6.[36]

Urea

Urea or carbamide is a nitrogenous compound which is an effective fertiliser for certain soils and climatic conditions. Its chemical formula is H_2NCONH_2 and it is made by reacting carbon dioxide and ammonia at about 200 °C and 3000–3300 psi (about 20 N/mm^2) pressure. The product of this reaction is ammonium carbamate $H_2NCO_2NH_4$ which is converted into urea by removal of water. The solution is concentrated and extruded into a prilling tower where granules or prills of dried urea are formed.

The welding problem in urea manufacture is due to the fact that ammonium carbamate is highly corrosive. It has a strong reducing action such that where shielded or stagnant conditions occur the protective oxide film on austenitic stainless steel may be damaged. It is necessary to inject air or oxygen to ensure the maintenance of a passive film when stainless steel is used for lining critical vessels. Moreover any ferrite in plate or weld metal is selectively attacked, so that the welds must be, so far as possible, fully austenitic.

The two critical vessels are the reactor and the stripper. These are illustrated, together with some structural details, in Fig. 5.59. The vessel and stainless liner shown here are as specified by Stamicarbon. The stainless material originally used for the liner was type 316L, ferrite-free. However, in recent years the general preference has been for a 25Cr22Ni2Mo type, which has better resistance to carbamate attack, increases service life and gives better operational flexibility.

The problem of obtaining a ferrite-free weld deposit that is free from serious cracking has been considered earlier. The main requirements are:

> Low carbon (below 0.1%);
> Low silicon (0.3–0.4%);
> Low phosphorus and sulphur (below 0.015%);
> Molybdenum between 2 and 3%;
> Nitrogen between 0.1 and 0.2%;
> Freedom from impurities and trace elements.

The nature of the coating is also said to play a part in avoiding fissures. The compositions of plate and weld metal are listed in Table 5.17. Stringent tests are applied to both strip and coated electrode overlays, as shown in Fig. 5.60. This is the testpiece used by Kellogg Continental[37] and from it samples are taken for microexamination, mechanical testing and corrosion testing. It is accepted that a fissure-free deposit is not obtainable, but subsurface cracks are limited to a length of 0.6 mm and the number is limited, for example in one bead to not more than eight. Cracking is also indicated by a low elongation in the tensile test, and a minimum acceptable value is 30%.

The corrosion test used for urea is the Huey test, in which a specimen is immersed for five consecutive periods of 48 hours each in boiling nitric acid. The average weight loss must not be greater than that corresponding to a wastage of 1.5 microns/48 hour. The samples are examined for selective attack, and this is specified to be not greater than 100 microns deep. The results obtained with the BM 310 MoN electrode were maximum 0.74 microns/48 hour for the Huey test and selective

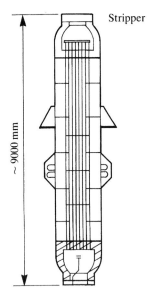

Stripper

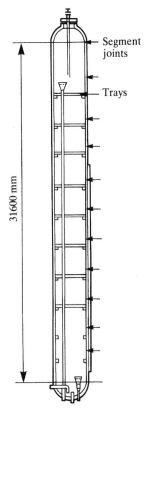

Reactor
Lined with 8 mm thick sheets
in 25-22-2, joined to side
wall at segment joints

Segment
joints

Trays

31600 mm

Stripper
dome

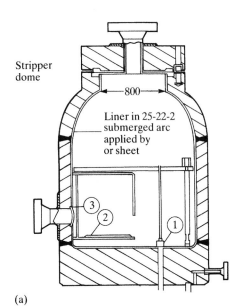

~9000 mm

800

Liner in 25-22-2
submerged arc
applied by
or sheet

(a)

5.59a) Urea reactor and stripper (Stamicarbon type) showing details
of cladding and nozzle connection.[37]

① 2800 tubes in 25-22-2 material
   installed in stripper

② Cladding transition in dome bottom

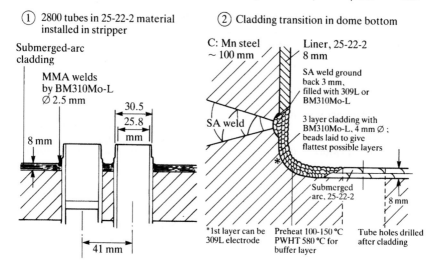

Submerged-arc
cladding

MMA welds
by BM310Mo-L
Ø 2.5 mm

30.5

25.8

mm

8 mm

41 mm

C: Mn steel
~ 100 mm

Liner, 25-22-2
8 mm

SA weld ground
back 3 mm,
filled with 309L or
BM310Mo-L

3 layer cladding with
BM310Mo-L, 4 mm Ø ;
beads laid to give
flattest possible layers

SA weld

Submerged
arc, 25-22-2

8 mm

*1st layer can be
309L electrode

Preheat 100-150 °C
PWHT 580 °C for
buffer layer

Tube holes drilled
after cladding

③ Example of welding HP nozzles
   with C:Mn steel body, clad or lined

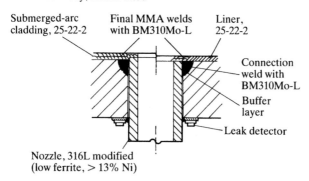

Submerged-arc
cladding, 25-22-2

Final MMA welds
with BM310Mo-L

Liner,
25-22-2

Connection
weld with
BM310Mo-L

Buffer
layer

Leak detector

Nozzle, 316L modified
(low ferrite, > 13% Ni)

Welding of loose lining sheets
at side joints of reactor segments

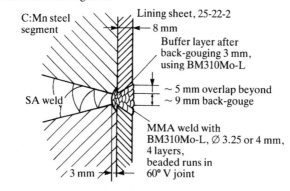

C:Mn steel
segment

Lining sheet, 25-22-2

8 mm

Buffer layer after
back-gouging 3 mm,
using BM310Mo-L

~ 5 mm overlap beyond
~ 9 mm back-gouge

SA weld

MMA weld with
BM310Mo-L, Ø 3.25 or 4 mm,
4 layers,
beaded runs in
60° V joint

(b)

3 mm

5.59b) Details of connections in the head of a urea stripper.

Table 5.17 Fully austenitic chromium-nickel alloys for cladding urea reactors

Material		Composition % max or range								
		C	Cr	Ni	Mo	Mn	Si	N	P	S
316L	(UHB 724L)	0.03	17.5 nom	13.5 min	2.6 nom	2.0	1.0	–	0.045	0.03
310MbN	(UHB 725LN)	0.02	24.5– 25.5	21.5– 22.5	1.9– 2.3	1.5– 2.0	0.4	0.1– 0.14	0.02	0.015
Weld deposit	FILARC BM 310 MoN	0.04	24– 26	19– 23	1.9– 2.4	3.0– 5.5	0.5	0.1– 0.2	0.015	0.015

attack was less than 50 microns.[37] In general service behaviour with stainless cladding is good but not perfect; pitting or wastage may occur from time to time due for example to the build-up of deposits on the metal surface.

Although the pressure in a urea reactor is high the temperature is modest and there is no hydrogen. Therefore a solid wall carbon-manganese shell may be used. Longitudinal welds in the liner are made in the shop and the rings are inserted into the shell. Circumferential welds in the liner are made as shown in Fig. 5.59. The heads are clad by a combination of strip submerged-arc welding and manual metal arc.

In the stripper the process fluid flows through the tubes, and cladding is required only for the heads. Details of seal welding of tubes, cladding of dome and welding of nozzles are shown in Fig. 5.59b).

Titanium is also used for the corrosion-resistant linings of urea reactors. Most vessels lined in this manner have been of multilayer construction. In one such type a core tube is formed in carbon steel but not welded. The layers of thin quenched and tempered steel are then longitudinally welded successively until the required thickness has been obtained. Weep holes are bored through the layered cylinder to the surface of the core tube, to drain off or to detect any leakage through the liner. In some cases a large hole is bored in the location where the weep hole is required; this is filled with weld metal and the final bore is made through the weld metal. In this way any liquid that escapes is not allowed to penetrate between the layers. The holes are connected to a piping system which is purged and may incorporate detectors (Fig. 5.61). Detectors may also be installed in the weep holes.

Problems with titanium liners have been discussed by Krystow.[38] Corrosion and localised cracking has occurred due to iron contamination. Corrosion of the iron results in hydrogen embrittlement of the titanium, with the formation of star shaped cracks and, ultimately, leakage. The iron contamination was thought to have occurred during fabrication, and it is recommended that such work be carried out

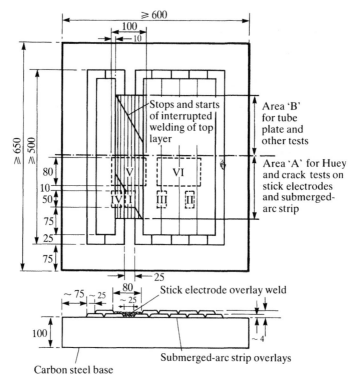

5.60 Testplate for fully austenitic stainless steel cladding of urea reactor. Submerged-arc strip and coated electrode weld deposits.[37]

under the same type of clean conditions that are used for nuclear work.

Liners have been inserted in the same manner as for stainless steel, described earlier, with the circumferential seams made inside the vessel. Titanium strip was then welded over the joint, in the manner shown in Fig. 5.39. Anodising in 1% ammonium sulphate has been used as a means of removing any traces of iron and strengthening the oxide film.[38]

Power boilers and steam plant

General

Steam boilers are required for the generation of high pressure steam to drive turbogenerators in central power stations or they may be used to supply steam for process requirements, for example in steam-methane reforming. In some cases, as at the Sasol plant in South Africa, the boilers supply steam for both process requirements (primarily coal gasification) and for power generation. The general design and layout

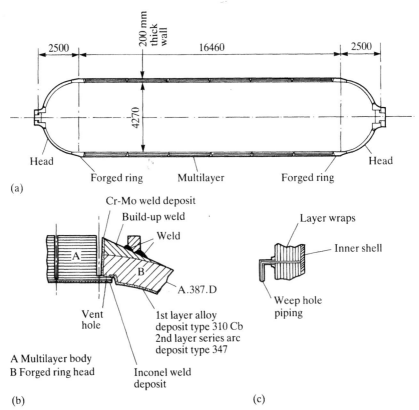

(a)

(b)

(c)

5.61 Multilayer pressure vessel, showing details of head to body connection and weep holes: a) Typical multilayer reactor; b) Shell to head connection, vented to avoid hydrogen build-up in weld metal; c) Typical weep hole connection for multilayer vessel.[39]

of boilers for process steam and power generation is similar but may differ in detail; for example, in ammonia plant steam is generated also in other pieces of equipment and fed to the steam drum; consequently the number of connections to the drum is greater and this may increase the build-up of stress during fabrication.

A steam power station contains many thousands of welds and the failure of just one can result in a shutdown of 2–3 days, so that good weld quality is of great importance.

The maximum attainable efficiency η of a steam engine is:

$$\eta = \frac{T_i - T_o}{T_i} = 1 - T_o/T_i \qquad [5.1]$$

where T_i is the steam inlet temperature and T_o the exhaust temperature, all in K. Therefore, since T_o is necessarily close to atmospheric temperature, high efficiency is obtained by increasing the steam inlet temperature to the turbines. The degree to which such an increase is economic depends upon the price and availability of fuel. High temperatures and pressures make heavy demands on materials of construction and increase operating risks. Thus, in the 1960s, when oil was cheap, power stations planned in the UK were oil-fired and at the Littlebrook 'D' station the turbine inlet temperature was 538 °C at a pressure of 16.6 N/mm².[40] However the US Department of Energy is sponsoring development work on materials suitable for steam temperatures up to 650 °C and pressures up to 35 N/mm² on the grounds that more efficient power plant will be required in the 1990s and beyond.[41]

Plant layout

A cross-section of a typical oil-fired power station is shown in Fig. 5.62, whilst Fig. 5.63 is a section of the boiler house. Starting with condensed steam below the low pressure turbine the circuit is as follows: condensate is first de-aerated and then pumped through the high pressure preheater and economiser to the steam drum. From here it flows by downcomers to the lower headers of the 'water walls'. These are banks of parallel tubes that surround the combustion chamber. Here the water is partially vaporised, forming a steam-water mixture. This passes to cyclones in the steam drum which separate the two phases; water is recycled and steam passes to the superheaters and thence to the high pressure end of the turbines. In the case illustrated, which is the Littlebrook 'D' power station near Dartford, UK, the pressure is now 17.5 N/mm² and the temperature 541 °C. Steam exhausts from the low pressure turbines at 61 mbar absolute (6.1 × 10⁻³ N/mm²) and is condensed. Between turbine stages the steam passes back to the furnace to be reheated (Fig. 5.64 shows the reheat coils). Combustion air is taken from the top of the boiler house and is preheated by exhaust flue-gas, which then flows to the stack. In the case illustrated residual oil from the tank farm is also preheated before burning in the combustion chamber.

A more frequent cause of shutdown than weld failure is corrosion. The boiler water is of high purity and free from dissolved gases, so it should be non-corrosive. Upsets can occur however. Also, there may be corrosion from the outside – fireside corrosion. In such a case it may be necessary to shutdown, cut out a short length of one of the water wall tubes, and weld in a replacement. To avoid cooling down the whole boiler and erecting scaffolding on the inside the repair can be made from the outside by welding, through a 'window'. Figure 5.64 illustrates the principle of the technique. A lozenge shape hole is cut in the insert length. Then by a suitable edge preparation it is possible to make a GTA butt weld of the back part of the tube by working through the window, whilst the front is welded from the outside. Finally an

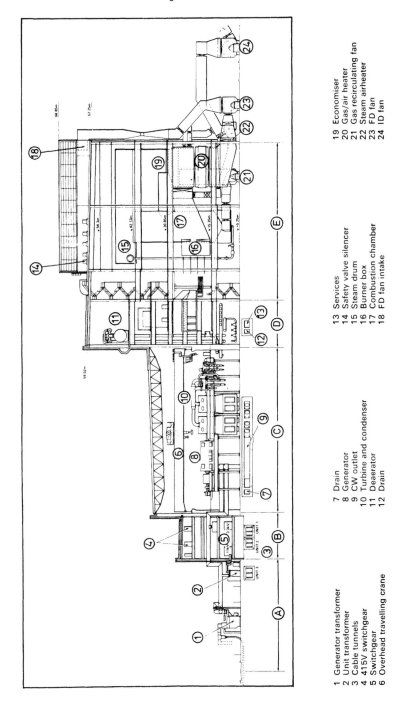

5.62 Cross-section of the Littlebrook 'D' oil-fired power station.[40]

1 Generator transformer
2 Unit transformer
3 Cable tunnels
4 415V switchgear
5 Switchgear
6 Overhead travelling crane

7 Drain
8 Generator
9 CW outlet
10 Turbine and condenser
11 Deaerator
12 Drain

13 Services
14 Safety valve silencer
15 Steam drum
16 Burner box
17 Combustion chamber
18 FD fan intake

19 Economiser
20 Gas/air heater
21 Gas recirculating fan
22 Steam airheater
23 FD fan
24 ID fan

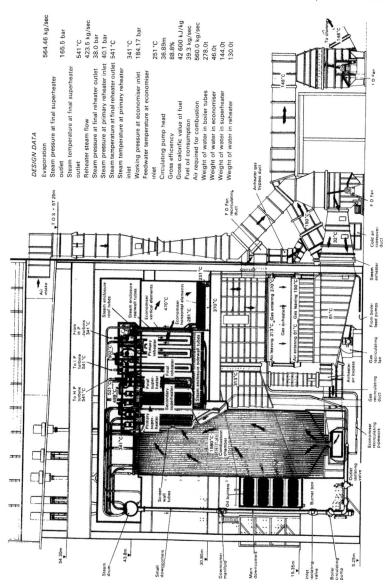

DESIGN DATA	
Evaporation	564.46 kg/sec
Steam pressure at final superheater outlet	165.5 bar
Steam temperature at final superheater outlet	541°C
Reheater steam flow	423.5 kg/sec
Steam pressure at final reheater outlet	38.0 bar
Steam pressure at primary reheater inlet	40.1 bar
Steam temperature at final reheater outlet	541°C
Steam temperature at primary reheater inlet	341°C
Working pressure at economiser inlet	184.17 bar
Feedwater temperature at economiser inlet	251°C
Circulating pump head	36.89m
Gross efficiency	88.8%
Gross calorific value of fuel	42 600 kJ/kg
Fuel oil consumption	39.3 kg/sec
Air required for combustion	560.0 kg/sec
Weight of water in boiler tubes	278.0t
Weight of water in economiser	46.0t
Weight of water in superheater	144.0t
Weight of water in reheater	130.0t

5.63 Cross-section of the boiler for the Littlebrook 'D' power station.[40]

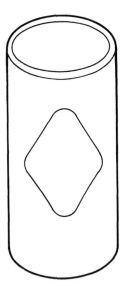

5.64 Tube repair by welding through a window.

insert piece is butt welded to close the window. A specially skilled welder is required for this job.

Materials

The bulk of the tubing is used in the furnace walls, and is carbon steel. For higher temperatures in the superheaters and reheat coils 1Cr½Mo and 2¼Cr1Mo steel is specified. Where temperatures exceed 550–600 °C (the actual breakpoint depends upon pressure and other considerations) it becomes necessary to employ austenitic chromium-nickel steel and a transition joint between 2¼Cr1Mo and stainless steel is required. Main steam lines and furnace headers up to about 560 °C are 2¼Cr1Mo steel except that in the UK the ½Cr½Mo¼V steel is used. Early supercritical power stations required austenitic stainless steel for main steam lines and these were constructed from 18Cr12Ni1Nb steel, which suffered badly from strain relaxation cracking. This material was replaced by type 316H (18Cr10Ni2.5Mo). The steam drum is fabricated from an ASTM or proprietary alloy such as one of those listed in Table 5.11.

9Cr1Mo and 12Cr1Mo steel may find a place in superheater or reheater coils for positions between 2¼Cr1Mo and austenitic chromium-nickel steel. Table 5.18 lists typical piping steels.

Table 5.18 Specifications for power piping

Material	ASME specification		BS specification	
	Tubing	Pipe	Tubing	Pipe
Carbon steel	SA53 SA192 SA210 SA178	SA83 SA106 SA135	BS3059 p⁺ 2 CFS440	BS3602 p⁺ 1 HFS410
1-1¼Cr½Mo	SA213T11	SA335P11 SA369FP11	BS3059 p⁺ 2 CFS620	BS3604 HFS620–440
2¼Cr1Mo	SA213T22	SA335P22 SA369FP22	BS3059 p⁺2 CFS622-440	BS3604 HFS622-440
9Cr1Mo	SA213T9	SA335P9 SA369FP9		
18Cr8Ni	SA213 TP304H SA249 TP304H	SA312TP304H SA376TP304H SA430TP304H SA312TP304H		
18Cr8NiMo	SA213TP3164 SA249TP316H	SA376TP3164 SA430TP316H SA412TP316H		

Experimental alloys that remain to be fully evaluated include the 2¼Cr1Mo¼V-B type noted earlier, and a 9Cr1Mo alloy with vanadium addition (ASME TP 91). TP 91 was developed by Oak Ridge National Laboratory, who found that at 593 °C it gave a 10^5 hour rupture stress of over 100 N/mm², as compared with just under 50 N/mm² for 2¼Cr1Mo. Other research has determined a rupture stress of 80 N/mm² for the vandium treated steel under the same conditions.[41]

Welding and non-destructive testing

The furnace membrane walls (water walls) are made from tubes with an integrally extruded fin at three o'clock and nine o'clock. The fins are edge-welded together in the shop by submerged-arc to form large panels. These panels are welded together on-site to form the combustion box. Tubes are butt welded by GTA welding; where practicable this is done using an orbital machine, otherwise manually. Likewise tube bundles for the economiser, superheat and reheat coils are shop fabricated to minimise site work.

Piping sub-assemblies are welded on-site with a GTA root and weld out by MMA. Figure 5.65 shows weld procedure specifications for 2¼Cr1Mo tubes and ½Cr½Mo¼V pipe that were used for the Littlebrooke 'D' power station. The filler rod and coated electrodes for CrMoV pipe deposit a 2¼Cr1Mo weld metal.

Site butt welds are examined primarily by radiography, usually gamma radiography with Iridium 192. Tube welds are radiographed by the double wall single image technique; that is to say, the isotope source is positioned outside the tube offset from the weld with the film on the opposite side. The complete weld then appears as an oval ring on the developed film. With small diameter tubes two shots may be sufficient but usually multiple shots are required. When the pipe diameter exceeds about 200 mm a panoramic radiograph is made by placing the isotope in the centre of the pipe and wrapping the film around the outside of the weld. In such cases it may be necessary to drill a hole in the pipe wall to insert the gooseneck that carries the isotope. The hole is closed by a threaded and seal welded plug. Lead intensifying screens are used to improve image quality.

Ultrasonic testing is specified for larger diameter pipe particularly in ½Cr½Mo¼V alloy, where there is a possibility of strain relaxation cracking. It may also be employed for tube welds where there is sufficient access. Larger pipe welds are also examined by the magnetic particle technique.

Careful control is maintained over the storage and issue of electrodes but it may still be desirable to check alloy welds by means of a portable spectroscope.

Special problems

Transition joints
Where austenitic stainless steel is specified for part of the superheater and reheat coils, or for larger diameter pipe, it is necessary to make a ferritic-austenitic joint. The simplest technique is to butter the cross-section of the ferritic pipe with austenitic weld metal to match the austenitic pipe and then weld the two together with an electrode of similar composition. Current practice however is to prepare a transition joint as a separate item, and to weld this to the ferritic steel (usually 2¼Cr1Mo) on the one side and the austenitic steel on the other.

Such joints may be produced in a number of ways. Most commonly the two parts are simply welded together using either an austenitic or a

5.65 Welding procedure specifications for a) 2¼Cr1Mo tube; b) ½Cr½Mo¼V pipe (source: NEI).[40]

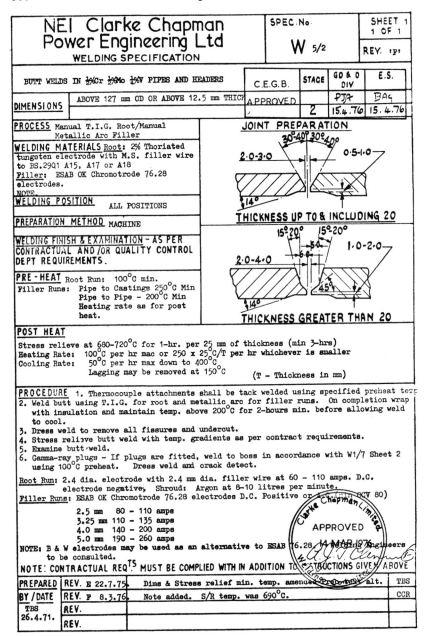

NEI Clarke Chapman Power Engineering Ltd
WELDING SPECIFICATION

SPEC. No.		SHEET 1 1 OF 1
W 5/2		REV. 'F'

BUTT WELDS IN ½%Cr ½%Mo ¼%V PIPES AND HEADERS

	C.E.G.B.	STAGE	GO & O DIV	E.S.
DIMENSIONS ABOVE 127 mm OD OR ABOVE 12.5 mm THICK	APPROVED	2	PJA 15.4.76	BAC 15.4.76

PROCESS Manual T.I.G. Root/Manual Metallic Arc Filler

WELDING MATERIALS Root: 2% Thoriated tungsten electrode with M.S. filler wire to BS.2901 A15, A17 or A18
Filler: ESAB OK Chromotrode 76.28 electrodes.
NOTE.

WELDING POSITION ALL POSITIONS

PREPARATION METHOD MACHINE

WELDING FINISH & EXAMINATION – AS PER CONTRACTUAL AND/OR QUALITY CONTROL DEPT REQUIREMENTS.

PRE-HEAT Root Run: 100°C min.
Filler Runs: Pipe to Castings 250°C Min
Pipe to Pipe – 200°C Min
Heating rate as for post heat.

JOINT PREPARATION

30°-40° 30°-40°
2.0-3.0   0.5-1.0
14°

THICKNESS UP TO & INCLUDING 20

15°-20°   15°-20°
5.0
2.0-4.0   5.0   1.0-2.0
45°
14°

THICKNESS GREATER THAN 20

POST HEAT

Stress relieve at 680-720°C for 1-hr. per 25 mm of thickness (min 3-hrs)
Heating Rate: 100°C per hr mac or 250 x 25 °C/T per hr whichever is smaller
Cooling Rate: 50°C per hr max down to 400°C
Lagging may be removed at 150°C   (T – Thickness in mm)

PROCEDURE 1. Thermocouple attachments shall be tack welded using specified preheat temp.
2. Weld butt using T.I.G. for root and metallic arc for filler runs. On completion wrap with insulation and maintain temp. above 200°C for 2-hours min. before allowing weld to cool.
3. Dress weld to remove all fissures and undercut.
4. Stress relieve butt weld with temp. gradients as per contract requirements.
5. Examine butt weld.
6. Gamma-ray plugs – If plugs are fitted, weld to boss in accordance with W1/7 Sheet 2 using 100°C preheat. Dress weld and crack detect.

Root Run: 2.4 dia. electrode with 2.4 mm dia. filler wire at 60 – 110 amps. D.C. electrode negative, Shroud: Argon at 8-10 litres per minute.
Filler Runs: ESAB OK Chromotrode 76.28 electrodes D.C. Positive or A.C. (min OCV 80)

2.5 mm	80 – 110 amps
3.25 mm	110 – 135 amps
4.0 mm	140 – 200 amps
5.0 mm	190 – 260 amps

NOTE: B & W electrodes may be used as an alternative to ESAB 76.28 ... to be consulted.

NOTE: CONTRACTUAL REQ^{TS} MUST BE COMPLIED WITH IN ADDITION TO INSTRUCTIONS GIVEN ABOVE

(Clarke Chapman Limited APPROVED 14 MAR 1976 ... Engineers)

PREPARED BY / DATE	REV. E 22.7.75	Dims & Stress relief min. temp. amended. Procedure alt.	TBS
	REV. F 8.3.76	Note added. S/R temp. was 690°C.	CCR
TBS 26.4.71.	REV.		
	REV.		

(b)

nickel base electrode. Alternatively a joint with graded composition may be made by electroslag or vacuum arc remelting, to make an ingot that is subsequently bored or forged to match the required pipe size. Most experience is with the first two types. These have a limited life, in the region of 15 000 hours for the austenitic joint and 70 000 for the nickel base type. These figures are to be compared with the design life of the power station, commonly 200 000 hours.

An austenitic-ferritic weld exposed to elevated temperature, either by stress relief heat treatment at say 725 °C, or in service as a transition joint at say 565 °C, suffers carbon migration from the ferrite into the austenite. This has the effect of reducing the creep strength of the ferritic material. At the same time the ferritic side is subject to a tensile strain due to the higher expansion coefficient of austenitic. In a transition joint this strain relaxes in service but not completely; sufficient remains to generate cracking in the coarse-grained region of the heat affected zone on the ferritic side.

The nickel based electrode normally used for a transition joint is Inconel 182, which gives a deposit of normal composition 15Cr10Fe0.5Cu1.75Nb1.0Ti, the balance being nickel. Using this alloy conditions are more favourable because there is no carbon migration across the interface, and the differential expansion is lower. However there is a region of mixed composition close to the fusion boundary. This layer is martensitic in structure, and forms a band 0.5 to 50 microns in thickness. In service at 600 °C carbides ($M_{23}C_6$ and M_6C) precipitate within the martensitic region. Most frequently the precipitates form a single line with a mean size of 0.5 micron or more, and are classified as type I (Fig. 5.66). Where a larger band of martensite is present the precipitates are finer and more dispersed; these are designated type II. Transition joints that have failed in service usually have type I precipitation. The carbides grow until they are large enough to act as crack nuclei. This results in low ductility cracking along a line very close to the fusion boundary on the weld deposit side.[42]

This state of affairs is analogous to the disbonding that occurs in hydrocracker service, where improved resistance to failure is obtained by producing a relatively wide band of mixed composition adjacent to the interface. On this basis transition joints with a graduated composition would be expected to behave well in service.

Strain relaxation cracking

In the UK power generation industry this defect is generally known as 'reheat cracking'. The origin of the term is uncertain but it could refer to cracking in the reheat circuit of early supercritical power stations where the piping was made of the very susceptible 18Cr12Ni1Nb austenitic steel. The metallurgy of the problem is discussed in Chapter 3. In power station practice it is a particularly insidious problem

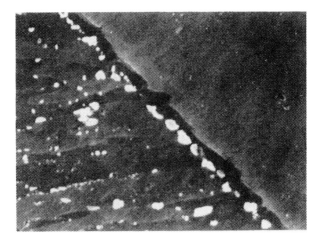

5.66 Precipates at the interface between Inconel 182 weld metal (left) and wrought 2¼Cr1Mo steel (right) after 1600 hours at 630 °C (reproduced by courtesy of The Institute of Materials).[42]

because the size of the cracks that form during fabrication (particularly post-weld heat treatment) may be below the level at which they can be detected ultrasonically. Such cracks may grow by a creep mechanism during service. Figure 5.67 shows the type of cracking defects found in a survey of welds in ½Cr½Mo¼V piping in some UK power stations after service. Cracks were found in about 10% of welds. The majority of these were thought to have propagated from pre-existing cracks.[43] Much work has been done to establish welding procedures that minimise the risk of cracking, and to determine which impurities are most harmful, but ½Cr½Mo¼V steel remains a material that is prone to weld cracking, and is best avoided.

Flue-gas desulphurisation
Sulphur is removed from flue-gas by treating with a slurry of lime in water. The components of this system are at first sight non-corrosive; however acid condensates form that are extremely corrosive to carbon steel and even to austenitic chromium-nickel steels. Similar conditions may occur in part of the air preheat system, where vitreous enamelled steel recuperative heaters or glass tube heat exchangers have been used. In flue-gas desulphurisers Hastelloy G is often specified for those parts of the ducting exposed to acid condensate. Hastelloy G is one of a family of nickel base and high nickel alloys that were developed for resistance to strong mineral acids, wet chlorine and other severely corrosive media. For convenience their compositions and uses are listed, together with Incoloy and Inconel in Table 5.19.

Electrostatic precipitation hoppers and bottom ash pits may also require protection of Hastelloy G, whilst other pollution control systems

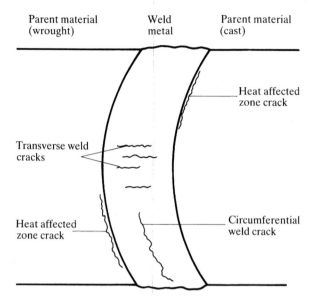

Parent material
(wrought)

Weld
metal

Parent material
(cast)

Heat affected
zone crack

Transverse weld
cracks

Heat affected
zone crack

Circumferential
weld crack

5.67 Location of cracks found in ½Cr½Mo¼V pipe welds after service in power stations. The weld metal is 2¼Cr1Mo.[43]

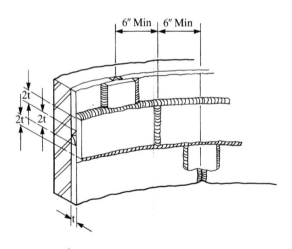

6″ Min   6″ Min

2t

2t   2t

t

5.68 Method of strip lining a steel vessel with a corrosion resistant nickel base alloy: a) Weld parallel strips to surface using 18Cr8Ni type electrodes; b) Fillet weld cover strips of the nickel base alloy over the 18/8 welds.

Table 5.19 Corrosion and heat-resistant nickel base and high nickel alloys

Name	Chemical composition, maximum or range %											Use
	C	Si	Mn	P	S	Cr	Co	Mo	W	Fe	Other	
Hastelloy B	0.05	1.0	1.0	0.025	0.03	1.0	2.5	26.0–30.0	–	4.0–6.0	V 0.20–0.40	Resists HCl at all concentrations and temperatures. Boiling H_2SO up to 60%
Hastelloy B2	0.02	0.10	1.0	0.04	0.03	1.0	1.0	26.0–30.0	–	2.0	–	As Hastelloy B but resistant to knifeline attack and weld decay
Hastelloy C-276	0.02	0.08	1.0	0.04	0.03	14.5–16.5	2.5	15.0–17.0	3.0–4.5	4.0–7.0	V 0.35 max	Strong oxidising conditions, mineral acids, wet chlorine in the as-welded condition
Hastelloy C4	0.015	0.08	1.0	0.04	0.03	14.0–18.0	2.0	14.0–17.0	–	3.0	Ti 0.70 max	As C-276 with improved high temperature stability
Hastelloy G	0.05	1.0	1.0–2.0	0.04	0.03	21.0–23.5	2.5	5.5–7.5	1.0	18.0–21.0	Cu 1.5–2.5 Nb 1.75–2.5	Resists hot sulphuric and phosphoric acids in as-welded condition
Hastelloy X	0.05–0.15	1.0	1.0	–	–	20.5–23.0	0.5–2.5	8.0–10.0	0.2–1.0	17.0–21.0	–	Good mechanical properties at elevated temperature
Haynes Alloy 20 mod.	0.05	1.0	2.5	–	–	21.0–23.0	–	4.0–6.0	–	Bal	Ni 25–27 Ti 4 × C	Resistant to pitting in chloride solutions
Haynes Alloy 625	0.10	0.50	0.50	0.015	0.015	20.0–23.0	1.0	8.0–10.0	–	5.0	Nb 3.15–4.15 Al 0.4 max Ti 0.4 max	Good mechanical properties at elevated temperatures. Resists chloride ion attack

Table 5.19 cont'd

Incoloy 825	0.05	–	–	–	–	20.0–24.0	–	2.5–3.5	–	Bal	Ni 40–45	Resists chloride attack and mineral acids
Inconel 600	0.08	0.50	1.0	–	0.015	14.0–17.0	–	–	–	6.0–10.0	Ti 0.3 max Cu 0.50	Oxidation resistant to 1175 °C. Resists stress corrosion cracking
Incoloy 800	0.10	1.0	1.5	–	0.015	19.0–23.0	–	–	–	Bal	Ni 30–35	Good mechanical properties at elevated temperature

Hastelloy is a trade name registered by the Cabot Corporation. Incoloy and Inconel are registered trade names of the International Nickel Company. The Table shows approximate compositions; for current figures the company in question should be consulted.

such as scrubbers for chemical waste incinerators may be made from this or other Hastelloys. Welding follows the same practice as for austenitic chromium-nickel steel. Thin sections may be welded by the GTA process or by using a GTA root pass and welding out with coated electrodes. Matching composition electrodes and filler rod are suitable. Earlier versions of Hastelloy were subject to intergranular precipitation in the heat affected zone and were liable to suffer intergranular corrosion or knifeline attack at the fusion boundary. Such problems have been largely overcome by reduced carbon content and stabilising additions.

Where practicable solid wall construction is preferred since the welds are more readily inspected and fabrication costs are lower. Alternatively ducts and other equipment items are lined or fabricated from clad steel. Figure 5.68 illustrates the recommended technique for applying strip cladding. Strips are welded to the backing steel using austenitic stainless steel electrodes: then a cover strip of Hastelloy is fillet welded over the joint. As in welding titanium clad steel, this overcomes problems of contamination from the base material. For clad plate a technique similar to that used for urea reactors and illustrated in Fig. 5.59 may be used.

REFERENCES

1 Shelton E, Rothwell A B and Coote R I: 'Steel requirements for current and future Canadian gas pipeline systems' *Metals Technology* 1983 Vol 10 234–241.
2 Sage A M: 'Physical metallurgy of high strength, low alloy line pipe steels' *Ibid* 224–233.
3 van Berkelom P J B, Dilthey U and Mursic M: 'Using the four-wire electrode process for the production of longitudinally welded pipe'. 2nd Int Conf on Pipewelding, 1979, The Welding Institute, Cambridge.
4 Matthews G T et al: 'Pipeline welding in the '80s' *Ibid*.
5 Fairhurst W et al: 'Weldability of low-carbon Mo-Nb and Mn-Mo-Nb $\times$ 70 pipeline steel' *Ibid*.
6 Rothwell A B, Dorling D V and Glover A G: 'Welding metallurgy and process development research for the gas pipeline industry'. In Advanced Joining Technologies (Ed T H North), Chapman and Hall, London, 1990.
7 Mathias R: 'Pipeline across the Zagros mountains' *Metal Construction* 1980 Vol 12 24–28.
8 Parlane A J A and Still J R: 'Pipelines for subsea oil and gas transmission' *Materials Science and Technology* 1988 314–323.
9 Still J R and Rae G: 'Laybarge inspection of submarine pipelines' *Metal Construction* 1984 Vol 16 268–276.
10 PD 6493 'Guidance on some methods for the derivation of acceptance levels for defects in fusion welded joints' British Standards Institution 1980.
11 de Sevry B and Bonnet C: 'Electron-beam welding of J-curve pipelines'. 2nd Int Pipeline Conference, The Welding Institute, Cambridge, 1979.
12 Eichorn F: 'High energy joining' in Advanced Joining Technology (Ed T H North), Chapman and Hall, 1990.

13 Masubuchi K, Gaudiano A V and Reynolds T J: 'Technologies and practices of underwater welding'. In Underwater Welding, Pergamon Press, Oxford, 1983.

14 Lancaster J F: 'Materials for the petrochemical industry' *Int Metals Reviews* 1978 (3) 101–148.

15 Gladman T: 'Developments in stainless steels' *Metals and Materials* 1988 351–355.

16 Anon: 'Learning from experience: fracture avoidance practice for large tanks' *Metal Construction*, 1988 Vol 20 16–19.

17 Lundin C D: 'Materials and their weldability for the power generation industry'. In Advanced Joining Technologies' (Ed T H North) Chapman and Hall, London, 1990.

18 API 941: 'Steels for hydrogen service at elevated temperatures and pressures in petroleum refineries and petrochemical plant'. American Petroleum Institute, Washington.

19 Bean A J: 'Low alloy steels in oil refinery service' *Metal Construction* 1984 Vol 16 671–677.

20 Macmillan W L and Gueld J C: 'Tube to tubesheet joints used in the chemical industry' in Welding and the Engineer: the Challenge of the 80s, South African Institute of Welding, 1983.

21 Shertz W R: 'Inbore welding of tubes to tubesheet on HP-heaters for power stations' *Ibid.*

22 Elliott S: 'Nooter thrives on quality' *Metal Construction* 1983 Vol 15 658–660.

23 Nichols R W: 'Reheat cracking on welded structures' *Welding in the World* 1969 Vol 7 (4) 244–261.

24 Ellis D J: 'Mechanised narrow gap welding of ferritic steel' *Joining and Materials* 1988 Vol 1 80–86.

25 Davis P: 'Pressurised fluid bed combustor vessels use flux-cored wire' *Ibid* 1989 Vol 2 55–62.

26 Breen A J: 'Low alloy steels in oil refinery service' *Metal Construction* 1984 Vol 16 671–677, 1985 Vol 17 23–29; 237–241; 293–296.

27 Cassie B A: 'Welding on to live gas pipelines' in 2nd Int Conf on Pipewelding. The Welding Institute, Cambridge, 1979.

28 Anon: 'Construction of Littlebrook "D" power station' *Metal Construction* 1981 Vol 13 106–111.

29 Anon: 'Learning from experience' *Metal Construction* 1987 Vol 19 699–704.

30 Towers O L: 'SCC in welded ammonia vessels' *Metal Construction* 1984 Vol 16 479–485.

31 de Wit J: 'New EEMUA recommendations for liquified gas storage tanks' *Metal Construction* 1987 Vol 19 204–205.

32 Suzuki H: 'Weldability of modern structural steels'. 1982 Houdremont Lecture, International Institute of Welding.

33 Schillmoller C M: 'Solving high-temperature problems in oil refineries and petrochemical plant' Chemical Engineering 1986 (1) 83–87.

34 Tucker A J P: 'Reformer tube and weld inspection, replacement and repair' *Ammonia Plant Safety* AIChE Vol 15 75–83.

35 Hale C C: 'Ammonia storage design practice' *Ammonia Plant Safety* 1974 Vol 16 23–27 AIChE.

36 Jarecki A and Lancaster J: 'Heavy-wall reactors for hydrotreating refinery

applications and for coal liquifaction plants' South African Institute of Welding.

37 Kastelein P and Verburg S J: 'A new electrode for urea plant fabrication' *Metal Construction* 1986 Vol 18 222–226.

38 Krystow P E: 'Materials and corrosion problems in urea plants' *Ammonia plant safety* 1971 Vol 13 93–102 AIChE.

39 Lancaster J F and Nichols R W: 'Fabrication of pressure-resistant holders'. Commission of the European Communities Congress 1968, 'Steel in the Chemical Industry'.

40 Anon: 'Construction of Littlebrook "D" power station' *Metal Construction* 1980 Vol 12 440–445 and 588–597.

41 Denys R M: 'Materials and their weldability for the power generation industry' in Advanced Joining Technologies (Ed T H North), Chapman and Hall, London, 1990.

42 Nicholson R D: 'Effect of ageing on interfacial structures of nickel-based transition joints' *Metal Technology* 1984 Vol 11 115–124.

43 Taft L H and Yeldham D E: 'Weld performance in high pressure steam generating plant' in Welding Research related to Power Plant, Marchwood Engineering Laboratories, Southampton, 1972.

6 The reliability of welded structures and process plant

General

Regardless of how it is made, a weld constitutes a discontinuity in a structure and as such may diminish its integrity. The weld itself may well be, and often is, stronger than the parts that it joins, but there are many secondary effects such as reduced resistance to fatigue or to corrosion which must be countered in one way or another. Much of this book has been concerned with the nature of these effects and the means of overcoming them. It is appropriate at this stage to take a look at the general problem of ensuring the reliable behaviour of welded equipment. Four main aspects will be covered; firstly, the means of making a quantitative measurement of reliability; secondly, the character and incidence of failures; thirdly, the available methods of assessing the reliability of flawed equipment, and finally the means of assuring reliability during construction and in service.

Fundamentals

The bathtub curve

According to classical reliability theory the behaviour of a large number of identical working parts in service can be described by the 'bathtub' curve illustrated in Fig. 6.1. When first put into operation this group suffers a relatively high rate of startup failures. Such failures are due to defective material or a lack of quality control, and the components in question are eliminated during a relatively short period of operation. At the end of the operating period there is again a relatively high rate of failure due to wear because the parts have exceeded their normal operating life. In between there is a period where the failure rate is relatively low and constant; this is due to the occurrence of chance failure.

Start-up failures may in principle be avoided by running the components in question for a trial period long enough to eliminate defective

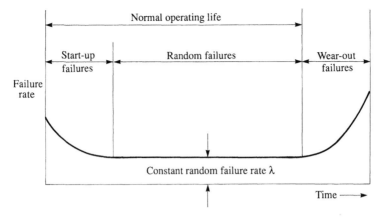

6.1 Bathtub curve.

parts. In pressure equipment the pressure test performs such a function by showing up leaks and, on rare occasions, causing the fracture of weak material. Rotating equipment such as pumps and compressors are run in and tested before installation. However process plant is, by its nature, complex and it is rarely possible to achieve perfection in all parts. The start-up of such plant is usually attended by a higher than normal rate of failures and plant shutdowns often follow the course indicated by the first part of the bathtub curve. With structures, of course, this is not the case. Rarely, there have been catastrophic brittle failures, associated with weld defects or embrittlement, after a short period of service.

In process plant wear-out failures are normally avoided by the replacement of equipment before the design life is exceeded, and by regular inspection and maintenance.

In the intermediate period, chance failures may result from a multitude of causes, including mechanical breakdown and human error. Of these, the human factor is probably the most important. Studies of failure rates in similar items of process plant (to be discussed later) show differences that cannot be accounted for by mechanical defects, and which are more likely to be due to differences in the operational skills of the organisations concerned.

Failure probability distributions

The probability that any event will occur is usually expressed as a number lying between 0 and 1. For example, in tossing a coin, the probability that it will fall showing a head is 0.5 which is equal to the probability that it will show a tail. The probability that it will show either a head or tail is 0.5 + 0.5 = 1; that is to say, it is a certainty.

The probability of failure of a number of components having a limited life is a function of the time interval to be considered. For example, if there are 100 components and the probability of failure during a ten hour period is 0.05, it would be expected that five components will fail during this period. The failure probability may be represented by the letter P. The reliability R of a component is then:

$$R = 1 - P \tag{6.1}$$

In a group of components the proportion that survive after a given period is a measure of the reliability. In the case quoted earlier, after a ten hour period there have been five failures and 95 survivals out of an original population of 100, so the reliability is 0.95.

The distribution of failures as a function of time depends on the type of environment in which the components function. For example, if failures are due to chance, the failure probability is the same for any given period during the life of the part. If, however, as in the case of a furnace tube, failures are due to creep rupture then the failure rate increases with time to a maximum at the mean rupture life and then falls.

The exponential distribution

The failure probability for a system exposed to chance hazards may be expressed as:

$$P = 1 - e^{-\lambda(t - t_o)} \tag{6.2}$$

where λ is the chance failure rate, t is time and t_o is the time at the start of the period under consideration. An alternative expression is:

$$P = 1 - e^{-(t - t_o)/m} \tag{6.3}$$

where m is the mean interval between failures. It is evident that:

$$R = e^{-\lambda (t-t_o)} = e^{-(t - t_o)/m} \tag{6.4}$$

Where failures are purely by chance, λ and m are constants. Consider the case where $t = 1.1\ t_o$, then:

$$R = e^{-0.1\ t_o/m} \tag{6.5}$$

which is also constant. Thus, independent of elapsed time the reliability or failure probability for any given time interval is the same. This is one way of defining chance failures.

The quantity P represents the proportion of the original population that has failed. Thus, if in [6.2] we put $(t-t_o) = m$, then:

$$P = 1 - e^{-1} = 0.632 \tag{6.6}$$

In other words 63.2% of the population that existed at time t_o will have failed at a time equal to the mean interval between failures. This apparent anomaly arises because the failure rate is proportional to the remaining population. If the original population was N_o, and at time t the number of failures is N_f, then the failure rate is:

$$\lambda = d/dt \left(\frac{N_f}{N_o - N_f} \right) = 1/m \qquad [6.7]$$

Thus the time interval between failures is equal to:

$$\Delta t = m \left(\frac{N_f}{N_o - N_f} \right) \qquad [6.8]$$

and this quantity increases with the number of failures; i.e. with time. Likewise the absolute number of failures per unit time decreases with time, whilst the proportion of failures per unit time remains the same. The absolute failure rate for the case where $t_o = 0$ is:

$$dP/dt = \frac{1}{m} e^{-t/m} \qquad [6.9]$$

This is the probability density function and is plotted in Fig. 6.2a, whilst Fig. 6.2b shows the cumulative proportion of failures as a function of time.

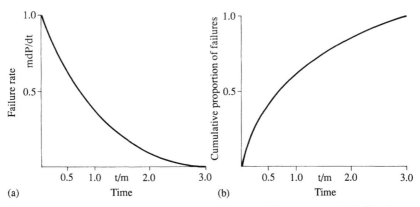

6.2 Exponential failure distribution: a) Failure rate (probability density distribution); b) Cumulative proportion of failures.

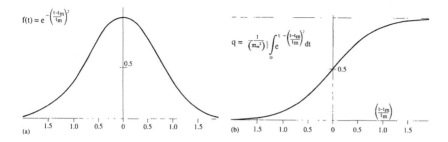

6.3 The normal distribution: a) Failure rate (probability density distribution); b) Cumulative proportion of failures. In both cases t = time and t_m = mean time of failure.

The normal distribution

The normal, or Gaussian distribution represents the random distribution of quantities around a mean value. For example, the velocities of gas molecules have a normal distribution about the mean value corresponding to the gas temperature. Likewise it gives a good representation of the distribution of times of wear-out failures. For example, in the case of metals, the times to failure by creep rupture have a normal distribution when a large number of identical specimens are tested.

Unlike the exponential distribution the probability of failure in any given period changes with age in the case of the normal distribution, and the number of failures is evenly distributed about the mean value. The normal probability density function and the cumulative proportion of failures are plotted in Fig. 6.3a and b respectively.[1]

The Weibull distribution

In real life failure rates do not as a rule conform to the simple patterns considered above, and it is often convenient to use the expression developed by Weibull:[2]

$$P = 1 - e^{\left(\frac{t - t_o}{t_m}\right)^b}$$

[6.10]

where b is the Weibull slope or form factor and t_m is the mean time for failure when b = 1. t_m is also known as the scale parameter. The mean time to failure for any value of b is given by:

$$\text{mean} = t_m \, \Gamma \, (1 + 1/b) + t_o$$

[6.11]

where $\Gamma(1 + n)$ is the gamma function and:

$$\Gamma(1 + n) = n \int_0^\infty x^{n-1} e^{-x} dx \qquad [6.12]$$

The failure rate is:

$$\lambda = \left[b (t - t_0)^{b-1} \right] / t_m^b \qquad [6.13]$$

When $b = 1$ [6.10] represents the exponential distribution and in this case since $\Gamma(2) = 1$, the mean time to failure is t_m, and the failure rate is $1/t_m$. When $b = 2$ the failure rate increases in a linear manner with time; this is the Rayleigh distribution.

Weibull probability charts are available on which the cumulative percentage of failures may be plotted as a function of time. Figure 6.4 shows such a plot for the failure of HK 40 catalyst tubes in steam-methane reformer furnaces. These were in ammonia plants located in Europe, and covered the decade up to 1974. They represent early experience in the operation of such plant and greater reliability has been achieved subsequently.

The failures were of all types, and were due for example to flame impingement, bending stresses, weld failures and so forth; in other words they were random failures. The results fall into two groups, group 1 having a much higher failure rate than group 2. Both sets of data yield a reasonably straight line, and in both cases the slope lies between 1 and 2. Salot[3, 4] surveyed US reformer furnaces for tube failures during the same period and obtained similar results; in particular different Weibull slopes were obtained for sets of tubes in furnaces operated by different companies. Clearly, the higher the slope the lower the reliability of the operation. As noted earlier, there is no material difference that could account for such results, and the only reasonable explanation is that in some plants operators were more skilled than others. Such differences would of course be expected, but not to the degree observed here. In this case at least, the human factor is dominant in determining reliability.

Also shown on Fig. 6.4 is the Gaussian distribution for the failure rate assuming a mean life of 3.3×10^5 hours (similar to that obtained by extrapolating the group 1 curve to 63.2% failure). Such a distribution might be obtained if the failures were all due to creep rupture; evidently this is not the case and the failures were of a random type, although increasing in frequency with age.

The Weibull plot is a useful tool for those concerned with monitoring and control of multiple failures in process plant, and indeed in other operations where numbers of identical components are exposed to severe conditions. One way of expressing the Weibull distribution is:

$$\frac{1}{1 - P} = e^{[(t-t_0)/t_m]^b} \qquad [6.14]$$

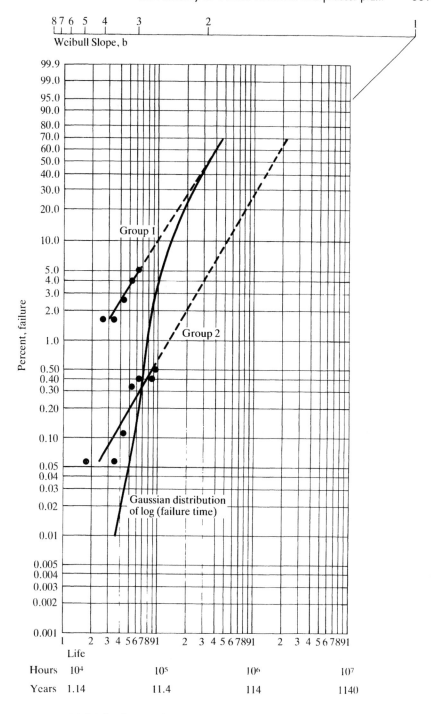

6.4 Weibull plot of steam-methane reformer catalyst tube failure.

where, taking logarithms, and putting $t_o = 0$:

$$\log \left[\ln \left(\frac{1}{1 - P} \right) \right] = b \, (\log t - \log t_m) \qquad [6.15]$$

P is the cumulative proportion of failures at time t. Thus, a plot of t against $\ln \left(\frac{1}{1 - P} \right)$ on log-log graph paper may be used as an alternative to plotting on a Weibull chart.

The reliability of boilers and pressure vessels

Historical background

Power for industry in the nineteenth century, particularly in the second half, was mainly provided by steam, and the population of boilers increased remarkably. So, unfortunately, did the number of deaths and injuries from boiler explosions. Design was empirical, material of variable quality, and maloperation was commonplace. In the USA an explosion that resulted in several deaths caused the ASME to formulate a boiler code, now part 1 of the Boiler and Pressure Vessel Code. In the UK the Manchester Steam Users Association was formed in the middle 1850s and this organisation undertook the periodic inspection of land based boilers, whilst Lloyds Register mainly inspected ships' boilers. In 1882 the Boiler Explosion Act required users to report all boiler explosions, and at a later stage, periodic inspection became a mandatory requirement. Records of failures have been maintained by the Associated Offices Technical Committee, which is an inspection organisation for a group of insurance companies. In Germany it is compulsory to report failures of boilers and pressure vessels to TUV (Technischen Uberwachungs Vereine). There is no legal requirement for reporting of failures in the USA but a need to establish figures for reliability of nuclear plant led to the collection of extensive data, reported by Bush[5] in 1975. There is therefore information about the reliability of boilers and pressure vessels in Germany, the UK and the USA.

Failure statistics

Boilers in the present context are shell boilers, fired-tube boilers or boiler drums, including nozzles but not including piping. The power boilers used in central power stations fall into the same category as pressure plant so far as failure statistics are concerned; in other words they are complex systems in which numerous items of plant contribute to the reliability or unreliability of the whole.

The failures in question are of the random type and the failure rate λ is numerically very small, such that λt for a normal vessel life of 10–20 years is much less than unity. Therefore the probability of failure is:

$$P = 1 - e^{-\lambda t} = 1 - \left(1 - \lambda t + \frac{(\lambda t)^2}{2} \cdots \right) \simeq \lambda t \qquad [6.16]$$

i.e. it is equal to the failure rate times life. Failure rates are usually expressed as number per vessel year. Since they are of the order of 10^{-5} to 10^{-4} it is of little practical use to express the reliability in numerical terms.

Failure statistics for boilers from about 1860 to 1960 are shown in Fig. 6.5.[6] The decline from an initially high rate of casualties and equipment failure will be evident. This fall is almost certainly associated with a steady improvement in construction, operation and inspection.

Potentially serious and catastrophic failures of pressure vessels surveyed by Phillips and Warwick[7] in 1968 are shown in relation to the age of the vessel in Fig. 6.6. This is a classical bathtub curve with a peak failure rate at two years' service. Such early failures point to deficiencies in material and/or workmanship so that although, from the boiler explosion record, there had been a very significant improvement up to 1960, there was still room for further improvement in 1968. No doubt the same remains true to this day.

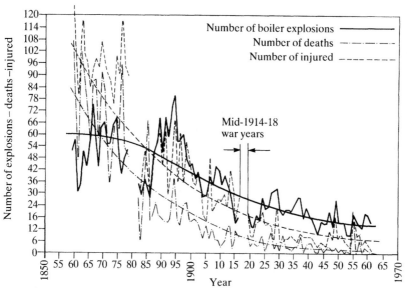

6.5 Boiler explosions and deaths and injuries therefrom 1860–1960[6] (reproduced by courtesy of the Council of the Institution of Mechanical Engineers).[6]

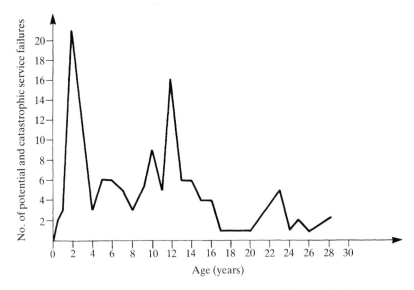

6.6 Potentially dangerous or catastrophic failures of boilers surveyed by AOTC during the period 1962–1967 (reproduced by courtesy of UKAEA).[7]

Like the Bush survey referred to earlier, and one by Kellerman[8] about the situation in the Federal Republic of Germany, the Phillips and Warwick report was directed towards obtaining an assessment of the reliability of nuclear vessels. It covered welded or forged boilers and pressure vessels manufactured to Class 1 standards (similar to ASME Sections I and VIII) and less than 30 years old. The Bush survey covered only vessels in central power stations, whilst the German survey, from TUV records was comprehensive. The reliability figures obtained from these and other sources are shown in Table 6.1. The probability of disruptive failure given for the FRG was estimated from observations of 'severe vessel damage'. That for the USA was obtained by a series of extrapolations and statistical techniques that make the final result questionable. Of the nine failures reported in the UK survey, seven were due to maloperation. This reflects the fact that the survey covered a wide variety of service applications, where the operating standards would be quite different to those applicable in nuclear operations. Thus the figures given in Table 6.1, whilst of interest in the present context, cannot be taken as accurate indicators of nuclear vessel reliability.

The causes of all failures in the UK survey are listed in Tables 6.2 and 6.3. Cracking, due mainly to fatigue and corrosion-assisted fatigue, predominates. Most of the cracking was at branches and at fillet welds. In some instances the cracking caused leaks; in others the crack was detected before leakage and repaired. Fatigue failure is more likely to be a problem with boilers, air receivers and other vessels that operate intermittently, than with process plant vessels.

Corrosion is another major cause of failure according to this survey and indeed this is in accordance with general experience. Even in power boilers, where it is possible to maintain close control over water quality, corrosion may occur on the fire side. For small boilers such control may not be practicable and, where for example steam is used for food processing, may be unacceptable. In these cases there is bound to be a corrosion problem.

Periodic inspection is an important safety measure in the operation of boilers and pressure vessels, and in many countries there are statutory requirements for such inspections. For example, in the USA boilers must be inspected annually and static pressure vessels once every two years. In the UK there is a 14 month interval between inspections of boilers, which may be extended to 26 months for large water-tube boilers. It is implicit that any defects found will be repaired or that the equipment will be replaced. The way in which periodic inspection can affect reliability is illustrated in Fig. 6.7.[9] In this diagram the calculated failure probability of a vessel subject to cyclic loading at 7 cycles/day, with a maximum stress of 7.5 kgf/mm² (73.6 N/mm²) and minimum stress zero is shown as a function of time for various inspection frequencies. It is assumed that the welds contain cracks that are just below the limit of detectability (a depth of 0.6 mm) and that they grow according to the equation:

$$da/dn = C(\Delta K)^m \qquad [6.17]$$

Table 6.1 Failure statistics for pressure vessels: number per vessel year[5]

Country	Vessel years operation	No. of failures		Estimated failure probability	
		Non-critical	Disruptive	Non-critical	Disruptive
FRG	1.7×10^6	596	0	1.5×10^{-4}	2.7×10^{-6} to 4×10^{-5}
UK	2.057×10^5	111	9	5.4×10^{-4}	4.4×10^{-5}
USA	7.25×10^5		0		6.3×10^{-6}

Table 6.2 Causes of failure of pressure vessels and boilers[7]

	No of cases	%
Cracks	118	89.3
Corrosion (stress corrosion, corrosion-fatigue and wastage)	2	1.5
Maloperation (e.g. no water)	8	6.1
Manufacturing defect	3	2.3
Creep failure	1	0.8

Table 6.3 Causes of cracking of pressure vessels and boilers[7]

	No of cases	% of total
Fatigue	47	35.6
Corrosion (including stress corrosion, corrosion- fatigue and wastage)	24	18.2
Pre-existing from manufacture	10	7.6
Not known	37	27.9
	118	89.3

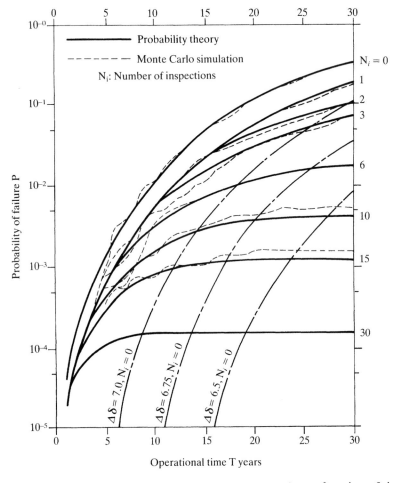

6.7 Probability of failure of a pressure vessel as a function of time.[9] Interval between inspections is $30/N_i$ years.

where a is crack depth, n is number of cycles, ΔK is the stress intensity factor range and C and m are material constants (in this instance 10^{-10} mm/cycle and 3 respectively). Repairs are assumed to be carried out after inspection. The conditions are intended to simulate a storage sphere 13 m in diameter, 30 mm thick fabricated from HT 60 steel and subject to a cyclic pressure of 7 kgf/cm². Also shown are reliability curves for lower peak stresses.

These calculations indicate that periodic inspection, and in particular annual inspection, can have a very beneficial effect on reliability. Other models[5] have predicted similar advantages for periodic inspection.

All such models implicitly assume that the quality of repair welding is equal to that of the original seam. It must always be remembered, however, that repair welding conditions are fundamentally less favourable than those obtained originally. In particular, the heat flow regime is more three dimensional; cooling rates are higher and the maintenance of preheat more difficult. It is desirable, wherever practicable, to make a separate procedure specification and qualification for repair welding and to subject the testpieces to both ultrasonic and magnetic particle examination in addition to radiography. Where stress relief is part of the original weld procedure it must also be applied after repair, bearing in mind that the residual stress field due to localised welding is more severe than that, say, of a circumferential seam.

The reliability of process plant

Failure statistics

Summaries of the results of two sets of investigations into types of failure are listed in Tables 6.4 and 6.5. The first[10] was carried out at Du Pont chemical and petrochemical plants in the USA during a period prior to 1973, and the second[11] relates to failures in Britoil North Sea production and utilities systems during the ten years 1978–1988. In spite of the difference between the types of process plant and the different periods of the survey, the results are remarkably similar. One outstanding feature is the predominance of fatigue as a cause of failure. This has already been noted in the case of boilers and pressure vessels. It also applies to other fields of industry. Figure 6.8 shows the relative incidence of major failure mechanisms in aircraft accidents as determined by the Royal Aerospace Establishment, Farnborough.[12]

Fatigue failures in process plant occur mainly in rotating equipment such as pumps and compressors, and for the most part in those components directly subject to alternating stress such as crankshafts. However a significant proportion occurs in pipework associated with pumps, and in such cases the crack is usually initiated at the fusion boundary of the weld, either internally or externally. Short branches

Table 6.4 Causes of non-disruptive failure in process plant[10]

Corrosion	%	Mechanical	%
General	15	Fatigue	15
Stress corrosion		Abrasion and	
cracking	13	wear	5
Pitting	8	Overload	5
Intergranular	6	Poor welds	4
Other	13	Other	16
Total corrosion	55	Total mechanical	45

Table 6.5 Causes of non-disruptive failure in offshore processing plant[11]

Corrosion	%	Mechanical	%
CO_2 related	9	Fatigue	18
H_2S related	6	Mechanical	
Preferential		damage/overload	14
weld corrosion	6	Brittle fracture	9
Pitting	4	Poor fabrication	
Erosion	3	(excluding welds)	9
Galvanic	2	Poor welds	7
Crevice	1	Other	10
Impingement	1		
Stress corrosion	1		
Total	33	Total	67

attached to a main line and terminating in a flange or valve are vulnerable features. The Britoil survey gives a typical example. This was a branch in the water injection piping system used for fitting corrosion monitoring equipment. It was attached to the line by a 100 mm to 50 mm reducing T, and terminated in a blank flange. Fatigue failures originated at the toe of the internal weld joining the branch to the reducer, and propagated outwards through the weld metal. Failures of this type are commonly due to a pendulum type oscillation of the branch, set up by relatively minor vibrations in the main line; at worst, the branch oscillates at its natural frequency and fails very quickly. Proper design can eliminate such failures.

Less frequently failures may occur in main circumferential welds in pipework that is subject to vibration due, for example, to cavitation in a pump or to organ-pipe vibrations set up in a tubular heat exchanger. Again, the cracking occurs in welds between a rigidly held section and a length of piping that is free to vibrate or oscillate.

Corrosion would be expected to be a major cause of failures in chemical plant so that the figures for the Du Pont survey are not surprising. Offshore plant, on the other hand, is primarily devoted to separation of crude oil from water and gas, and to providing means of

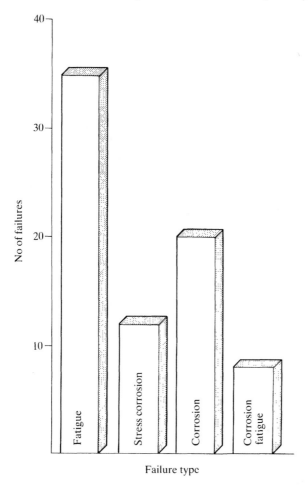

6.8 Relative incidence of various modes of failure in aircraft (reproduced by courtesy of The Institute of Materials).[12]

injecting sea water into the formation. Figure 6.9 is a diagram of the
separation system, and at first sight this would seem a relatively non-
corrosive process. However in the well itself the aqueous phase contains dissolved CO_2, and this combines with iron:

$$Fe + H_2CO_3 \rightarrow FeCO_3 + H_2$$

The carbonate film so formed is relatively soft and is removed in areas
of high turbulence, resulting in localised attack. 13%Cr steel has good
resistance to CO_2 corrosion at temperatures up to 150 °C; above this
higher chromium ferritic/austenitic steels are required. Welds in carbon
steel may be preferentially attacked, and the use of a ½Ni½Cu or similar
low alloy filler metal, which is used in sea water injection lines, does

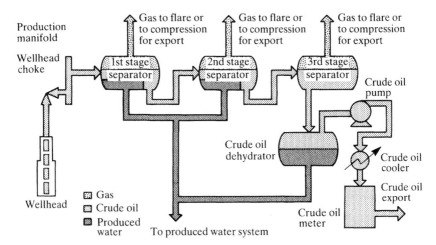

6.9 Typical offshore production unit (reproduced by courtesy of The Institute of Materials).[11]

not appear to solve the problem. This is not surprising, since CO_2 corrosion is a simple chemical attack, whereas the sea water corrosion is primarily electrochemical and can be inhibited by making the weld metal slightly more noble than the surrounding material.

The failures due to H_2S corrosion are likewise surprising at first sight, since North Sea crude is sweet. In fact the H_2S is produced by sulphate reducing bacteria in parts of the system that are relatively stagnant. H_2S has caused brittle failures of high tensile wire and corrosion of carbon steel and cupronickel but hydrogen-induced cracking of welds does not appear to be a problem.

Preferential corrosion of welds in sea water injection systems, of which a sample is shown in Fig. 6.10, has already been discussed. A similar type of attack has also been observed in sections of production pipework, including overboard dump water piping, and it has occurred on subsea structural parts where the cathodic protection has been inadequate. However in the latter case, and particularly where a 2½% Ni filler was used, corrosion occurred preferentially along the fusion line. In some cases, fatigue cracks developed at the base of the grooves so formed.[11]

Brittle fracture is mentioned as a cause of failure in Table 6.5; in fact these failures occurred during the hydrostatic testing of defective spheroidal iron castings, and in a defective nickel base alloy bolt. They could equally be placed under the heading of 'poor fabrication'.

One of the notable features of these surveys is the relatively small percentage of failures that are attributed to weld defects. In one case, typical of general experience, there were no weld failures whatsoever in pressure equipment. Those that did occur were in secondary and

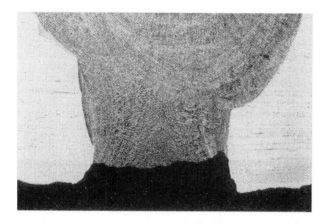

6.10 Laboratory simultation of preferential corrosion of carbon steel weld metal in sea water injection system (courtesy of The Welding Institute).[11]

structural parts that were originally considered non-critical and were not subject to NDT.

Brittle fracture

General

The avoidance of catastrophic brittle fracture remains a major concern for engineers and for welding engineers in particular. This is a problem that can affect any metallic fabrication, but the material that is most affected is steel, and the structures which experience has indicated to be most at risk are thick-walled pressure vessels (other than the multilayer type), storage tanks, ships, offshore structures and bridges. The precautions that are currently taken have reduced the probability of failure by brittle fracture to a low level. Nevertheless it is useful to review the historical record.

The brittle failure of pressure vessels

A list of disruptive brittle failures of pressure vessels reported in the literature between 1960 and 1970 is given in Table 6.6. The wall thickness in all these cases was 50 mm or greater, and so far as is known, all but the 1970 and possibly the 1964 and 1967 failures occurred during hydrostatic testing. The two storage spheres were fabricated from quenched and tempered microalloyed steel. As noted, earlier welds in this type of fabrication are prone to microcracking if not adequately

Table 6.6 Disruptive failures of pressure vessels and boiler drums 1960–1970[13]

Date	Incident	Material	Cause
1960	Steam drums, FRG failed on hydrotest	Mo-Ni-Cu	Embrittlement due to intergranular copper precipitation
1962	Pressure vessel, France		Notch-ductility of alloy steel impaired by prolonged heat treatment
1963	Nuclear power boiler, Sizewell, UK	Ducol W30	Failure of support causing shock loading during hydrotest
1964	High pressure heat exchanger channel section, USA	ASTM A 302B	Hard zone cracking and low notch-ductility in forging
1965	Ammonia converter, UK	Ducol W30	Hard zone cracking and embrittlement due to defective heat treatment
1966	Steam drum, Cockenzie, Scotland	Ducol W30	Large pre-existing crack formed during heat treatment
1967	High pressure heat exchanger channel section, USA	ASTM A 302B	Probably similar to 1964 failure
1968	2000 m^3 storage sphere, Japan	HT 80 Q and T	Not known but weld cracking found in similar vessel
1968	1000 m^3 storage sphere, Japan	HT 60 Q and T	As above
1970	Ammonia plant heat exchangers, Finland		Pre-existing crack and embrittlement of forging due to incorrect heat treatment

preheated. The remainder were made of low alloy steel. Ducol W30 (BS 1501-271, MnCrMoV type) was used in no less than three of the failures, but investigations showed that in each case the fault lay with the fabricator and not with the steel. For example, the steam drum at Cockenzie power station in Scotland failed when it was being hydrostatically tested for the seventh time, and examination of the fracture surface showed that it had been initiated by a pre-existing crack 330 mm long and 90 mm wide which ran between an angle bracket and an economiser nozzle (this is the black area in Fig. 6.11). The vessel had been ultrasonically examined before, but not after,

6.11 Failure of steam drum during hydrotest. Blackened area is a crack which formed during stress relief heat treatment (reproduced by courtesy of Elsevier Applied Science Publishers Ltd).[13]

final stress relief so evidently the cracking occurred during this operation. Current practice is to inspect both before and after post-weld heat treatment.

The disruptive brittle failure rate for pressure vessels and boiler drums indicated by Table 6.6 is one per year. The rate of installation of boiler drums in central power stations in the USA for the years 1959–1968 averaged 3300 per year. Assuming that the numbers of heavy pressure vessels and steam drums were about equal, and that world production (other than the USSR) was three times US production, this corresponds to a failure rate of about 5×10^{-5}. A previous estimate[13] gave a rate of 1×10^{-5}. These figures represent the probability of a brittle failure of a vessel during hydrotest. Service failures are rare, in part because the operating pressure is lower than the hydrotest pressure, but also because in most cases heavy pressure vessels, and all steam drums, operate at elevated temperature. An exception is found in ammonia plants, where some high pressure heat exchangers may operate at subzero temperatures. The 1970 failure in Finland given in Table 6.6 represents such equipment.

Two ammonia converters are known to have failed in service. One of these was forged in two halves, joined together by an interrupted thread, and then seal welded around the outside of the joint. However the threads did not fully engage and as a result there was a stress concentration at the end of one of them. Ammonia converter shells operate typically at 100 °C, within the range where hydrogen embrittlement and cracking is possible. In the vessel in question a hydrogen crack initiated at the point of stress concentration and propagated slowly around the circumference until the stress intensity factor was high enough for an unstable brittle fracture. The vessel then split in two, the top half flying some distance whilst the bottom half was driven into the ground.

In the other case a forging was used to make the top closure of a Casale (high pressure) converter. This part was also threaded and seal welded, but with a continuous thread. It happened that there were some relatively large inclusions in the forging just below the thread, and these provided a point of initiation for a hydrogen crack. Eventually the crack propagated far enough for the closure to be blown off.

These two examples, and some others quoted earlier, underline the fact that welding is not the only cause of brittle fracture. It is also of interest to note that hydrogen can play a part in the initiation and propagation of very large cracks. Thus, in making an engineering critical assessment of a vessel in hydrogen service the fracture toughness to use in calculation is not K_{IC} but K_H.

The brittle fracture of marine structures

The US Government's requirement in World War II for the rapid construction of cargo ships and, later, oil tankers could only be met by use of welding in place of riveting. By introducing pre-fabrication and other techniques that were not customary in shipbuilding at that time, the construction period for a Liberty ship was reduced to an average of 17 days. Likewise a T2 tanker could be produced in just over 40 days. Inevitably the quality both of design and fabrication was not the best, so that the casualty rate became serious in the period 1942–1944. Between March 1943 and March 1944 seven Liberty ships had broken in two or had been abandoned because of structural brittle fracture. At a later date these failures were the subject of an extensive review by the US Ship Structures Committee. The casualties were designated Class 1 if the fracture endangered the vessel or resulted in its loss, Class 2 if the fracture did not endanger the vessel but occurred in a critical location, whilst Class 3 included minor fractures.

Most of the fractures were initiated by fatigue cracks, and most of the fatigue cracks originated at unfavourable design features, particularly the square hatch corners (52%) and cutouts in shear strakes (24%). The ships were redesigned to eliminate such stress concentrations. At the same time four or more horizontal runs of riveted joints were inserted in the hull as crack arrestors. Not surprisingly, these crack arrestors did not work, and by 1948 at least 18 Liberty ships having this feature had become major casualties.

Although design details were mainly responsible for the fractures, welding defects did play a part. In 1943 the tanker 'Schenectady' had completed sea trials and was moored overnight in the fitting-out basin. During the night the air temperature fell to −5 °C, and in still water the tanker broke in two. The origin of the fracture was a defective weld in combination with a design detail at the deck. The 'Esso Manhattan' also broke in two in fine weather and a slight sea; in this case the fracture originated in a defective butt weld. The highest casualty rate (about 1.3×10^{-2} per vessel year) occurred in the early vessels built at the Swan Island yard.

The results of the Ship Structures Committee Survey are summarised in Table 6.7.[14] The casualty rates are about two orders of magnitude higher than those given in Table 6.1 for land based pressure vessels. This is not surprising; the ships in question were hastily built using plate of low notch-ductility (transition temperatures ranged from 15 °C to 60 °C) and operated under severe wartime conditions. Comparable figures for the present day are not available, but it is safe to say that the casualty rate is much lower. Radical improvements have been made in standards of design and fabrication, whilst steel quality, and in particular notch-ductility, has been greatly improved.

This improvement started, as noted previously, in the early post-war years, and at the same time work on the brittle behaviour of steel plate intensified. In particular, the wide plate test developed by Wells made it possible to reproduce a low stress brittle fracture in the laboratory. Wells' test provides a reference for validating small scale tests, and is so used to this day.[15] On the negative side, the notion that welding is primarily responsible for initiation of brittle fracture is widespread in industry, and this derives in part from the failure of the Liberty ships and T2 tankers. In fact, poor design which resulted in fatigue cracking was a much more important cause of these failures. On the other hand, half the pressure vessel failures listed in Table 6.6 were probably initiated

Table 6.7 Casualties to all-welded Liberty ships and T2 tankers up to 1953[14]

Liberty ships	Ship Years	Casualty rate per ship year	
		Class 1	Classes 1 and 2
Original design	2100	4.2×10^{-2}	0.19
Improved details	2600	5.4×10^{-3}	3.6×10^{-2}
T2 tankers			
Original design	1483	1.9×10^{-2}	6.5×10^{-2}
With riveted crack arresters	2064	1.2×10^{-2}	5.8×10^{-2}
Overall record for T2 tankers*	8600	6.7×10^{-4}	

* excluding the first fifty built at Kaiser Swan Island Yard.

by welding cracks, and weld cracking has been responsible for the failure of at least three bridges.

Brittle failures do not necessarily have catastrophic results. After the failure of the ammonia converter in 1965 (Table 6.6) the vessel was reordered from another fabricator and was delivered in sufficient time for the plant to be completed without significant delay. Only two of the ten T2 tankers that broke in two were a total loss. The 'Schenectady' for example was repaired, fitted with crack arresters, and completed 20 years' service at sea. The 'Esso Manhatten' was also repaired and did a further 16 years of service before being converted into a dry cargo ship.[14]

Catastrophes

General

From time to time a large scale disaster such as a major fire, the sinking of a passenger ship, an explosion and fire in a chemical plant, or an aircraft crash causes human fatalities and financial loss. Such incidents receive widespread publicity. There is public concern and there may be an enquiry following which there are changes in laws and regulations. Such events may be precipitated by a mechanical failure, but in many cases the failure is the result of human error.

For the present purposes it is of interest to examine those catastrophic incidents where welding may have been in some degree responsible for the mechanical failure. Table 6.8 lists a selection of catastrophic incidents that occurred during the period 1960–1980,[16] together with an indication of whether or not welding contributed to the failure. This list is selective and no quantitative conclusion can be drawn from it. Four of the incidents are described below.

Case histories

Kings Bridge, Melbourne[17]

The high level section of this bridge consisted of a series of alternate cantilever and suspended spans, each carriageway being supported on four lines of main girders. The girders were I beams with flange plates welded to the steel web plate. To accommodate the higher force to which the centre of a span was subject, a second or cover plate was welded on the underside of the flange. The material specified was BS 968-1941 with additional impact properties. BS 968 was a niobium treated carbon-manganese steel having augmented yield and tensile strength. At the time in question it was produced in open hearth furnaces and was semi-killed. The bridge was completed in August 1961, and in

Table 6.8 Catastrophes[16-20]

Date	Incident and location	Description	Welding a contributory cause?
1962	Kings Bridge (Melbourne, Australia)	Brittle failure of main girders in service. No death or injuries	Yes
1965	'Sea Gem' (North Sea)	Brittle fracture of tie bars caused the rig to collapse and sink. 13 killed	Possible
1970	West Gate Bridge (Melbourne, Australia)	One span of a box girder bridge under erection collapsed when bolts were removed in an attempt to remedy buckled plates. 35 killed	No
1973	Markham Collery (Derbyshire)	Fracture of a threaded rod in the mechanical brake for the pithead winding gear allowed the cage to strike the bottom of the shaft. 18 killed	Yes
1973	Ammonia bullet (South Africa)	Brittle fracture of 50 ton ammonia storage vessel, causing an ammonia spill. 18 killed	Yes
1973	Summerland (Isle of Man)	Fire started by vandals spread to interior of building. 50 killed	No
1974	Flixborough (NE England)	Failure of temporary pipework led to explosion and fire. 28 killed	No
1980	'Alexander L Kielland' (North Sea)	Fatigue failure of main brace caused this accommodation platform to capsize. 123 killed	Yes

July 1962 a span close to the southern end collapsed suddenly under the load of a truck weighing 47 tons, which was well within its load capability. Fortunately the span was supported by the concrete deck and by vertical concrete walls below, so that it did not fail completely, but merely sagged by about one foot.

Examination of the fractures showed that they were brittle failures initiated by pre-existing cracks in the heat affected zone of the transverse weld attaching the cover plate to the flange. This was a manual weld and the inquiry committee considered that the cracking was probably hydrogen-induced and associated with inadequate preheat. Examination of other girders indicated that all were probably in a similar condition.

In 1961 BS 968 was not used to any significant extent in Australia and both welders and inspectors were unfamiliar with the required welding procedures. Where such a situation exists it is self-evident that all suppliers should carry out procedure testing in accordance with the relevant Code rules. The records did not show that this was done, nor did they confirm that such procedures as were laid down in the specifications were followed. In other words there were serious deficiencies in the control of welding operations that resulted in avoidable weld cracking. It is very improbable that such weld cracking would occur under present conditions.

The 'Sea Gem'

This vessel started life as a pontoon which underwent a series of conversions and finally became a self-elevating drilling barge, fitted with ten legs which could be jacked down to the sea bottom so as to raise the deck to a height above the sea suitable for drilling operations. In 1965 it was towed to a location in the North Sea in preparation for drilling. Once in position it was jacked up to about 50 ft above sea level, and drilling commenced. The drilling was completed and the well head secured by 18 December 1965. Preparations were then made to lower the barge until it was waterborne with the object of towing it to a new area.

The jacks were pneumatically operated and were each connected to the platform by four tie-bars. During the jacking operations the platform was therefore suspended from the legs via these tie-bars. This was the situation on 27 December when, prior to lowering the barge it was decided to raise it by one stroke of the jacks to ensure that they were in good order. In fact the barge came up unevenly, and whilst it was being returned to its original position there was a bang and the structure lurched violently to port. It then righted itself and fell into the sea in a horizontal position. However the bottom plating was severely damaged and the vessel sank rapidly.

The wreckage was later described as being 'like a heap of broken glass'; almost all the fractures having occurred in a brittle manner, in particular the tie-bars that were recovered had suffered brittle failures. These tie-bars were flame-cut from plate, and in a number of places had been patched by welding and then grinding flat, presumably where the flame-cutting was irregular. Cracks had initiated at weld defects, at fatigue cracks and at the sharply radiused fillets between the spade end and the shank.

The inquiry tribunal concluded that the loss was due to brittle failures of the tie-bars. These could have been initiated by shock loading due to movement of the legs or of the platform but there was no direct evidence of such movement and it was thought that a fall in temperature, which was 3 °C at the time, was sufficient to trigger off the failures. It was evident that the whole structure was in a notch-brittle condition, whilst the tie-bars had fatigue cracks and defects in welds that had not been stress-relieved.[18] Clearly these items should have been forgings; even without weld repair and other defects, square cut sections were quite unsuitable for such a duty.

This design of rig is unlikely to be used in the future, nor will such notch-brittle steel be used for North Sea structures. It is impossible to say how far the weld repairs contributed to this disaster but they could very well have played a part.

There were 32 men on board the rig, and 13 lost their lives.

Ammonia tank failure, South Africa
The tank in question was one of four bullets (horizontal cylindrical tanks) 2.9 m in diameter and 14.3 m in length, with a capacity of 50 metric tons of liquid ammonia. They operated at atmospheric temperature (19 °C at the time of the accident) and a pressure of 90 lb/in² gauge (0.62 N/mm²). The vessels were designed and constructed to BS 1515 from steel to BS 1501–151–28 A. They were not stress-relieved, since this was not required by the Code. The dished ends were fabricated from two plates 23 mm thick, cold formed at the major radius and hot flanged at the knuckle. After commissioning in 1967 they remained in service until 1971, when they were taken off-line for statutory inspection. Radiography and ultrasonic testing of the dished end in No 3 tank disclosed some defects in the weld between the two plates. One fault was ground out and repaired successfully. The second, which appeared to be lack of sidewall fusion or slag entrapment, was ground out at least twice before being passed. Radiography then disclosed a new crack 100 mm long a short distance from the repair; this was removed and rewelded. The final repair was about 200 mm long. No stress relief was carried out, but the tank was hydrostatically tested to 374 lb/sq in (2.58 N/mm²).

After these operations the tanks were returned to service. On July 13 1973 two tanks, No 3 and 4, were being filled simultaneously from a railway tank car at a pressure of 90 lb/in² gauge (0.62 N/mm²), which corresponds to an equilibrium temperature of 15 °C for liquid ammonia. At 16.15 hours No 3 tank exploded (without any known triggering incident) releasing an estimated 30 tons of anhydrous ammonia. A further 8 tons were released from the rail car. A gas cloud about 150 m in diameter and 20 m deep was formed. A slight breeze blew this cloud in a westerly direction. All the men working to the west of the blast eventually died, whilst most of those in other areas escaped, some by using wet cloths as masks.[19]

The explosion was due to the failure of the recently repaired dished end on tank No 3. Figure 6.12 is a sketch of the rupture. A brittle failure had initiated at the point marked C and an irregular area, including the more severe weld repair (marked A), was blown out. There was no thinning at the edge of the fracture, which showed the typical chevron markings of a brittle crack. Impact testing indicated that both plates were in a brittle condition, the large plate having a transition temperature of 115 °C and the small one 20 °C. The hardness was in the region of 200 VPN. It was concluded that the plate had suffered strain-age embrittlement as a result of cold forming and possibly hydrostatic testing. Stress-relieving of samples at 620 °C reduced the hardness level to a mean of about 160 VPN.

The failure did not originate in the repair weld and indeed appeared to skirt around it. Nevertheless the repair could have generated a residual stress field of sufficient intensity to drive the crack in the observed direction. As noted earlier, no triggering event was observed. However metal objects exposed to the sun can become quite hot and part filling the tank with liquid at 15 °C could well set up a temperature gradient and a corresponding stress gradient.

The investigators concluded that stress relief was required for bullets and was also required after any weld repair, and that the use of cold forming for the dished heads of pressure vessels was undesirable.

This incident caused the death of 18 people, and at least 65 others required hospital treatment.

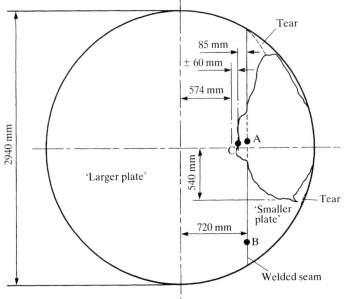

6.12 Brittle failure of dished end of an ammonia storage tank (reproduced by permission of the American Institute of Chemical Engineers. © 1974 AIChE).[19]

The Alexander L Kielland accident

The platform in question was semi-submersible; that is to say, the deck was supported by columns mounted on pontoons, and it was moored in position by wires or chains attached to anchors. This design of rig originated in the USA but the Alexander L Kielland was of a type known as Pentagone developed in France.[20] Pentagone rigs have five pontoons, each supporting a column, with two anchors attached to each column. The Alexander L Kielland was intended to be a drilling rig and its original layout is shown in Fig. 6.13. However from the start it was used as an accommodation platform, equipped with living quarters, mess rooms and a cinema. It was delivered to the owners in 1976 and from that time operated in the Ekofisk field. In 1980 it was anchored close to the production platform Edda 2/7 C, the two platforms being connected by a walkway. In bad weather the walk bridge was lifted on board the Alexander L Kielland, which was then winched away from the production area. Such a move was made on 27 March 1980, in poor visibility with a wave height of 6–8 m. About half an hour after completing the move a main brace fractured and one of the columns broke off. Other braces then failed and the platform keeled over to 35 °. It continued to list and sink but was retained more or less in position by one anchor wire. Eventually this wire broke and the platform overturned, trapping many of those on board.

Figure 6.14a shows the layout of the brace, (D6), that failed initially. It was fitted with a drainhole, required for stability, and a hydrophone fitting, details of which are given in Fig. 6.14b. The hydrophone is an instrument that picks up acoustic emissions from a source located in the well and enables the rig to be positioned accurately when drilling.

The failure was due to fatigue fracture of the brace, initiated on opposite sides of the hydrophone fitting, as shown in Fig. 6.15. The fitting itself had suffered extensive lamellar tearing below the 6 mm fillet welds attaching it to the brace. It is considered that the presence of such tears resulted in augmented strain at the twelve o'clock and six o'clock positions around the fitting, and this in turn initiated and helped to propagate the crack. Figure 6.15 shows how, as the crack progressed, the crack length per cycle is increased until finally it propagated as a fast ductile tear.

Samples from the failed brace were subject to extensive testing in Norwegian laboratories. The fatigue crack growth rate was measured as a function of ΔK and was found to obey the usual law:

$$da/dN = C(\Delta K)^m \text{ and}$$
for da/dN in mm/cycle and ΔK in $N/mm^{3/2}$
$$C = 6.8 \times 10^{-18} \text{ and } n = 4.47$$

These figures are characteristic of a steel containing a high proportion of inclusions, and imply a low initial and high final rate of growth.

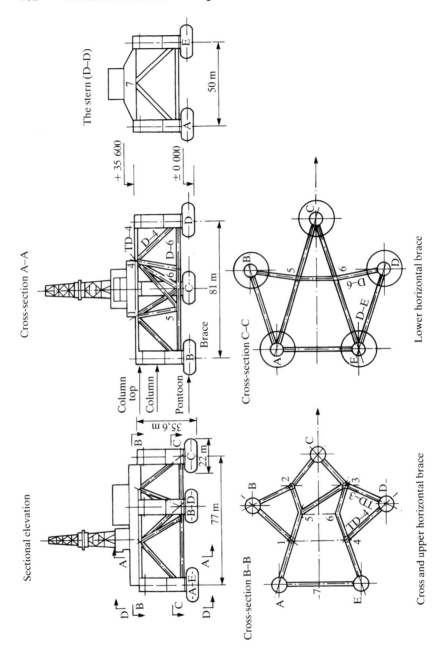

6.13 Layout of the Alexander L Kielland.[20]

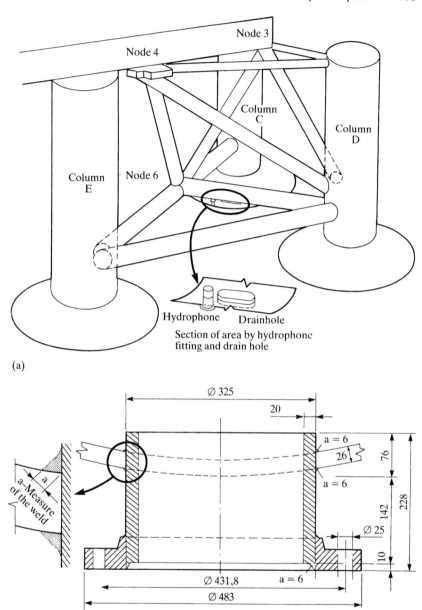

(a)

(b)

6.14 The hydrophone fitting on brace D-6 of the Alexander L Kielland: a) Location of the hydrophone fitting on brace D.6; b) Sectional elevation of the hydrophone fitting with nominal measures.[20]

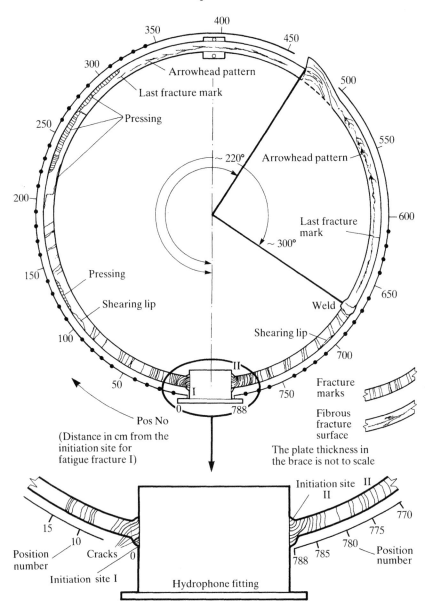

6.15 The surface appearance of the fracture in brace D-6 of the Alexander L Kielland.[20]

This is not thought to have affected the final result significantly. Other properties showed slight but not significant deviations from specification. Through-thickness tests on the hydrophone fitting gave a reduction of area of 8.3%. Although there were no requirements for minimum through-thickness properties in the job specification this figure is a long way below the value of 20% recommended by Det Norske Veritas for North Sea structures.

The commission of inquiry concluded that the failure was due to a combination of poor design with inadequate through-thickness ductility in the hydrophone fitting material. The fact that the hydrophone was classified as a fitting and not as a structural component contributed to the acceptance of substandard design features. It is ironic that the fitting was only required for drilling operations and was redundant whilst the rig was being used for accommodation.

212 men were on board when the Alexander L Kielland capsized and overturned; 123 were killed because of the accident.

Failures: summary and comment

A consistent feature for all records and for all the types of structure surveyed is the prominence of fatigue as a cause of failure, with corrosion coming a close second. Brittle fracture has always been a relatively rare failure mode since it primarily affects steel items that are either thick-walled or have exceptionally high tensile properties. Another class, of course, comprises those structures such as the 'Sea Gem', the ammonia tank and early German steam drums which were fabricated from steel that was brittle in the first place or embrittled in service. The rapid improvement in steel properties that has been achieved in recent years is such that a repetition of such failures is extremely unlikely. There is no record of the impact properties of the steel used for the Sea Gem but its behaviour suggests a bottom-shelf value of 5–15 J at 0 °C. By contrast, tests on the Alexander L Kielland after the accident gave longitudinal impact strengths at the same temperature typically in the range 100–200 J. Moreover, even in the presence of a crack about 3 m in length and with a stress in the range 100–300 N/mm, the final failure was ductile. Bearing in mind this sort of trend, brittle fracture is likely in the future to be increasingly rare as a mode of failure.

One feature of the catastrophic incidents reviewed here is that in two cases the weld that may have contributed to the failure was a repair weld. The same applies in the case of the fire on a Boeing 737 aircraft at Manchester Airport in 1985.[21] This was caused by thermal fatigue cracking on a pressure chamber, which led to fuel ignition and the resulting fire, in which 55 people died, mostly from inhaling toxic smoke. A weld repair had been carried out on the fuel chamber and this was then rated as having the same life expectancy as a new component. This was not the case, nor was it for the tie-bars on the Sea Gem or the dished head of the ammonia tank. It was pointed out earlier that repair welds are exposed to a more severe thermal environment than a normal butt weld, whereas they often receive the least amount of attention from management. Clearly, the opposite should be the case, and any weld repair on a critical component or vessel should be subject to a careful appraisal. In the first place a reliability analysis should be carried out to determine whether or not a repair is really necessary. If this proves to be the case, every alternative should be explored before considering repair, and a repair should only be undertaken if the material is known to be in good condition, and then only after qualifying a procedure that simulates the actual conditions.

In process plant defective welds rate low as a cause of failure; 4% in the case of the 1973 review[10] and 7% for the decade up to 1988.[11] The difference could be due to many factors and is probably not significant; nevertheless it would have been more satisfactory to record a positive improvement. As it is, there is no cause for complacency and every effort needs to be made to improve weld quality, not only for special work such as oil rigs, but for all welding operations.

In doing so it must be borne in mind that improved welding quality alone may not improve reliability. In the case of the Sea Gem designers should have required the use of forgings for the tie-bars and should not have accepted material flame-cut out of plate. It should have been appreciated that welding a hydrophone into a main brace of the Alexander L Kielland generated a fatigue hazard and it should have been located where there was no such risk. Such examples can be multiplied indefinitely; there is almost always an element of human error in any catastrophic event. The quality assurance systems to be discussed later represent a means of reducing the risk of such errors.

Reliability analysis

General

Codes and standards detailing methods for assessing the probable life of a structure containing flaws, particularly weld discontinuities, have been established in a number of countries, and these are reviewed in the IIW Document SST-1157-90 'Fitness for purpose'[23]. Those based on linear elastic fracture mechanics follow the technique originally established in the USA by Paris and others. Most structural materials however fall outside the scope of the linear elastic analysis and require a treatment that takes account of plasticity. These include crack tip opening displacement, J integral and R curve methods, as described in

Chapter 3. The most commonly used technique is CTOD, which has the advantage that the testing is relatively simple and that it is possible to locate the crack tip in a particular region to be treated such as the heat affected zone of a weld. CTOD has also been adopted in a modified form for the reliability of pipelines containing discontinuities, for example in API 1104 Appendix A and BS 4515 Appendix H.

A reliability analysis may be carried out at the design stage to establish how far the structure is vulnerable to discontinuities that may develop in service, or to the effect of embrittlement. More often it is used to assess the serviceability of a part which has been found to contain cracks or inclusions, such as the subsea pipeline described in Chapter 5. The IIW document allows the use of such analysis during the course of manufacture. Normally, however, the acceptability of defects found during fabrication is governed by the applicable construction code.

For nuclear work there are code requirements covering reliability calculations, but for other applications the procedure, if used, must be agreed by the parties concerned. Nuclear vessels are subject to periodic non-destructive examination, and any flaws that are disclosed may be subject to an analysis in accordance with the relevant codes.

Techniques for calculating reliability

Before discussing the available alternatives it is worth recapitulating some of the basic features of fast running (sometimes called 'instantaneous') fractures.

In metals these fall into two categories. In the first the failure is ductile, and is characterised by a 45 ° shear failure surrounded by a region in which there has been significant plastic deformation and thinning. Such failures may occur due to the sudden overloading of a structure or over-pressurising a vessel. They can also occur in gas pipelines if, after initiation of a failure, the rate of energy release due to depressurisation is greater than the rate of energy absorption in the fracture. In such a case a long fast running fracture can develop. This problem is discussed in Chapter 5.

In the second type of running failure the fracture surfaces are more or less at right angles to the surface, with a relatively small width of ductile 45 ° shear close to the edge. The flat part of the fracture usually exhibits a mixture of cleavage failure and microvoid coalescence. The rate of energy absorption is much lower than for a fully ductile failure, and in the classic case the strain energy released from the surrounding plate is sufficient to keep the crack moving. Intermediate crack morphologies are possible but are rarely, if ever, observed. For the present purposes a distinction is drawn between elastic and elastic-plastic types of failure. In both conditions, however, the anticipated type of failure is the same, namely a quasi-brittle failure with a flat fracture surface.

The distinction is drawn because linear elastic fracture mechanics is based on the assumption of elastic behaviour, whereas the presence of a large plastic zone at the crack tip, or general yielding in the vicinity of the crack, invalidates this assumption. Likewise, the relationships established by fracture mechanics analysis and, for example, the stress intensity factor associated with particular forms of crack do not necessarily apply to the elastic-plastic condition.

There are a number of techniques that aim to overcome this problem; eleven of these are listed by Burdekin and Harrison.[22] All are open to criticism, and in practice only two are in general use; the crack tip opening displacement and the J integral methods. The technical background to CTOD and J integral testing is given in Chapter 3. There is a degree of empiricism in both cases, and ultimately the justification for their use comes from the accumulation of experience and correlation with the results of large scale tests such as the Wells wide plate test.

Standards and codes for reliability assessment

Various standards or codes for assessing the acceptability of defects are listed in Table 6.9. A number of these apply specifically to nuclear pressure vessels. For the most part these relate to ASTM A 533 and similar alloys, and are generally concerned with the use of linear elastic fracture mechanics to determine whether or not discontinuities found during routine inspection are tolerable. However the ASME procedures are also sometimes used for non-nuclear applications.

Other standards have general application and usually cover both elastic and elastic-plastic conditions. The EPRI method employs crack opening displacement and J integral measurements for an elastic-plastic analysis that is applicable to nuclear components.[24] The German Welding Society Merkblatt gives formulae for calculating the limiting crack size for brittle fracture and plastic collapse, and allows for crack growth due to fatigue. The Japanese standard is based on CTOD testing but utilises strain rather than stress as a means of calculating the probability of fracture. Whilst this is formally more realistic, it makes little difference to the final result.

The British Standard document PD 6493: 1980 relies primarily on CTOD testing. In its original form it embodied a design curve based on the strip yield model of crack opening discussed in Chapter 3, with a built-in factor of safety on crack size. The 1991 revision of PD 6493 is more complex, and includes three levels of assessment. The first corresponds generally with the original version. The second level is

Table 6.9 Methods of calculating the reliability of a flawed structure: national standards (data from Ref. 23)

Country	Standard or code	Application	Method	
			Type	Property
France	RCC-M rules	Nuclear	LEFM and	K_{IC}
	Annex ZG		EPFM	J_R
Germany	KTA rules	Nuclear	LEFM	K_{IC}
	DVS M 2401	General: including piping and offshore	LEFM and EPFM	K_{IC}, δ_C J_R
Japan	WES 2805	General	EPFM	δ_C
Norway	Norske Veritas Recommended Practice D404	Offshore	LEFM and EPFM	K_{IC}, δ_C J_R
UK	PD 6493	General	LEFM and EPFM	K_{IC}, δ_C
USA	ASME code Sections III and XI	Nuclear	LEFM	K_{IC}
	EPRI[24]	General	EPFM	J_R
Inter-national	IIW SST-1157-90[23]	General	LEFM and EPFM	K_{IC}, δ_C J_R

LEFM = Linear elastic fracture mechanics
EPFM = Elastic-plastic fracture mechanics

intended to give a more accurate calculation of failure risk and does not include built-in safety factors; users are expected to apply their own safety margins. Level 3 is appropriate to materials of high strain hardening capacity and is not intended for general structural steel applications.

IIW SST-1157-90 'Guidance on assessment of the fitness for purpose of welded structures' is not a code and in particular does not provide specific guidance as to calculations for avoiding brittle fracture. It does however give a procedure for calculating crack growth due to fatigue, together with quantitative information about the effect of misalignment and on the stress intensity factors associated with the particular types of discontinuity.

The procedure given in the ASME code Section III Division I Appendix G assumes the presence of an elliptical surface crack of width five times its depth and of depth ¼ × wall thickness of ¼ inch whichever is greater. The stress intensity associated with this crack is calculated, and the value so obtained is compared with a fracture toughness/temperature curve obtained by J integral testing. In this type of plot the temperature is that in excess of the nil-ductility temperature, obtained either by the Pellini drop-weight test or by Charpy testing. Such a plot makes it possible to allow for embrittlement when the shift in nil-ductility temperature due to the embrittling effect is known.

ASME Section XI Appendix A details a procedure for determining the acceptability of flaws that have been found during routine inspection. It applies to ferritic steel four inches or greater in thickness and having a yield strength of 50 ksi (345 N/mm²) or more. The actual flaw is resolved into an idealised geometry and the stress intensity factor is calculated for all stress conditions. Crack growth is then determined from known stress fluctuations to obtain the final crack size. This is compared to the critical crack size obtained using a standard curve for crack arrest fracture toughness.

Procedure for risk assessment

The sequence of operations listed in the original version of PD 6493, which is an elaboration of ASME Section XI, is probably the most straightforward and simplest to follow. This is detailed below:

(1) Determine the dimensions (width and depth) of a crack that is equivalent to the observed defect, in accordance with procedures given in the relevant standard.
(2) Calculate the crack size $\bar{a}_m$ that would cause failure by the following mechanisms
    (a) Brittle fracture
    (b) Ductile fracture or bulk yielding
    (c) Leakage
    (d) Corrosion
    (e) Creep
(3) Assess the size $\bar{a}$ to which the initial defect will grow during the design life of the vessel or structure due to the following:
    (a) Fatigue
    (b) Corrosion
    (c) Creep crack growth
    (d) Creep/fatigue interaction

Then if $\alpha\bar{a}_m > \bar{a}$ the defect is acceptable. α is a safety factor, and in the PD 6493 level 1 assessment this is built in to the calculation, so that the defect is acceptable when $\bar{a}_m > \bar{a}$.

The treatment of brittle fracture risk may be in accordance with an elastic or elastic-plastic analysis, depending on the stress level. According to PD 6493 if the total stress (applied plus local stress) is below yield point linear elastic fracture mechanics applies. Others use the formula:

$$B > 2.5 \left[\frac{K_{IC}}{\sigma_{yp}} \right]^2 \qquad\qquad [6.18]$$

If the plate thickness B conforms to this equation, an elastic analysis is used. Otherwise either J integral or CTOD methods must be employed.

Brittle failure

If condition [6.18] is met, the first step is to calculate the stress intensity K for the type of discontinuity in question. For example, in the case of a surface of embedded elliptical crack:

$$K = \frac{M_m \sigma_m + M_b \sigma_b}{\Phi} (\pi a)^{\frac{1}{2}} \qquad\qquad [6.10]$$

where σ_m = membrane stress
σ_b = bending stress
a   = crack depth
M_m, M_b = correction factors
Φ = complete elliptic integral of the second kind

Formulae for M_m, M_b and Φ may be obtained from Ref. 23 and 25. K is then plotted as a function of αa, where α is the desired safety factor. The permissible maximum value of a is given when $K = K_{IC}$. It is assumed that K_{IC} is known for the relevant material. Alternatively K is plotted as a function of a, and (in line with PD 6493) the critical value of a is given when:

$$K = 0.7 \, K_{IC} \qquad\qquad [6.20]$$

This is equivalent to applying a safety factor of 2 to the crack size.

IIW SST-1157-90 gives a useful case study illustrating this procedure. The material is assumed to be a steel plate 50 mm thick subject to a tensile stress of 100 N/mm². The yield and ultimate stress are 400 N/mm² and 550 N/mm² respectively and K_{IC} = 1600 N/mm$^{3/2}$ (50.6 MN/m$^{3/2}$). It is required to determine the allowable size for a through-thickness discontinuity. Equation [6.18] requires:

$$B > 2.5 \left[\frac{1600}{400} \right]^2 = 40 \text{ mm} \qquad\qquad [6.21]$$

so LEFM plane strain conditions are met. For a through-thickness crack width 2a:

$$K = \sigma(\pi a)^{\frac{1}{2}} \qquad\qquad [6.22]$$

At fracture

$$a = \frac{1}{\pi} \left[\frac{K_{IC}}{\sigma} \right]^2 = 81.5 \, \text{mm} \qquad [6.23]$$

Applying a safety factor of 2 the total permissible discontinuity length is 81.5 mm.

Alternatively, partial safety factors may be used for the various quantities in this calculation. Partial safety factors are currently used in some design codes. They are intended to allow for uncertainty in measurement of a quantity, its scatter (e.g. the spread of tensile properties) or to reflect the relative hazard (for example, if failure would result in collapse the partial safety factor would be greater than if it resulted in a redistribution of stress). In the present case partial safety factors may be assigned to stress, to discontinuity size and to the value of K_{IC}. Assuming these to be 1.4, 1.2 and 1.2 respectively.

$$1.2a = \frac{1}{\pi} \left[\frac{K_{IC}}{1.2} \right]^2 \left[\frac{1}{1.4\sigma} \right]^2 \qquad [6.24]$$

giving a maximum value of 2a equal to 48 mm.

Where the conditions for LEFM are not met, the allowable discontinuity size may be calculated using the formula developed by Burdekin and Dawes:[26,27]

$$\bar{a}_m = \frac{\delta_c E \sigma_{yp}}{2\pi\sigma^2} \quad \text{for} \quad \frac{\sigma}{\sigma_{yp}} \leqslant 0.5 \qquad [6.25]$$

$$\bar{a}_m = \frac{\delta_c E}{2\pi(\sigma - 0.25\sigma_{yp})} \quad \text{for} \quad \frac{\sigma}{\sigma_{yp}} > 0.5 \qquad [6.26]$$

where σ = sum of applied $\div$ local stresses

δ_c = critical CTOD

The flaw is acceptable if

$$\bar{a}_m > a$$

The 'design curve' in PD 6493 is based on equations [6.25] and [6.26]. This curve was validated by practical tests.[26]

Procedures for calculating the formalised defect size from observed defect types and sizes are given in various standards including PD 6493. Likewise the value of σ must be calculated or estimated, allowing for welding residual stress, localised stress and peak stress.

Evaluating techniques for the prediction of brittle fracture

In 1984 the Netherlands Institute of Welding started an investigation into the reliability of the various methods currently available for assessing the risk of brittle fracture in welded structures, and van Rongen[28] has summarised the results up to 1990. In the first phase of this work, where wide plates and pipes were tested, it was determined that all analysis techniques based on CTOD and J integral tests gave safe predictions of fracture. The second phase (1987–1990) used a variety of full scale tests on pressure vessels and piping and obtained the same result. One of the tests (on a pressure vessel) is of particular interest and is recorded below.

The vessel in question was a heat exchanger shell from an ammonia plant which had been rejected because of the presence of planar flaws. Figure 6.16 shows a section of the shell, which was fabricated from 76 mm thick carbon-manganese steel. The actual location of the defects, which were situated in the transition region between shell and tubesheet, is indicated in Fig. 6.17. The crack sizes and location were determined after the burst test. Prior to testing however ultrasonic examination was not capable of resolving individual defects and it was necessary to assume that there was a 15 mm deep crack that ran around the circumference. A reliability assessment based on this assumption and on the known material properties indicated the vessel to be unsafe; hence its rejection.

The burst test was carried out at 9 °C. Plastic deformation of the cylindrical part commenced at 30 N/mm² pressure and the vessel failed by a circumferential brittle fracture at 53 N/mm². The failure was initiated by a planar surface defect in the transition region, 6 mm deep and less than 6 mm long. The material properties were deter-mined from Charpy impact tests, which gave 15 J at 15 °C, correspond-ing to K_{IC} $(C_V) = 1600$ N/m$^{3/2}$. In addition J integral tests were made on three samples of material from the vicinity of the failure and these gave a minimum of 2070 N/m$^{3/2}$ and a maximum of 2700 N/m$^{3/2}$. Based on these and other known properties, fracture analyses were carried out to ASME XI, PD 6493-1980 and the revised PD 6493. The results are shown in Fig. 6.18. Only the lowest value of measured fracture tough-ness, represented by the middle points on the curves, would be used in a reliability analysis, so that all methods gave conservative results, with the ASME XI value being closest to actual behaviour.

The design curve used in PD 6493 is based on [6.25] and [6.26] in which σ_{yp} is taken to be the yield strength as measured in a uniaxial tensile test. It has been suggested however that the true yield stress under the conditions of restraint that exist in the propagating crack is increased by a factor 'm', the value of which lies between 1 and 2. It fol-lows that:

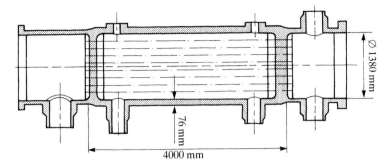

6.16 Heat exchanger shell (DSM pressure vessel) used by Dutch Institute of Welding for assessing the validity of various methods of fracture analysis.[28]

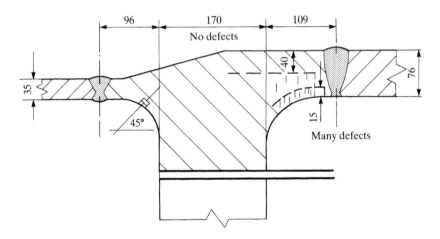

6.17 Defective region in the shell to tubesheet transition of the DSM pressure vessel.[28]

$$J = m(1 - v^2) \, \delta\sigma_{yp}$$

Rongen concludes that since there remains an uncertainty about the value of the restraint factor, J should be used in preference to the crack tip opening displacement for reliability analyses.

Fatigue crack extension

Calculations of fatigue crack growth are, in principle, based on the growth rate law given by [6.17].

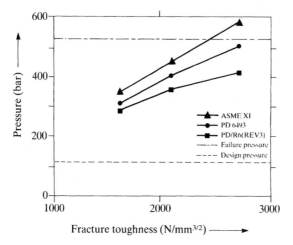

6.18 Predicted failure pressure of the DSM pressure vessel as a function of the fracture toughness assumed in the analysis.[28]

Based on examination of a large number of S-N curves for steel the constants in this equation are given in IIW SST-1157-90 and in PD 6493. The IIW values are:

$$m = 3$$
$$C = 3 \times 10^{-13}$$

in air and

$$m = 3$$
$$C = 2.3 \times 10^{-12}$$

in a marine environment.

In both cases da/dn is expressed in mm/cycle and ΔK in N/mm$^{3/2}$. The figures correspond to the upper bound of published data for crack growth by a striation mechanism. Where crack growth is by cleavage or microvoid coalesence PD 6493 gives the following:

$$m = 4$$
$$C = 7.4 \times 10^{-16} \text{ for } 97.5\% \text{ probability of survival}$$
$$C = 1.7 \times 10^{-15} \text{ for } 99.5\% \text{ probability of survival}$$

Using such figures it is in principle possible to calculate the growth of any initial crack of known size. For example for a through-thickness crack:

$$da/dn = C(\pi a)^{m/2} (\Delta\sigma)^m \tag{6.27}$$

and for m = 3

$$2 \left[\frac{t}{a_i^{1/2}} - \frac{1}{a_N^{1/2}}\right] = C(\pi^{1/2}\Delta\sigma)^m N \tag{6.28}$$

where a_i is the initial crack size and a_N is the size after N cycles.

Unfortunately the shape factors M_m and M_b for surface and buried cracks, together with the secondary stress contribution to the total stress, change with crack size. So the calculation must be made cycle by cycle, which may not be practicable. Therefore simplified methods are used. These are essentially similar in PD 6493 and in the IIW document. A required 'quality category' is determined by the applied stress range to which the joint is subject. The quality category for the discontinuity is then obtained using a series of charts in which governing factors are the initial crack size $\bar{a}_i$ and the maximum tolerable crack size $\bar{a}_m$. If the discontinuity quality category is higher than the required quality category the discontinuity is acceptable. There are formulae to convert variable amplitude loading to an equivalent constant amplitude loading.

In the case of pressure vessels or pipelines it may be of interest to calculate the probable crack size that will cause leakage, and to compare this with the crack size for unstable brittle failure. If the crack size for leakage is the smaller, the condition ('leak before fracture') is a relatively safe one. On the other hand leakage is not acceptable so that a crack that penetrates the wall represents an extreme limit for fatigue crack growth.

Other modes of crack extension

None of the documents surveyed give specific procedures for calculating crack growth for stress corrosion cracking, creep or creep plus fatigue, although the user is urged to consider these possibilities. At the present time there are not enough quantitative data available on these mechanisms to make it possible to set up such a procedure. Crack growth rates for creep are known in some cases but it is uncertain as to whether these would be applicable to weld imperfections. It is also not known how defects will affect the risk of stress corrosion cracking; in most cases they would be irrelevant, the major predisposing factor being the level of residual stress (or weld hardness, which amounts to the same thing so far as carbon steel is concerned).

As noted earlier, the IIW document proposes a crack growth rate in marine exposure about eight times that in air. This figure is based on published data relating to North Sea environments. Crack growth rates are also higher in water at elevated temperature, and are sensitive to the ratio R, equal to K_{min}/K_{max}. The ASME code section XI plot of typical crack growth rates under these conditions is shown in Fig. 6.19. The IIW growth rate curve for $m = 3$ and $C = 3 \times 10^{-13}$ is plotted on the same figure.

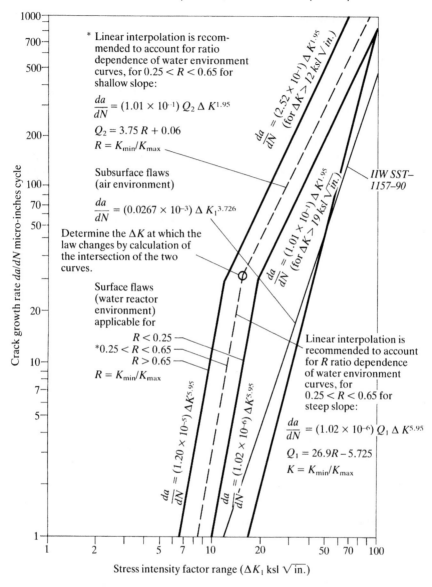

The graph contains the following labels and equations:

Vertical axis: Crack growth rate da/dN micro-inches cycle (1, 5, 7, 10, 20, 50, 70, 100, 200, 500, 700, 1000)

Horizontal axis: Stress intensity factor range (ΔK_1 ksl $\sqrt{in.}$) (1, 2, 5, 7, 10, 20, 50, 70, 100)

* Linear interpolation is recommended to account for ratio dependence of water environment curves, for $0.25 < R < 0.65$ for shallow slope:

$$\frac{da}{dN} = (1.01 \times 10^{-1}) Q_2 \, \Delta \, K^{1.95}$$

$$Q_2 = 3.75 R + 0.06$$

$$R = K_{min}/K_{max}$$

Subsurface flaws (air environment)

$$\frac{da}{dN} = (0.0267 \times 10^{-3}) \, \Delta \, K_1{}^{3.726}$$

Determine the ΔK at which the law changes by calculation of the intersection of the two curves.

Surface flaws (water reactor environment) applicable for

$R < 0.25$
$*0.25 < R < 0.65$
$R > 0.65$
$R = K_{min}/K_{max}$

$\frac{da}{dN} = (2.52 \times 10^{-1}) \, \Delta \, K^{1.95}$ (for $\Delta K > 12$ ksl $\sqrt{in.}$)

$\frac{da}{dN} = (1.01 \times 10^{-1}) \, \Delta \, K^{1.95}$ (for $\Delta K > 19$ ksl $\sqrt{in.}$)

IIW SST–1157–90

Linear interpolation is recommended to account for R ratio dependence of water environment curves, for $0.25 < R < 0.65$ for steep slope:

$$\frac{da}{dN} = (1.02 \times 10^{-6}) Q_1 \, \Delta \, K^{5.95}$$

$$Q_1 = 26.9R - 5.725$$

$$K = K_{min}/K_{max}$$

$\frac{da}{dN} = (1.20 \times 10^{-5}) \Delta K^{5.95}$

$\frac{da}{dN} = (1.02 \times 10^{-6}) \Delta K^{5.95}$

6.19 ASME fatigue crack growth curves for air and boiling water exposure compared with IIW SST-1157-90 curve for air.

Case study: hydrocracker

The problems that may affect hydrocracker reactors, particularly those fabricated in the early 1970s, are discussed in Chapter 5. These include cracking and disbonding of the cladding. Disbonding, although it could result in a collapse of the catalyst trays and other internals, has never in practice had such an effect, and does not otherwise pose any threat to the integrity of the vessel. Cracks in the cladding are not a problem during operation because the temperature (about 450 °C) is above the brittle-ductile transition and above that for hydrogen assisted crack growth. At atmospheric temperature however, during shutdown, hydrogen cracking is possible. The driving force is residual tensile stress in the austenitic chromium-nickel steel weld deposit, which, as noted earlier, remains at yield strength level after post-weld heat treatment.

Such cracks were found in hydrocracker reactors at the Natref oil refinery in South Africa.[29] The vessel had a wall thickness of 259 mm with weld overlay cladding 8 mm thick. It operated under an internal pressure of 22.5 N/mm^2 and at a maximum temperature of 460 °C. The cylindrical part of the shell consisted of forged rings of 2¼Cr1Mo steel, clad with one layer of type 309 and a second layer of type 347 stainless steel. In most cases the cracking affected only the type 347 layer but in places the cracks penetrated to the austenite-ferrite interface.

Values of the stress intensity associated with such cracks oriented circumferentially and longitudinally were computed by The Welding Institute (now TWI) and the results are shown in Fig. 6.20. The values of K fall initially with increased crack depth due to the diminishing effect of the cladding residual stress but with deeper cracks the pressure stress takes over as the driving force and the stress intensity increases. For cracks that just penetrate the cladding the value of K_I is between 45 and 65 MN/m$^{3/2}$ (1423-2056 N/mm$^{3/2}$). Values of K_H are shown as a function of yield strength in Fig. 3.30, and for the hydrocracker shell (with a yield strength of 350 N/mm^2) the predicted value of K_H is 50 MN/m$^{3/2}$ (1581 N/mm$^{3/2}$). Temper embrittlement has the effect of reducing K_H still further. Figure 3.31 shows this effect and indicates that for higher degrees of embrittlement the value of K_H may fall to about 30 MN/m$^{3/2}$ (950 N/mm$^{3/2}$). Therefore hydrogen assisted crack growth during shutdown periods was judged to be a possibility. For this and other reasons the vessels concerned were replaced.

The material for the replacement vessels was subject to J integral testing to ASTM E 813-81 over a range of temperatures and transition curves K_c (J) were obtained, as plotted in Fig. 6.21. The minimum pressurisation temperature for the reactors was specified to be −5 °C, and in all cases this lay in the upper shelf region. Step cooling (simulating long term temper embrittlement) had a relatively small effect on the

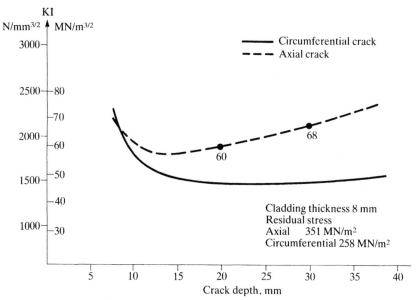

6.20 Stress intensity factor for internal surface cracks in a hydrocracker reactor.[29]

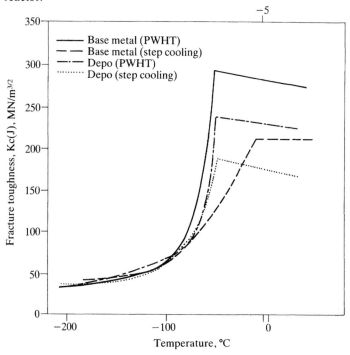

6.21 Fracture toughness transition curves for 2¼Cr1Mo forged ring material and weld deposits.[29]

transition temperature, but quite a substantial effect on the upper shelf level of fracture toughness. Critical crack depths are shown in Fig. 6.22. For all stress conditions the crack size corresponding to the measured fracture toughness is large and would be readily detectable by ultrasonic testing. However the stress intensity due to a crack increases continuously with crack depth, and the values of K_I at the full thickness was such that a 'leak before break' condition could not be guaranteed for the vessels in question.[29]

The assurance of reliability

General

Assuring reliable behaviour of welded structures and plant requires that satisfactory procedures be followed in all phases of construction and operation; in the production of the material of construction, during design and construction, and during operation. In this book we have been primarily concerned with the construction phase and in particular with welding. To achieve optimum reliability in this type of activity increasing reliance is placed on formal systems of job organisation and records. Before considering this aspect however it will be convenient at this point to review the effect of proof testing.

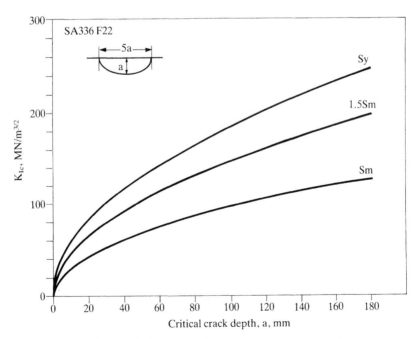

6.22 Relationship between critical crack depth and fracture toughness for 2¼Cr1Mo forged rings.[29]

Proof testing

The proof test is as old as metalworking itself. So are the means of passing such tests. One of the most convenient metallurgical tools was the old type of tensile testing machine, where a weight was moved along the lever arm until the tensile testpiece broke. This weight was moved by turning a handwheel, and deft manipulation of this wheel often made the difference between a pass and a fail. Hand operated tensile test machines have long since gone, but it must not be forgotten that the human operator is still there, and the human factor may still be important in all matters affecting quality.

Experience has shown that one of the best guarantees of reliability is the periodic overload test. Most vessels or structures suffer some deterioration in the course of service through fatigue, cracking or corrosion but this is rarely so rapid that it results in failure within a period of say twelve months. So if a test is carried out annually, there is a good chance that the defect will be disclosed by the test. For critical items of equipment the physical test is accompanied by non-destructive testing such as radiography, ultrasonic testing and acoustic emission, and the probability of failure is still further reduced.

It is also generally considered that an overload test pre-conditions a structure in a favourable manner. For example, if a crack is present the metal at the tip of the crack will deform plastically when the load is applied. On removing the load the surrounding volume contracts elastically and puts the deformed region into compression. When a lower service load is applied this compressive stress may be sufficient to reduce or even eliminate the tensile stress at the crack tip. Likewise areas of generalised stress concentration may be affected in a beneficial manner. Against this view it has been argued that metal subject to plastic deformation in the cold may suffer strain-age embrittlement in service, which would result in the worst possible condition: brittle material at the tip of a notch. However the majority of experts would consider the reduction of stress at the crack tip to be the more important factor. Moreover the current generation of steel plate has a low susceptibility to strain-age embrittlement and no doubt this improvement will be maintained in the future.

A further benefit may be obtained by warm prestressing. If the proof test can be carried out at a slightly elevated temperature plastic yielding at stress concentrations is increased and the induced compressive stress on completion of the test is correspondingly greater. There is also a greater possibility of blunting any crack tips that may exist. Warm prestressing combined with periodic inspection can make a substantial contribution to the reliability of pressure equipment.

Quality assurance: general

Quality assurance systems are generally considered to have originated with the problems encountered by the US Navy during the early phases of nuclear submarine construction. These were highlighted by Admiral Rickover in an address to ASME in 1963. For example:

> 'there are 99 carbon steel welds in one particular nuclear plant steam system. The manufacturer stated that all these welds were radiographed and met specifications. Our own (US Navy) re-evaluation of these welds – using correct procedures and proper X-ray sensitivity – showed however that only 10% met ASME standards; 35% had defects definitely in excess of ASME standards and the remaining 55% had such a rough external surface that the radiographs obtained could not be interpreted with any degree of assurance'.

During the same period there were serious delays in the UK power industry due to defective welds, mostly in boiler tubes.[30] The need for numbers of hydrostatic tests of boilers has already been noted in connection with the failure of the steam drum at Cockenzie power station; one witness has described these hydrotests as like a rainstorm. As a result of such problems CEGB set up a quality assurance group and issued a QA document in 1969.

It is the intention of quality assurance programmes to minimise defects by applying surveillance to the whole production process, rather than making inspection the last step of the operation. In an ideal system, the shopfloor operatives are responsible for quality and correct or reject any deficiencies in the course of production, rather than to be subject to an external test which may or may not disclose defects. For example, in the case of the piping fabrication shop discussed in Chapter 5, the supervisor in charge of tackers receives pipe cut to length and bevelled for assembly in jigs. He is responsible for checking that pipe lengths and bevels are correct and signs a document to that effect. The parts are then set up in jigs and tack welded, and the sub-assemblies passed on to the supervisor in charge of the 2G welders, who ensures that they are correct and signs the job document. This process continues to the end of the line, when non-destructive tests are applied and any required repairs are made. The documentation is then reviewed and if it is in order the shop manager or deputy certifies completion and passes the item to the field for erection. Any serious or persistent non-conformity is reported to the welding engineer or other responsible technician, who investigates the problem and takes corrective action.

Similar procedures may be applied to a wide variety of industrial operations; for example, they may be used in a design office. It is customary for major purchasers to require the supplier to submit evidence that such a quality assurance programme is in use.

Quality assurance standards

There is an unusual degree of international agreement so far as QA standards are concerned. There are five ISO standards and these are produced both as European (EN) and British Standards. The ISO standards are:

ISO 9000-1987: Quality management and quality assurance standards: guidelines for selection and use;

ISO 9001-1987: Quality systems – Model for quality assurance in design, development, production, installation and servicing;

ISO 9002-1987: Quality systems – Model for quality assurance in production and installation;

ISO 9003-1987: Quality systems – Model for quality assurance in final inspection and test;

ISO 9004-1987: Quality management and quality system elements – Guidelines.

The corresponding European Standards are EN 29000–29004 whilst BSI designations are BS 5750 parts 0–3. In addition BSI publishes Handbook No 22 which reproduces all the BS documents relating to QA, including BS 5750.

In the UK the Department of Trade and Industry publishes a Quality Assurance Register which lists companies that have been assessed and certified as complying with the relevant part of BS 5750/ISO 9000. Assessments are carried out by third party certification bodies such as BSI QA, Lloyds Register QA Ltd, Bureau Veritas QA Ltd, etc. The certification bodies are themselves subject to scrutiny; they are accredited by the Secretary of State on the advice of the National Accreditation Council for Certification Bodies. The main requirements for their acceptance are that they should have a representative and independent board of directors and that they should employ lead assessors approved by the Institute of Quality Assurance. Companies that qualify under the scheme are subject to periodic surveillance by a certification body. Some organisations like British Gas may conduct their own assessment but will generally require certification to BS 5750 as a first step. Some industries, for example ready mixed concrete, have their own systems but for welded fabrication BS 5750 is used.

There is a similar accreditation scheme in Holland (Raad voor de Certificatie). Internationally, large purchasers operate their own systems; for example NATO quality assurance standards are based on the original US Military Specification Mil-Q-9858a, which in turn was developed following the deficiencies noted earlier.

The use of quality assurance systems is therefore widespread, and in the UK, in particular, it is virtually obligatory for any but the smallest organisation concerned with the design or fabrication of welded structures to be certified to BS 5750.

Operating a QA system

Since it applies to a wide range of industries BS 5750 is of necessity written in very general terms, and details of QA procedures must be developed by the companies concerned. The basic document is the quality assurance manual, which lays down procedures and details the various steps during design, manufacture or quality control that must be subject to checking. Based on such check lists documentation is formulated; this is signed by a responsible person on completion of each stage. It then becomes part of the job documentation. The resulting paperwork is voluminous and one of the tasks of QA personnel is the control of these records, and ensuring that they are complete.

Two other functions are essential to the QA operation; corrective action and audit. Persistent or serious non-conformities are referred to a corrective action group, which investigates the problem and proposes a solution. This may result in the amendment of procedures. The audit group operates at specified intervals of time and determines whether or not the QA procedures are being followed.

Berg and Finsnes[31] have given a useful outline of how Det Norske Veritas applied QA to the construction of LNG storage tanks. These were shipborne spherical aluminium tanks 36.5 m in diameter with plate thickness 36 to 72 mm. Each tank carried 25 000 m³ liquefied natural gas at a temperature of −163 °C, and five or six were mounted on each carrier. The consequences of failure of such tanks would be extremely serious and every effort was made to ensure sound fabrication.

Plates were received from the mill, formed and cut to size. These were mounted in jigs and tacked on the inside. The welding groove was then milled on the outside by a rail-mounted milling machine. Next, a metal inert gas welder was mounted on the same rails and three weld runs were made. A weld groove was milled on the inside and a single weld bead laid down. Subsequent runs were made alternately from each side to minimise distortion. Welds were ground and dye-checked between runs. Finally the completed weld was ground flush and dye-checked. The complete procedure is shown in Fig. 6.23, which indicates major check points. Figure 6.24 gives details of the procurement and storage phase. The main items checked on receipt were:

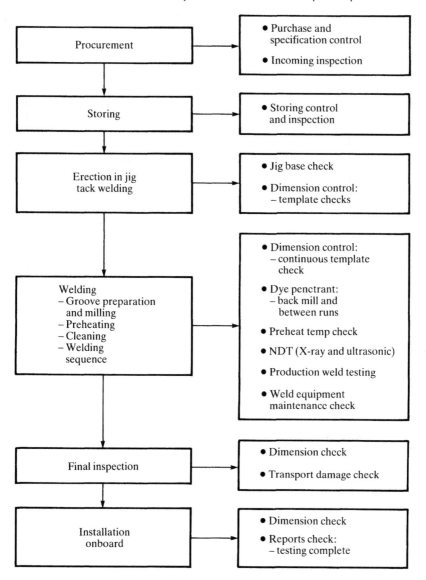

6.23 Manufacturing flow chart and major check points for LNG tanks.[31]

- Conditions of packing and protection;
- Check of identification and number of plates;
- Check of surface condition;
- Check of dimensions and form.

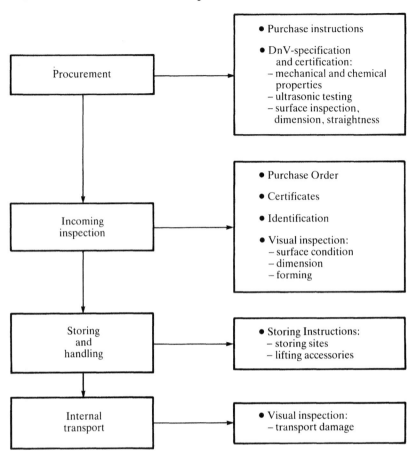

6.24 Checks for incoming inspection, storage and handling of formed aluminium plate for LNG tanks.[31]

Early identification of any defects enables replacements to be obtained in good time. Test certificates and other paperwork are checked at this stage. Storage of aluminium plate requires care: plates must be raised from the ground and spaced one from another to avoid condensation and potential corrosion problems.

The control of incoming welding consumables and their storage is equally important and the sequence and check points are shown in Fig. 6.25. One essential is to ensure that packets and reels are clearly marked and stacked in separate areas so as to minimise the chance of mix-up. The other important item with aluminium welding wire is to keep it dry: moisture contamination of any item, wire, gas leads, guns and other parts can lead to porosity.

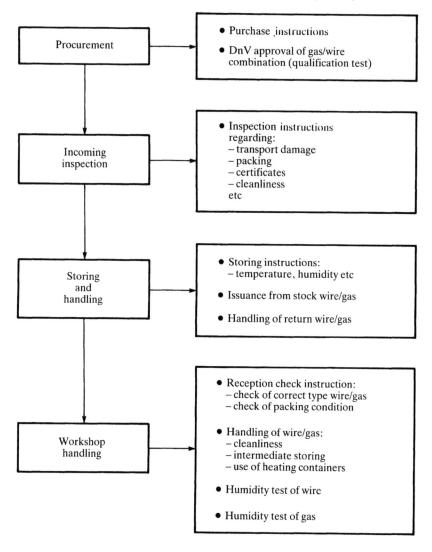

6.25 Checks for welding consumables for LNG storage tanks.

This and other requirements are also covered in a welding maintenance programme, listed in Fig. 6.26. Nozzle cleaning is an important item on this list. Even with inert gas shielding spatter builds up inside the nozzle and if not removed, disturbs the gas flow and causes weld defects. In this case it was intended that the welders should be responsible for items 1 and 2 and that the maintenance workshop should look after 3 and 4. This arrangement was not successful and eventually the maintenance shop took over all items. The failure of welders to

1. After each weld and when changing wire
   - Wire condition: dry and clean
   - Wire preheating function check
   - Gas nozzle cleaning
   - Gas flow check

2. After each shift
   - Cooling of weld gun, water flow check
   - Wire feeding: check of hub brake setting
   - Feed roller cleanliness and spring load

3. Monthly maintenance
   - Weld gun cleaning
   - Inspection of cables, hoses and connections
   - Change of parts (preventive programme)

4. Six monthly maintenance
   - General mechanical and electrical overhaul. Strip down and change of sub-assemblies according to programme

6.26 Maintenance programme for gas metal arc welding of aluminium plate.[31]

carry out maintenance was held to be due to lack of education and motivation.[31]

This description covers only a small part of the quality assurance procedures adopted for this job, but it is illustrative of the methods used. For the most part the QA manual simply describes good practice in the industry concerned. It does however impose a discipline and, when properly applied, ensures that work is not done in a slipshod manner and that corners are not deliberately cut. It must always be remembered, however, that the human factor is of the greatest importance. Poor workmanship at CEGB construction sites motivated that organisation to introduce QA systems. However, during the same period power boilers were being constructed in other countries without a QA programme and without significant weld failures. Earlier, the different failure rates of reformer furnace tubes between different operating companies were noted; a difference that could only have been due to differences in the competence and motivation of the operators. Even in the case just described, the system failed at one important point – torch maintenance – because of the human factor. Systems are no substitute for a well trained and well motivated workforce.

REFERENCES

1 Thoft-Christensen P and Baker M J:' Structural Reliability Theory and its Applications'. Springer-Verlag Berlin, Heidelburg 1982.

2 ASTM Special Technical Publication 91A.

3 Salot W J: 'Catalyst tube performance'. *Ammonia Plant Safety* AIChE Vol 15 1–5.

4 Salot W J: 'High pressure reformer tube operating problems' *Ammonia Plant Safety* AIChE Vol 17 34–40.

5 Bush S H: 'Pressure vessel reliability'. *J Pressure Vessel Technology* 1975 Vol 97 Series J 54–70.

6 Eyers J and Nesbitt E G: Proc Inst Mech Engrs 1964/5 Vol 179 part 31 paper 9.

7 Phillips C A G and Warwick R G: 'A survey of defects in pressure vessels'. UKAEA report AHSB (S) R162, HMSO 1968.

8 Kellermann O: 'Present views on recurring inspection of reactor pressure vessels'. Technical Reports series No 81, Int Atomic Energy, Vienna 1968.

9 Kitegawa H and Hisada T: 'Reliability analyses of structures under periodic non-destructive inspection'.

10 Collins J A and Monack M L: 'Stress corrosion cracking in the chemical process industry'. *Materials protection and performance* 1973 Vol 12 11–15.

11 Nelson P and Still J R: 'Metallurgical failures on offshore oil production installations' *Metals and Materials* 1988 Vol 4 559–564.

12 Peel C J and Jones A: 'Analysis of failures in aircraft structures'. *Metals and Materials* 1990 Vol 6 496–502.

13 Lancaster J F: 'Failures of boilers and pressure vessels' *Pressure Vessel and Piping* 1973 Vol 1 155–170.

14 Smedley G P: 'The integrity of marine structures'. Welding Institute Conference, Paper 26.

15 van Rongen H J M: 'Prediction of fracture behaviour of welded steel structures' *Lastechniek* 1991 Vol 57 183–187.

16 Bignell V F: 'Human aspects of catastrophes'. *Metallurgist and Materials Technologist* 1979 338–342.

17 Report of Committee of Investigation, Kings Bridge, Melbourne and Metropolitan Board of Works, Australia 1962.

18 Report of the inquiry into the causes of the accident to the drilling rig Sea Gem. Cmnd3409 HMSO 1967.

19 Lonsdale H: 'Ammonia tank failure – South Africa', *Ammonia Plant Safety* AIChE 1974 126–131.

20 The 'Alexander L Kielland' accident. Norwegian Public Reports Nov 1981: 11, 1981.

21 Everitt N M: 'Failures in aircraft'. *Metals and materials* 1991 Vol 7 522.

22 Burdekin F M and Harrison J D: 'Alternative elastic-plastic fracture mechanics concepts'. IIW Colloquium on the Practical Application of Fracture Mechanics 1979.

23 IIW Document SST-1157-90 'Guidance on assessment of the fitness for purpose of welded structures'.

24 Kumar V, German M D and Shih C F: 'An engineering approach for

elastic-plastic fracture analysis' Topical Report No EPRI Np 1931 GEC Schenectady NY 1981.

25 Newman J C and Raju I S: 'Stress intensity fractor for cracks' NASA Technical Memoranda 85793 NASA Virginia USA 1984.

26 Burdekin F M and Dawes M G: 'Practical use of linear elastic and yielding fracture mechanics'. Proceedings I Mech E Conference London, May 1971 pp 28–37.

27 Dawes M G: 'Fracture control in high yield strength weld metals' *Weld J* 1974 Vol 53 369-S to 379-S.

28 van Rongen H J M: 'Prediction of fracture behaviour of welded steel' *Lastechniek* 1991 Vol 57 179–187.

29 Jarecki A and Lancaster J: 'Heavy-wall reactors for hydrotreating refinery applications and for coal liquifaction plant' South African Institute of Welding.

30 Burgess N T and Levine L H: 'The control of quality in power plant' I Mech E Conf Sussex University 1969 Paper 12.

31 Berg A and Finsnes O F: 'Quality assurance in the building of aluminium tanks' in 'Quality assurance in welding construction', IIW, 1980.

Appendix: Nomenclature

	Operating parameter	nd
n, N	Number of cycles	nd
P	Probability	nd
q	Rate of heat or energy flow	W
	Cumulative number of failures	nd
Q	Quantity of heat or energy	J
r	Radius	m
R	Reliability	nd
	Ratio min/max stress	nd
s	Solubility (gas in metal)	ppm, g tonne^{-1}
t	Time	s
T	Temperature	K, °C
v	Velocity, welding speed	m s^{-1}
V	Voltage	V
w	Thickness	m
α	Diffusivity of heat	m^2 s^{-1}
	Safety factor	nd
γ	Surface tension	N m^{-1}
	Surface energy	J m^{-2}
δ	Longitudinal displacement	m
	Crack-tip opening displacement	m
δ$_c$	Critical crack-tip opening displacement	m
Δ	Increment	nd

Index

Printed and bound by CPI Group (UK) Ltd, Croydon, CR0 4YY

08/05/2025

01864847-0001